2021台达杯国际太阳能建筑设计竞赛获奖作品集

Awarded Works from International Solar Building Design Competition 2021

阳光·低碳社区
SUNSHINE & LOW-CARBON COMMUNITY

中国建筑设计研究院有限公司　编
Edited by China Architecture Design & Research Group

执行主编：张　磊　鞠晓磊
Chief Editor: Zhang Lei　Ju Xiaolei

中国建筑工业出版社
CHINA ARCHITECTURE & BUILDING PRESS

图书在版编目（CIP）数据

阳光·低碳社区：2021台达杯国际太阳能建筑设计竞赛获奖作品集 = SUNSHINE&LOW-CARBON COMMUNITY Awarded Works from International Solar Building Design Competition 2021/中国建筑设计研究院有限公司编；张磊，鞠晓磊执行主编. —北京：中国建筑工业出版社，2022.8
 ISBN 978-7-112-27739-1

Ⅰ.①阳… Ⅱ.①中… ②张… ③鞠… Ⅲ.①太阳能住宅－建筑设计－作品集－中国－现代 Ⅳ.①TU241.91

中国版本图书馆CIP数据核字（2022）第142903号

本次作品集围绕我国2030碳达峰和2060碳中和的战略目标，贯彻落实国家实现能源利用转型、推动低碳社会发展、建设绿色健康的生活环境，将低碳、绿色、可持续理念融入社区建设。以低碳社区和零碳社区为赛题，面向全球征集作品，本书中收录了竞赛的一、二、三等奖及优秀奖获奖作品，全面反映了我国及全球设计师在低碳建设方面的理念和技术。本书可供高校建筑设计相关专业本科生、研究生以及从事相关专业的设计公司、建筑师等人员参考阅读。

责任编辑：唐　旭　吴　绫　张　华
责任校对：王　烨
编　　辑：夏晶晶　张星儿　孙　畅

阳光·低碳社区
SUNSHINE & LOW-CARBON COMMUNITY
2021台达杯国际太阳能建筑设计竞赛获奖作品集
Awarded Works from International Solar Building Design Competition 2021
中国建筑设计研究院有限公司　编
Edited by China Architecture Design & Research Group
执行主编：张　磊　鞠晓磊
Chief Editor: Zhang Lei　Ju Xiaolei
*
中国建筑工业出版社出版、发行（北京海淀三里河路9号）
各地新华书店、建筑书店经销
北京雅盈中佳图文设计公司制版
临西县阅读时光印刷有限公司印刷
*
开本：787毫米×1092毫米　1/12　印张：$31\frac{1}{3}$　字数：972千字
2022年8月第一版　2022年8月第一次印刷
定价：228.00元
ISBN 978-7-112-27739-1
（39748）

版权所有　翻印必究
如有印装质量问题，可寄本社图书出版中心退换
（邮政编码 100037）

为实现我国2030碳达峰和2060碳中和的战略目标，建设绿色健康的生活环境，是将低碳、绿色、可持续理念融入社区建设的必要途径之一。如何在新时期建筑方针的指引下，设计出"零碳社区"、"低碳社区"，利用可再生能源实现社区低碳化，成为本次竞赛的关注点。本届竞赛首次设置了概念设计项目"零碳社区"与实际建设项目西藏班戈县低碳社区两个赛题，参赛团队施展才华，呈现出精彩纷呈的竞赛作品，为绿色低碳、节能环保、可持续的发展理念插上梦想的翅膀飞得更高、传播得更广。

感谢台达集团资助举办2021台达杯国际太阳能建筑设计竞赛！

谨以本书献给致力于低碳建设与绿色发展的同仁们！

To achieve China's strategies of peaking carbon dioxide emissions before 2030 and achieving carbon neutrality before 2060, building a green and healthy living environment is one of the necessary ways to integrate low carbon, green and sustainable concepts into community building. This competition focused on how to design "zero-carbon communities" and "low-carbon communities" and how to decarbonize communities with renewable resources to meet the requirements of architecture guidelines in the new period. Given this, the competition set two subjects—the Field Construction Project of low-carbon community in Baingoin county, China's Tibet Autonomous Region and set the Concept Design Project of zero-carbon community for the first time. All teams showcased their talents and created brilliant designs. They injected their dreams and passion into these works and spread the green and low-carbon development idea featured by energy saving, environmental protection and sustainable growth.

Thank Delta Electronics for the sponsorship of the International Solar Building Design Competition 2021!

This book is dedicated to all our colleagues for your efforts in low-carbon construction and green development!

目 录
CONTENTS

阳光·低碳社区　SUNSHINE & LOW-CARBON COMMUNITY

2021台达杯国际太阳能建筑设计竞赛过程回顾
General Background of International Solar Building Design Competition 2021

2021台达杯国际太阳能建筑设计竞赛评审专家介绍
Introduction to Jury Members of International Solar Building Design Competition 2021

获奖作品　Prize Awarded Works　001

综合奖·一等奖　General Prize Awarded · First Prize

藏街隐屋（实地建设项目）　Sinking Street and House (Field Construction Project)	002
归源谷（概念设计项目）　Back to the Original (Concept Design Project)	008

综合奖·二等奖　General Prize Awarded · Second Prize

风下旅人（实地建设项目）　Step on the Wind (Field Construction Project)	012
光的致意（实地建设项目）　Greeting of Light (Field Construction Project)	018
与海共舞（概念设计项目）　Dancing with Sea (Concept Design Project)	024
能源控制实验站（概念设计项目）　Energy Control Experimental Station (Concept Design Project)	026

综合奖·三等奖　General Prize Awarded · Third Prize

光·窖（实地建设项目）　The Cellar of Light (Field Construction Project)	030

格桑社区（实地建设项目）	Gesang Community (Field Construction Project)	036
故土·亲尘（实地建设项目）	The Gift of the Land (Field Construction Project)	042
向阳（实地建设项目）	Towards to Sun (Field Construction Project)	048
疆域零城（概念设计项目）	Uyghur-Zero (Concept Design Project)	054
巨构社区：律之海（概念设计项目）	Mega Community: Sea of Melody (Concept Design Project)	058
生生不息（概念设计项目）	Loop (Concept Design Project)	062
衣聚·宜居（概念设计项目）	Clothing in Living·Living in Comfort (Concept Design Project)	066

实地建设项目
Field Construction Project
综合奖·优秀奖
General Prize Awarded · Honorable Mention Prize

暖巢	Warm Nest	070
寨心·光·聚	The Heart of the Village· Light·Gather	074
伴日——低碳社区综合体设计	Accompanying the Sun	080
引璨流光	Welcome, Miss Sunshine	084
念青	Together to an Ecological Habitat	088
向光	Embracing Sunshine	094
藏进太阳里	Hide in the Sun	100
沐光	Bathe in Light	106
藏纹新译	Revival of Tibetan Patterns	112

藏时风光	Tibetan Scenery	116
阳起风落	Low Carbon Community Which Use Solar and Wind Energy	122
"沐"色·"阳"幡	Bathe in Light Born to the Sun	128
行寻圣湖，游居光筑	Green and Blue—A Design of Dongga Low-Carbon Community as the Gate of Holy Lake Namtso	134
坛城·融光	Glory City	140
光环·藏腔	Light Ring· Hot Cavity of Tibetan	146
回聚	Cycling and Converging	152
匡郭/织轨	Outer-Internal Weaving	156
聚·星空之下	Tribe Under the Stars-Low Carbon Community Design in Tibet	160
廊桥日暖渡游人	Ferrying Corridor	166
风拂经轮	The Wind Blows the Wheel	172
风回路转	Wind Back	176
暖厅·聚能	Warming Hall · Gathering Energy	182
"盒""光"同尘——西藏班戈县青龙乡东嘎村低碳社区设计 Sunshine, Box, Single Building, Group, Community		188
风涌墙阻	A Wall to Keep Out the Cold Wind	194
万物生	Growth of All Things	198
碳固·能转	Carbon Sequestration · Energy Circulation	204
融合社区·阳光市集	Convergence Community· Sunshine Town	208

斯贝廓罗	Samsara	214
轮回之径	The Path of Reincarnation	218
向阳而生，逐光而行	Facing the Sun, Chasing the Sun	222

概念设计项目
Concept Design Project
综合奖·优秀奖
General Prize Awarded · Honorable Mention Prize

田园"彝"居——以水为介质的太阳能和风能循环系统社区	Rural Yi Settlement-Community of Solar and Wind Energy Circulation Systems Using Water as the Medium	228
风曦·满居	Wind And Sunshine · MAN Design	232
归园·田居	Farming Community	236
三"R"零碳太阳能社区	Three "R" Community-Zero Carbon Community	240
零碳社区	Zero Carbon Community	242
零碳·侗寨	Zero Carbon·DONG Village	246
浅草托岳	The Marsh Prop Up Huge Mountain	250
后疫情垂直社区	Vertical Garden	254
城间一绿	Green Live Hood in the City	258
芯·社区	The Community Chips	262
零荷博益	Zero-Carbon Lotus Community	266
零社模方	M-Cube Zero Carbon Community	270

城市地衣·零碳社区　Urban Lichen·Zero Carbon Community	274
流动之"环"——基于智能化平台的意识型低碳社区设计　Mobile Loop	278
缝合·社区——大理低碳社区　Urban Farming Community Stitching City	280
森之谷　City Valley	284
生土楼"埂与生"　The Rebirth of Fujian Earth Building	288
共享办公　Office Community	292
不期而"寓"　Zero Carbon Community Design	296
江风摇巷　River Wind Shaking Alley	300
绿意·环生　Green Germination	304
纳兰住集·向阳而居　Living in the Naran Resistance Towards the Sun	308
叠榭·回环——基于空间重组理念的能量自循环低碳社区设计　Pavilion Loop—Design of Energy Self-Circulation Zero Carbon Community Based on the Concept of Spatial Reorganization	312
生长工厂　Growing Factory	316
旧城新衣——基于岭南湿热气候的广州老旧小区零碳改造探索　Exploration of Zero Carbon Transformation of Old Communities in Guanghzhou Based on Lingnan′S Hot and Humid Climate	320
何不归故里　Why Not Return Home	324
阡陌交通　Crisscross of Paths in Fields	326
"厂·能"低碳社区　Factory Energy	328
告别孤独的低碳社区——基于"共享"理念的定制化社区　Join the We Share Say Goodbye to the Loneliness—Customized Community Based on "Sharing" Concept	332
"芯"院里——基于碳中和视角下的社区设计　Ismart Community	336

有效作品参赛团队名单
Name List of All Participants Submitting Valid Works 340

2021台达杯国际太阳能建筑设计竞赛办法
Competition Brief on International
Solar Building Design Competition 2021 344

2021台达杯国际太阳能建筑设计竞赛过程回顾
General Background of International Solar Building Design Competition 2021

主题：阳光·低碳社区

2021台达杯国际太阳能建筑设计竞赛由国际太阳能学会、中国建筑设计研究院有限公司主办，国家住宅与居住环境工程技术研究中心承办，台达集团冠名。在社会各界的大力支持下，竞赛组委会先后组织了竞赛启动、媒体宣传、云讲堂、作品注册与提交、作品评审、现场答辩等一系列活动。这些活动得到了海内外业界人士的积极响应和参与。

一、赛题设置

习近平主席在第七十五届联合国大会一般性辩论上表示，中国将提高国家自主贡献力度，采取更加有力的政策和措施，二氧化碳排放力争于2030年前达到峰值，努力争取2060年前实现碳中和。为贯彻落实国家实现能源利用转型，本届竞赛以"阳光·低碳社区"为主题，分别设置西藏班戈县青龙乡东嘎村低碳社区实地建设项目和"零碳社区"的概念设计项目，推动低碳社会发展、建设绿色健康的社区环境。本届赛题首次采用虚拟与现实结合的方式，不仅在西藏设置了可实地建设的赛题供大众观摩、体验，还设置了概念设计的赛题，探寻更加理想与超前的绿色建筑技术设计理念，将科学技术与理论实践相结合，推动可再生能源在建筑上的应用，让科技走进生活，惠及民生。

二、竞赛启动

2021年3月31日，2021台达杯国际太阳能建筑设计竞赛在北京启动。在疫情依然严峻的情势下，竞赛采用线上启动，高效、广泛的传播竞赛理念，让更多感兴趣的网友们参与到竞赛的活动中。中国建筑设计研究院有限公司董事、总经理、党委副书记马海，中国可再生能源学会理事长谭天伟，中国建筑设计研究院副总建筑师、中国可再生能源学会太阳能建筑专委会主任委员仲继寿，台达集团创办人暨荣誉董事长郑崇华作为嘉宾参加本次竞赛线上启动仪式并分别致辞。

竞赛围绕我国2030碳达峰和2060碳中和的战略目标，贯彻落实国家实现能源利用转型，推动低碳社会发展、建设绿色健康的生活环境，将低碳、绿色、可持续理念融入社区建设。本次竞赛不仅延续了前8届竞赛的惯例，设置可实地建设项目：西藏班戈县青龙乡东嘎村低碳社区，还增设了以"低碳社区"为主题

Theme: Sunshine & Low-Carbon Community

International Solar Building Design Competition 2021 was co-hosted by the International Solar Energy Society (ISES) and China Architecture Design & Research Group (CAG), with China National Engineering Research Center for Human Settlements (CNERCHS) as the organizer and Delta Electronics as the title sponsor. With the support from all sectors of the society, the competition organizing committee has held a series of activities such as the launch ceremony, publicizing, online lectures, registration and submission of entries, assessment and evaluation, and on-site inquiring. Practitioners at home and abroad positively responded and took part in these activities.

I. Competition Preparation

At the 75th General Debate of the United Nations General Assembly, President Xi Jinping said that China would increase its nationally determined contributions with more vigorous policies and measures, spare no efforts to peak CO_2 emissions by 2030, and achieve carbon neutrality by 2060. To implement the principle of energy transformation, the year's competition is themed with low-carbon communities. It set the Field Construction Project for Donggar Village Low-Carbon Community in Qinglong Township, Baingoin County, Tibet Autonomous Region, and a conceptual design project for a "zero-carbon community" to promote low-carbon social development and build a green and healthy community environment. This competition, for the first time, combines VR and reality. The Field Construction Project was built in Tibet Autonomous Region for observation and experience. A conceptual design project was also set to explore more ideal and advanced green building technology design concepts. The competition also combines science and technology with theory and practice, promoting the application of renewable energy in buildings to apply technology to our daily life and increase people's well-being.

国家住宅与居住环境工程技术研究中心主任张磊主持会议
Zhang Lei, Secretary General of China National Engineering Research Center for Human Settlements over the Conference

竞赛海报
Competition Poster

II. Launch of the Competition

On March 31, 2021, International Solar Architecture Design Competition 2021 was launched in Beijing. Despite the pandemic, the competition was still launched via video link, efficiently and widely spreading the concept of the competition and attracting more netizens to participate in the activities of the competition. Ma Hai, Director, General Manager, and Deputy Secretary of the Party Committee of China Architecture Design & Research Group (CAG); Tan Tianwei, Chairman of the China Renewable Energy Society (CRES); Zhong Jishou, Deputy Chief Architect of CAG and Chairman of the Technical Committee of Solar Building, CRES; and Zheng Chonghua, Founder and Honorary Chairman of Delta Electronics attended the online launch of the competition and delivered speeches respectively.

The competition focused on the strategic goals of China's 2030 carbon peak and 2060 carbon neutrality. It aimed to implement energy transformation, promote low-carbon social development, build a green and healthy living environment, and integrate low-carbon, green, and sustainable concepts into community construction. The competition followed the practices of the previous eight competitions, setting a field able construction project: Donggar Village Low-Carbon Community in Qinglong Township, Baingoin County, Tibet Autonomous Region. It also added a conceptual design competition themed with "zero-carbon community" to create a virtual community that meets the needs of human work and life in whatever geography, climate, and topography, and promotes clean energy and other green building technologies in architecture.

III. Promotion Publicity

Since the first competition was held in 2005, the competition organizing committee has visited architecture departments of more than 50 universities, such as Tsinghua University, Tianjin University, Southeast University, Chongqing University, and Shandong Jianzhu University, to promote International Solar Building Design Competition, which has received positive response and praises

的概念设计竞赛题目，在不限地域、气候、地形的条件下，打造满足人类工作生活需求的虚拟社区，推动清洁能源等绿色建筑技术在建筑上的应用。

三、竞赛宣传

自2005年第一届竞赛举办以来，竞赛组委会已先后前往了清华大学、天津大学、东南大学、重庆大学、山东建筑大学等50多所建筑院校开展国际太阳能建筑设计竞赛巡讲活动，受到了高校师生的积极响应和好评。

2021年6月，国际太阳能建筑设计竞赛组委会走进湖南省长沙市，与师生们围绕国际太阳能建筑设计竞赛进行了交流。竞赛巡讲团一行还前往了湖南大学、长沙理工大学、中南大学等高等院校进行校园巡讲。随后，竞赛巡讲团继续前往西藏开展校园巡讲。巡讲主讲人为国家住宅与居住环境工程技术研究中心低碳建筑研究所所长鞠晓磊，巡讲内容涵盖了太阳能建筑技术应用趋势和现状、历届竞赛获奖作品分析和本届竞赛介绍。通过巡讲，师生们对太阳能建筑设计竞赛和节能技术有了更深入的了解，激发了参赛团队的设计灵感，对太阳能建筑应用技术

from teachers and students.

On June, 2021, the organizing committee of the International Solar Architecture Design Competition visited Changsha, Hunan Province, and talked with teachers and students there about the competition. The promotion delegation also went on a campus tour to architecture colleges of universities, including Hunan University, Changsha University of Science &Technology, and Central South University. Then, they went to Tibet Autonomous Region. The lecturer Ju Xiaolei, Director of the Low Carbon Building Research Institute of China National Engineering Research Center for Human Settlements, talked about the trends and status of solar building technology applications and the analysis of previous winning entries, and then introduced this year's competition. After the lecture, students and teachers gained a deeper understanding of the solar architecture design competition and energy-saving technology. Participants were inspired to have more innovative ideas of technology applied to solar

竞赛宣讲会现场
Competition Presentation Site

网络宣讲海报
Poster on Online Publicity

用技术进行了创新思考。同时,组委会还邀请领域专家开展包括太阳能建筑一体化设计、建筑节能技术综合应用、建筑能效提升等内容的云上讲堂,提高了竞赛关注群体和观众的知识积累及相关技术素养的传播,使竞赛宣讲会成为太阳能建筑领域行业知识、技术、理念的重要科普和交流平台。

四、媒体宣传

组委会自竞赛启动以来通过多渠道开展媒体宣传工作,包括:竞赛官方网站(双语)实时报道竞赛进展情况并开展太阳能建筑的科普宣传;在百度设置关键字搜索,方便大众查询,从而更快捷地登陆竞赛网站。在新华网、腾讯网、新浪网、Bustler、Architectural Record 等 50 余家国内外网站上报道或链接了竞赛的相关信息;同时,组委会与多所国外院校和媒体取得联系并发布竞赛信息与动态。通过微信公众号、微博实时发布竞赛进展、云讲堂预告等动态,并提供竞赛相关资料下载与案例介绍等,有效地提高了竞赛的影响力及参赛团队的技术能力。

energy buildings. Besides, experts were invited to conduct lectures via video, covering integrated solar building design, comprehensive application of building energy-saving technologies, and improvement of building energy efficiency. These activities, therefore, improved people's knowledge and promoted relevant technological literacy. As a result, the lecture has become an important platform for science popularization and communication of knowledge, technology, and ideas about solar energy buildings.

IV. Media Publicity

Since the launch of the competition, the organizing committee has carried out media publicity through various channels. For example, it reported the progress of the competition in real-time on the official website of the competition (bilingual) and popularized knowledge of solar architecture. Keywords can be

竞赛信息网络发布
Online Competition Information Release

竞赛网站
Official Website

五、竞赛注册及提交情况

本次竞赛的注册时间为 2021 年 3 月 31 日至 2021 年 8 月 15 日，共 645 个团队通过竞赛官网进行注册，其中，包括来自澳大利亚、德国、荷兰等境外注册团队 8 个。截至 2021 年 9 月 30 日，竞赛组委会收到有效参赛作品 149 份。

六、作品初评

2021 年 10 月 1 日 -21 日，组委会组织评审专家对通过形式筛查的全部有效作品开展初评工作。专家根据竞赛办法中规定的评比标准对每一件作品进行评审，经过合规性审查及初期评审成绩分别排名，选择每个赛题的前 50 名，共计 100 份作品进入中期评审。

2021 年 10 月 26 日 -11 月 21 日，组委会组织评审专家对 100 份进入中期评审的作品开展评审工作。经过专家组的严格评审，组委会将根据赛题分别排名，得分前 74 名将获得综合奖，得分前 18 名将进入最终的现场答辩环节。

searched on Baidu, which facilitated the public to log in to the official website. Over 50 domestic and international websites, including Xinhua Net, Tencent, Sina, Bustler, and Architectural Record, reported or put a link to the competition. The organizing committee contacted many foreign universities and media and released the information and process about the competition. The organizing committee also contacted many foreign universities and media to release information and news about the competition. Processes of the competition and online lectures were posted on the WeChat official account and Weibo. The introduction of cases and competition files were also provided for downloading. These measures increased the influence of the competition and the technical ability of the participating teams.

V. Registration and Submission

Registration for this competition was open from March 31 to August 15, 2021. 645 teams registered through the competition website, including 8 teams from other countries such as Australia, Germany, and the Netherlands. By September 30, 2021, the competition organizing committee had received 149 valid entries.

VI. Preliminary Evaluation

From October 1 to 21, 2021, the Organizing Committee carried out the preliminary evaluation of all valid entries. Experts evaluated each entry according to the appraisal standard in the competition specification. After the compliance review, experts gave preliminary evaluations to those eligible entries and ranked them according to their scores. The top 50 entries for each competition subject, totaling 100 entries, were selected for the mid-term evaluation.

From October 26 to November 21, 2021, experts reviewed the 100 entries in the mid-term evaluation. After the mid-term evaluation, the organizing committee ranked the entries separately according to the subjects. The top 74 scorers were awarded the general prize, and the top 18 entered the final inquiry.

七、作品终评

本届竞赛终评会采用线上答辩的形式让学生更好地展示作品设计内容，也能够使评审专家综合评价竞赛作品。2021年10月10日上午，进入决赛的18个团队在线上通过图像、文字、语言与视频等多元方式展示、阐述作品，并回答终评会现场评审专家组提问。专家们在关注作品设计与技术应用创意的同时，注重设计与技术应用的适用性和可操作性。10月10日下午由崔愷院士、杨经文博士等十位国际评审专家通过视频会议的方式连线，历经3轮评选和讨论，最终两个赛题各选出一等奖1项、二等奖2项、三等奖4项、优秀奖30项，综合奖项共计74项。

VII. Final Evaluation

The final evaluation of this year's competition was held online in the form of inquiry. In this way, students better presented their designs, and experts could evaluate those entries comprehensively. On the morning of January 10, 2022, the 18 finalists presented their works online through images, texts, speeches, and videos, and answered questions from the judging panel. In the final evaluation, the feasibility and operability were given the same importance as the design and technological ideas. On the afternoon of January 10, 2021, ten experts, including Academician Cui Kai and Dr. Yang Jingwen, held a video conference. After three rounds of evaluation and discussion, based on the two subjects, the panel selected 1 first prize, 2 second prizes, 4 third prizes, and 30 honorable mention prizes for each subject, totaling 74 general prizes.

终评会现场
Scenes of Final Evaluation Conference

终评专家组与答辩师生合影
Members of Final Evaluation Juries and on-site Statement Teachers and Students

2021台达杯国际太阳能建筑设计竞赛评审专家介绍
Introduction to Jury Members of International Solar Building Design Competition 2021

评审专家
Jury Members

Peter Luscuere：荷兰代尔伏特大学建筑系教授
Mr. Peter Luscuere: Professor of Department of Architecture, Delft University of Technology

杨经文：马来西亚汉沙杨建筑师事务所创始人、**2016**梁思成建筑奖获得者
Mr. King Mun YEANG: President of T. R. Hamzah & Yeang Sdn. Bhd (Malaysia), 2016 Liang Sicheng Architecture Prize Winner

林宪德：中国台湾绿色建筑委员会主席、中国台湾成功大学建筑系教授
Mr. Lin Xiande: President of Taiwan Green Building Committee, China and Professor of Faculty of Architecture of Cheng Kung University, China

Deo Prasad：澳大利亚科技与工程院院士、澳大利亚勋章获得者、新南威尔士大学教授
Mr. Deo Prasad: Academician of Academy of Technological Sciences and Engineering, Winner of the Order of Australia, and Professor of University of New South Wales, Sydney, Australia

崔愷：中国工程院院士、全国工程勘察设计大师、中国建筑设计研究院有限公司总建筑师
Mr. Cui Kai: Academician of China Academy of Engineering, National Engineering Survey and Design Master and Chief Architect of China Architecture Design & Research Group (CAG)

仲继寿：中国可再生能源学会太阳能建筑专业委员会主任委员、中国建筑设计研究院有限公司副总建筑师
Mr. Zhong Jishou: Chief Commissioner of Special Committee of Solar Buildings, CRES, and Deputy Chief Architect of CAG

宋晔浩：清华大学建筑学院建筑与技术研究所所长、教授、博士生导师，清华大学建筑设计研究院副总建筑师
Dr. Song Yehao: Director, Professor and Doctoral Supervisor of Institute of Architecture and Technology, School of Architecture, Tsinghua University, and Deputy Chief Architect of Architectural Design and Research Institute of Tsinghua University

钱锋：全国工程勘察设计大师，同济大学建筑与城市规划学院教授，博士生导师，高密度人居环境生态与节能教育部重点实验室主任
Dr. Qian Feng: National Engineering Survey and Design Master, Professor and Doctoral Supervisor of College of Architecture and Urban Planning Tongji University (CAUP), Director of Key Laboratory of Ecology and Energy-saving Study of Dense Habitat (Tongji University), Ministry of Education

黄秋平：华东建筑设计研究总院总建筑师
Mr. Huang Qiuping: Chief Architect of East China Architectural Design & Research Institute (ECADI)

冯雅：中国建筑西南设计研究院顾问总工程师，中国建筑学会建筑热工与节能专业委员会副主任
Mr. Feng Ya: Chief Engineer of China Southwest Architectural Design and Research Institute Corp. Ltd, and Deputy Director of Special Committee of Building Thermal and Energy Effciency, Architectural Society of China (ASC)

刘泓志：AECOM 亚洲区高级副总裁，中国区战略与发展负责人，城市设计策略团队负责人
Mr. Liu Hongzhi, AECOM Asia Senior Vice President, Leader for Strategy & Development Cities Market Sector Leader, China

获奖作品
Prize Awarded Works

藏街隐屋 · 1
Sinking Street and House

综合奖 · 一等奖
General Prize Awarded · First Prize

注 册 号：100506
Registration No：100506

项目名称：藏街隐屋（实地建设项目）
Sinking Street and House (Field Construction Project)

作　者：李睿晨、何嘉琦、谢 冲、冉铖李、王文轩
Authors：Li Ruichen, He Jiaqi, Xie Chong, Ran Chengli, Wang Wenxuan

参赛单位：重庆大学
Participating Unit：Chongqing University

指导教师：张海滨、李臻赜
Instructors：Zhang Haibin, Li Zhenze

■ **Design Specification**

设计受窑洞式民居启发，将建筑下沉、回填覆土以利抗风、保暖；同时引入更新颖的天街式商业空间——一次化舒适度较低的室外空间，将主要内街与铺面引至阳光房、阳光暖廊，营造出丰富的步行体系与公共空间；技术层面上，设计版应用夯土山墙、夹土砖墙、卵石蓄热地面等当地材料，又将阳光房、彩色薄膜光伏板、深色太阳能晶硅板、蓄热天窗、彩色压顶百叶、相变蓄热墙等新式技术与藏式传统建筑形象紧密结合，对于藏式传统文化的回应，设计选取五彩经幡的意象，使阳光房顶面与建筑压顶延续以彩色太阳能光伏与五色压顶百叶，形成一条条连贯的"彩带"。

Inspired by the cave dwelling, the building is sunk and backfilled with earth to resist wind and keep warm.At the same time, introduce a more novel sky street style commercial space -- outdoor space with lower secondary comfort, lead the main inner street and shop surface to sunshine room and sunshine warm corridor, create a rich walking system and public space;On the technical level, the design not only applies local materials such as rammed earth gable, brick wall with earth, pebble heat storage floor, but also closely combines the new technologies such as sun room, color thin film PHOTOVOLTAIC panel, dark solar crystal silicon panel, heat storage skylight, color pressure top louvers, phase change heat storage wall and other traditional Tibetan architectural images.In response to the traditional Tibetan culture, the design selects the image of colorful prayer flags, so that the sun roof surface and the top of the building continue to color solar photovoltaic and five-color top louvers, forming a coherent "ribbon".

藏街隐屋·2
Sinking Street and House

Major Technical-economic Indices:
Total Site Area: 62892.3m²
Total Floor Area: 1340m²
Commercial Street Area: 6848m²

Site Plan 1 : 1000

专家点评：

作品规划布局规整，各功能分区相互独立且联系紧密，建筑设计吸取地方本土化元素，造型简约，针对当地恶劣的气候条件，采取下沉商业内街的设计手法巧妙，实现被动节能和提升室内温、湿度环境的效果，设计空间整合性高。将太阳能主、被动技术与富有地域特色的建筑构配件创新性结合，具有一定的实用性和可推广性。

The planning is clear and neat, with each functional sub-area being independent and closely linked. The building design draws on local elements with a simple and distinctive style. The design of the sunken commercial inner street is ingenious and highly integrated due to local harsh weather conditions. It is also a passive design, which can increase indoor temperature and humidity. The innovative combination of solar active and passive technologies and building components with regional features is practical and replicable.

■ Logic Generation

1. The total land area of the site is 75391.5 m²
2. The venue has four main functional areas
3. The two axes meet at the long-term core of the site
4. Determine the relationship between the four functional areas
5. Organize the flow of vehicles and divide the site
6. Organize pedestrian circulation and deepen the site
7. Determine the building based on the orientation and wind direction
8. Determine the location and type of landscape green space

■ Space Partition

1. Commercial pedestrian street
2. Outdoor country fair
3. Residential area supporting
4. Traffic Plaza
5. Residential area
6. Sea of Gesang Flowers
7. Carbon sink forest
8. Gesang Flower Stairs

The site consists of commercial pedestrian street, outdoor country fair, residential area supporting, traffic plaza, residential area, sea of gesang flowers, carbon sink forest and gesang flower stairs.

■ Site Planning Analysis

— Roadway
--- sidewalk
··· Roadway in the community

— bicycle lane
● Solar carport

Site streamline analysis:
The traffic lanes and pedestrian lanes in the site are designed separately.

Low-carbon transportation analysis:
There are several solar carports charging electric cars to reduce carbon emissions.

■ Carbon Sink

Carbon dioxide emissions: manufacturing, transportation, construction, operation and maintenance of materials and components for commercial buildings, public facilities, residential buildings, etc.
Residents and tourists' traffic emissions and daily life emissions.
Carbon sink:
Forest trees are calculated at 24.45 tons per mu per year (direct)
Energy saving and emission reduction (indirect)
Solar and wind energy use of clean energy
Active and passive building technology combines heating, heat preservation, lighting and ventilation

■ Climatic Simulation

Direct Evaporative Cooling | Indirect Evaporative Cooling | Natural Ventilation

Night Purge Ventilation | Passive Solar Heating | Thermal Mass Effects

Wind Frequency(Hrs) | Average Temperatures | Maximum Temperatures | Minimum Temperatures

Direct and diffuse solar radiation analysis:

■ Technology System

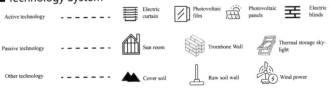

Active technology ----
- Electric curtain
- Photovoltaic film
- Photovoltaic panels
- Electric blinds

Passive technology ----
- Sun room
- Trombone Wall
- Thermal storage skylight

Other technology ----
- Cover soil
- Raw soil wall
- Wind power

藏街隐屋·3
Sinking Street and House

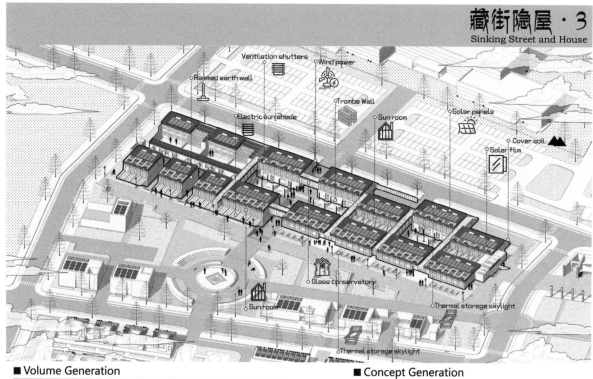

■ Volume Generation

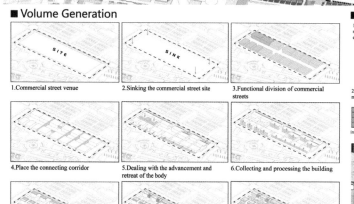

1. Commercial street venue
2. Sinking the commercial street site
3. Functional division of commercial streets
4. Place the connecting corridor
5. Dealing with the advancement and retreat of the body
6. Collecting and processing the building
7. Sink the roof
8. Built-in glass skylight system
9. Further deepen the treatment of buildings

■ Concept Generation

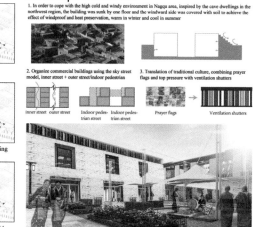

1. In order to cope with the high cold and windy environment in Nagqu area, inspired by the cave dwellings in the northwest region, the building was sunk by one floor and the windward side was covered with soil to achieve the effect of windproof and heat preservation, warm in winter and cool in summer

2. Organize commercial buildings using the sky street model, inner street + outer street/indoor pedestrian

3. Translation of traditional culture, combining prayer flags and top pressure with ventilation shutters

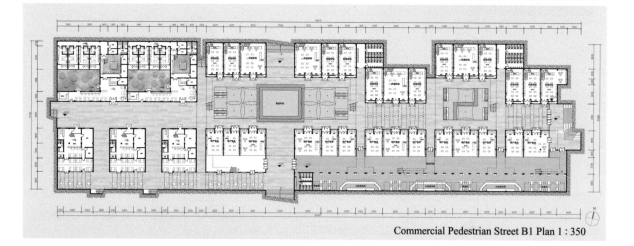

Commercial Pedestrian Street B1 Plan 1:350

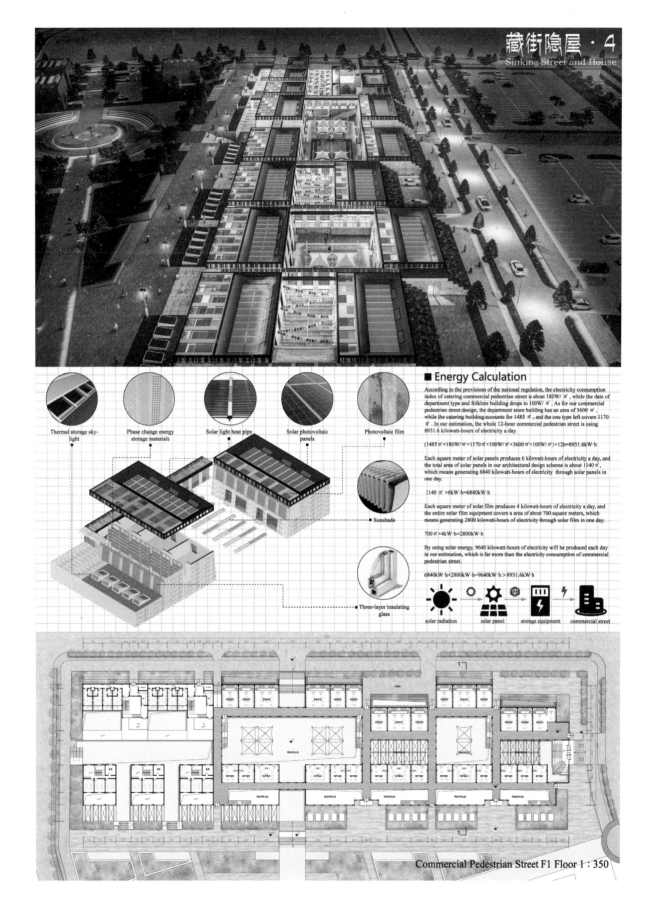

■ Energy Calculation

According to the provisions of the national regulation, the electricity consumption index of catering commercial pedestrian street is about 180W/㎡, while the date of department type and folklore building drops to 100W/㎡. As for our commercial pedestrian street design, the department store building has an area of 3600 ㎡, while the catering building accounts for 1485 ㎡, and the one type left covers 1170 ㎡. In our estimation, the whole 12-hour commercial pedestrian street is using 8931.6 kilowatt-hours of electricity a day.

$(1485㎡ \times 180W/㎡ + 1170㎡ \times 100W/㎡ + 3600㎡ \times 100W/㎡) \times 12h = 8931.6 kW \cdot h$

Each square meter of solar panels produces 6 kilowatt-hours of electricity a day, and the total area of solar panels in our architectural design scheme is about 1140 ㎡, which means generating 6840 kilowatt-hours of electricity through solar panels in one day.

$1140㎡ \times 6kW \cdot h = 6840kW \cdot h$

Each square meter of solar film produces 4 kilowatt-hours of electricity a day, and the entire solar film equipment covers a area of about 700 square meters, which means generating 2800 kilowatt-hours of electricity through solar film in one day.

$700㎡ \times 4kW \cdot h = 2800kW \cdot h$

By using solar energy, 9640 kilowatt-hours of electricity will be produced each day in our estimation, which is far more than the electricity consumption of commercial pedestrian street.

$6840kW \cdot h + 2800kW \cdot h = 9640kW \cdot h > 8931.6kW \cdot h$

Commercial Pedestrian Street F1 Floor 1:350

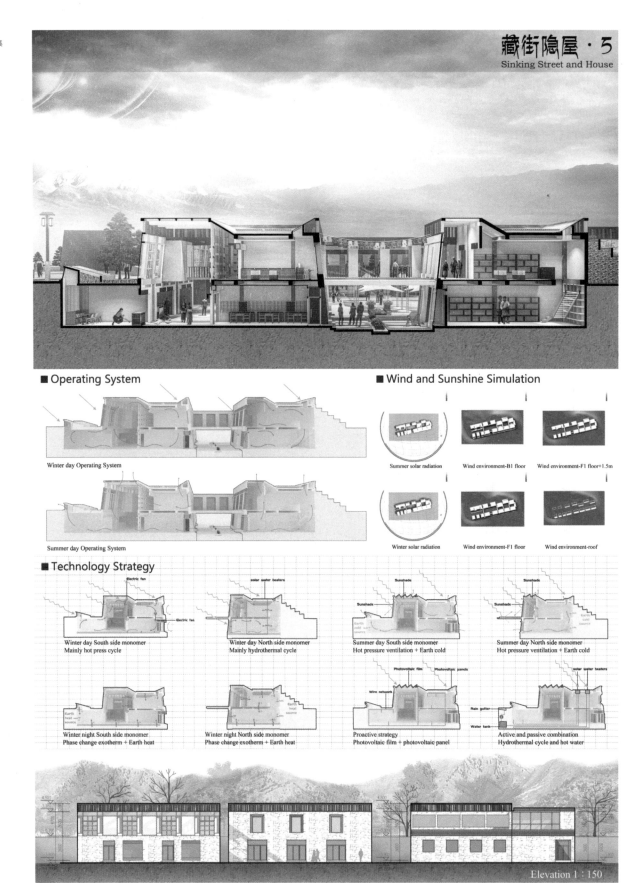

■ Construction Details

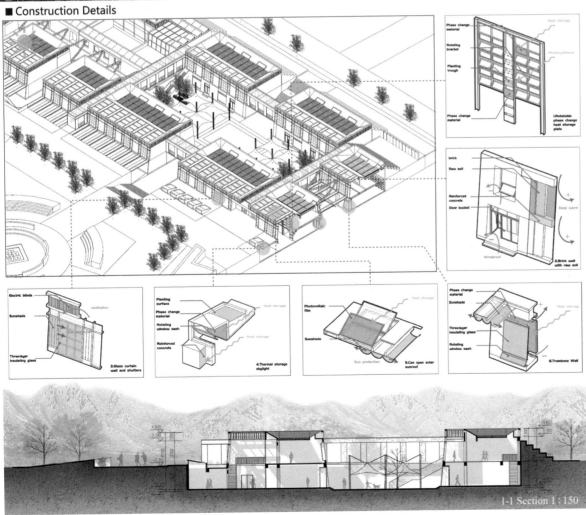

1-1 Section 1:150

归源谷 / BACK TO THE ORIGINAL

综合奖·一等奖
General Prize Awarded · Frist Prize

注 册 号：100373
Registration No：100373

项目名称：归源谷（概念设计项目）
Back to the Original (Concept Design Project)

作　　者：田琳琳、姜丹、王公睿、许烁钰
Authors：Tian Linlin, Jiang Dan, Wang Gongrui, Xu Shuoyu

参赛单位：厦门大学
Participating Unit：Xiamen University

指导教师：石峰
Instructor：Shi Feng

LOCATION

This design is located in Xiamen City, Fujian Province, China, in the center of the urban greenway river crossing base, which is organically integrated with the natural ecology.

随着城市化进程不断加快，环境问题越发突出，喧嚣和繁杂在带来极高碳排放量的同时也影响着都市人群的身心健康。在重压之下，人们对开放自然的乡村生活追求日益提升，城市与乡村的界限被逐渐抹除。为创造更适宜的生活环境，开拓新式城市生活，我们直接将乡村的自然人居体验带到城市中心，反述建筑与环境的关系，结合城市绿带，在城市建起峡谷"，将主动式与被动式节能技术有机融合，充分利用自然资源提升节能减排效率，为零碳生活创造新的可能。

With the accelerating urbanization process, environmental problems become more and more prominent. The noise and complexity not only bring high carbon emissions, but also affect the physical and mental health of urban people. Under the pressure, people are increasingly pursuing open and natural rural life, and the boundaries between cities and villages are gradually erased. In order to create a more suitable living environment and develop a new urban life, we directly bring the natural living experience in the countryside to the city center, reverse the relationship between architecture and environment, and build a "canyon" in the city in combination with the urban green belt, organically integrate active and passive energy-saving technologies, make full use of natural resources to enhance the efficiency of energy conservation and emission reduction, and create new possibilities for a zero-carbon life.

ANTIQUITY — SELF-SUFFICIENT NATURAL ECONOMY
MECHANICAL AGRICULTURE — THE FORMATION OF MODERN MACHINE SYSTEM AND THE WIDE APPLICATION OF AGRICULTURAL MACHINES
MODERN AGRICULTURE — UNIFICATION OF RURAL COMPLEX AND NEW URBANIZATION
COMMERCIALIZED AND SOCIALIZED AGRICULTURE — MODERN AGRICULTURE
AGRICULTURAL LIFE DEVELOPMENT
COMBINATION OF MODERN SCIENCE AND TECHNOLOGY WITH AGRICULTURAL PLANTING — WISDOM AGRICULTURE

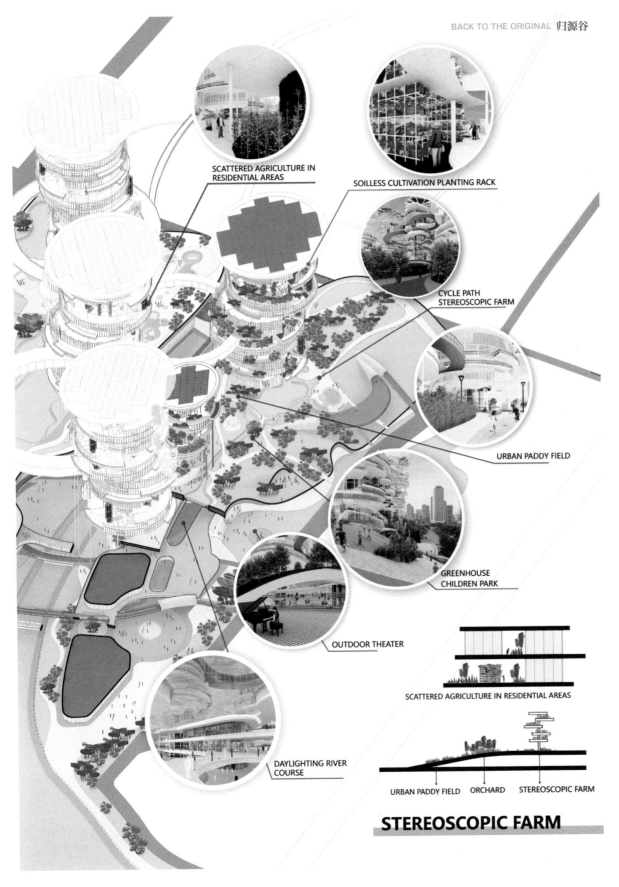

专家点评：

作品设计创新性高，形态表现突出。以乡村自然人居体验为主题塑造建筑与环境的关系，将建筑形态与低碳技术进行有机结合，对于太阳能主、被动技术利用以及其他技术的集成具有前瞻性。但总体方案氛围与归于本源的宁静仍有一定落差，形态的异化与地标感强于零碳建筑的具体效益或诉求。

The design is highly innovative, with an outstanding architectural image. The relationship between the building and the environment is shaped by the theme of the experience of rural and ecological habitat. The designer combined architectural form with low-carbon technologies, which is forward-looking in terms of the integration of active and passive solar systems and other technologies. However, the ambiance of the overall design still falls short of the required tranquility. Besides, because the design is too eye-catching and looks like a landmark, it will hide the specific benefits or needs of a zero-carbon building.

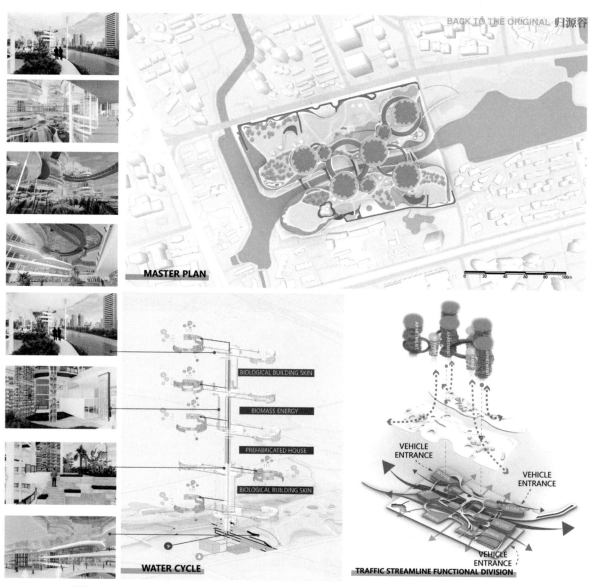

MASTER PLAN

WATER CYCLE

TRAFFIC STREAMLINE FUNCTIONAL DIVISION

BACK TO THE ORIGINAL 归源谷   2021 台达杯国际太阳能建筑设计竞赛获奖作品集

01 Rotating Roof Photovoltaic Panels
02 Alage-powered Dynamic Facades
03 Low Emissivity Glass
04 Ventilation and lighting Well
05 Daylighting Water
06 Overhead Ventilation
07 Air Vent
08 Cycle Path
09 PlantingFrame
10 Exhaust Ventilation

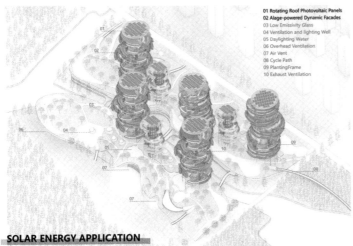
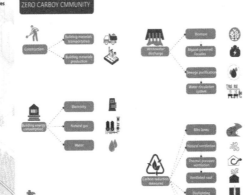

SOLAR ENERGY APPLICATION

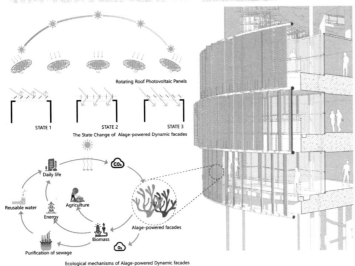

Rotating Roof Photovoltaic Panels

The State Change of Alage-powered Dynamic facades

Ecological mechanisms of Alage-powered Dynamic facades

		Area (㎡)	Carbon emission index	Life cycle carbon emissions (kg/㎡)
Hidden carbon emissions	Building materials production	107,047	40.36	671.46
	Building materials Construction			
Carbon emissions during operation	Residential carbon emission	10,325	0.15	1548.75
	Office	35572	0.4	14228.8
	Commerce	61150	0.35	21402.5
Total (kg/㎡)				37851.51
Carbon neutral	Photovoltaic panels	8,741	0.35	3059.28
	Algaed	6750	2.7	18225
	Greening			17000
Total (kg/㎡)				38284.28

VENTILATION AND LIGHTING   WIND SIMULATION

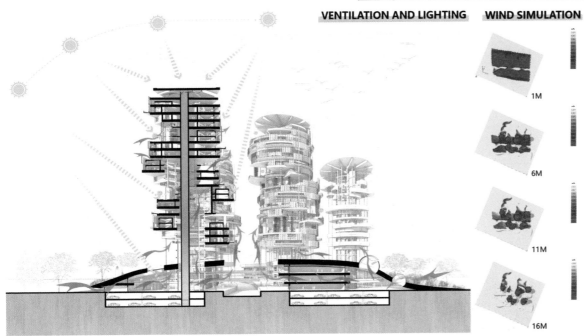

1M

6M

11M

16M

综合奖·二等奖
General Prize Awarded · Second Prize

注 册 号：100291
Registration No：100291

项目名称：风下旅人（实地建设项目）
　　　　　Step on the Wind
　　　　　(Field Construction Project)

作　　者：王远航、席 斌、蔡天舒、
　　　　　潘鹏宇
Authors：Wang Yuanhang, Xi Bin,
　　　　　Cai Tianshu, Pan Pengyu

参赛单位：南京工业大学
Participating Unit：Nanjing Tech University

指导教师：薛春霖、罗 靖
Instructors：Xue Chunlin, Luo Jing

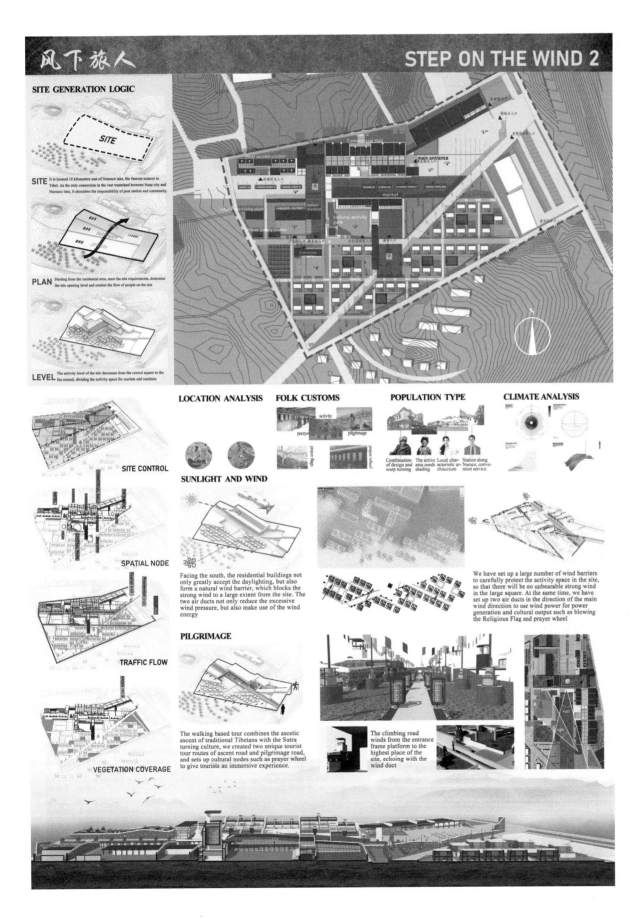

专家点评：

总体规划布局合理、形式感强，布局有特色、分区明确、功能流线组织合理。在建筑空间和建筑造型方面，地域特色浓郁又具有现代感。利用建筑设计与建筑构造的被动式节能，将传统建造方式与现代技术相结合，实现建筑采暖与蓄热。该方案应用了绿色、低碳、安全、健康技术，建筑单体设计功能合理，空间利用充分。作品具有可实施性，技术应用具有经济性和普适性，设计成果表达充分、完整，设计者具有较好的专业素质；居民区车行入口这条路考虑欠妥。

The overall planning layout is distinctive, with a strong architectural image, clear sub-areas, and a well-designed circulation. In terms of architectural space and form, the building has abundant regional features and a modern style. Through the architectural design and the passive architecture, the building combines traditional construction methods with modern technology to achieve building heating and heat storage. Green, low-carbon, safe, and healthy technologies have been applied, and the single building design is functionally sound, space-efficient, and implementable. The technology is both economical and universal. The architecture design is fully expressed and complete. However, the traffic routing of the residential vehicular entrance is inappropriate.

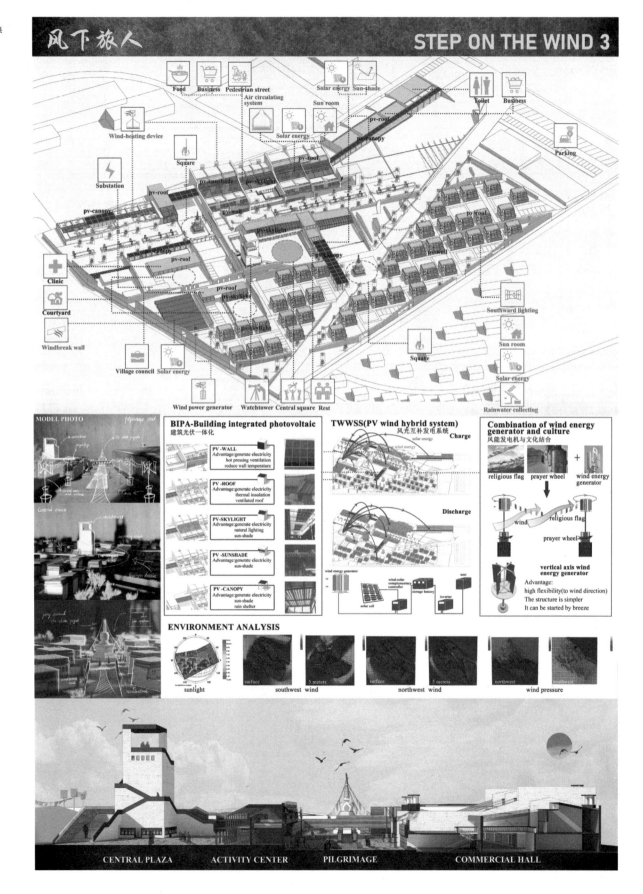

STEP ON THE WIND 4

风下旅人

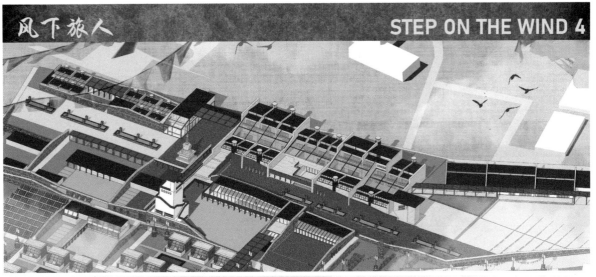

FIRST FLOOR PLAN OF THE HOTEL 1:300

SOUTH FACADE OF THE HOTEL 1:300

HOTEL SECTION 1-1 1:300

Roof costruction
AGA soil 阿嘎土 · gravel · rammer
Window contruction

1ST FLOOR PLAN 1:100　　2ND FLOOR PLAN 1:100

HOTEL ENERGY-SAVING TECH ANALYSIS

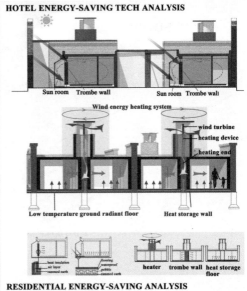

Sun room · Trombe wall

Wind energy heating system — wind turbine, heating device, heating end

Low temperature ground radiant floor · Heat storage wall

heat insulation / air layer / flooring / waterproof / pebble / rammed earth

heater · trombe wall · heat storage floor

RESIDENTIAL ENERGY-SAVING ANALYSIS

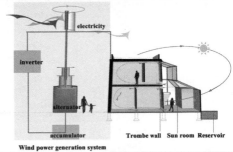

electricity · inverter · alternator · accumulator

Trombe wall · Sun room · Reservoir

Wind power generation system

SIDE ELEVATION 1:100　　FRONT ELEVATION 1:100　　SECTION 1-1 1:100

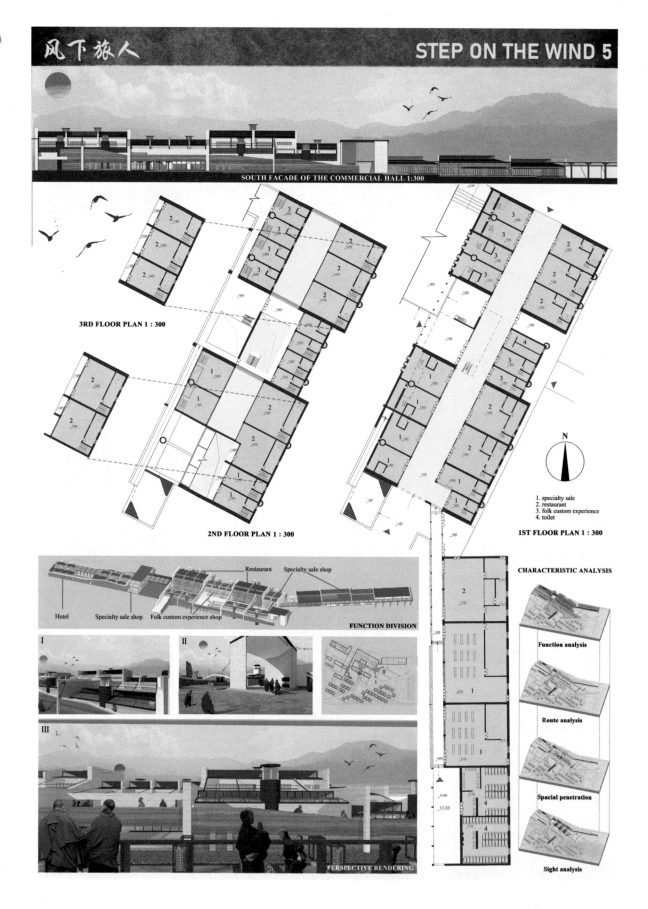

STEP ON THE WIND 6

NEW PHOTOVOLTAIC COMPONENT MATERIALS

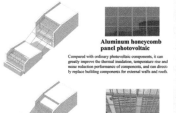

Aluminum honeycomb panel photovoltaic

Compared with ordinary photovoltaic components, it can greatly improve the thermal insulation, temperature rise and noise reduction performance of components, and can directly replace building components for external walls and roofs.

Vacuum glass photovoltaic

The heat transfer coefficient of vacuum glass photovoltaic components is lower than 1 W/m²·K, and the thermal insulation performance reaches level 10. Compared with insulating glass photovoltaic components, it can effectively reduce the energy consumption of buildings by more than 50%.

WIND-HEATING DEVICE ANALYSIS

- protection cover
- wind turbine
- heating end
- stirring rotor
- heating liquid
- thermal presentation crust
- climbing path
- room
- thermal presentation cavity

INDOOR ENVIRONMENT ANALYSIS

WINTER

SUMMER

DAY

NIGHT

VENTILATION

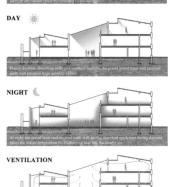

RAMMED EARTH MATERIAL

soil particles with different thickness

pebble | branches | earth | ramming
gravel | hey | salt | frame up

RAMMED EARTH MANUFACTURING

1. Build the 1st layer frame
2. Pour in the material
3. Compact material
4. Tamp a modular wall
5. Build the 2nd layer frame
6. Tamp the 2nd layer
7. Complete the construction
8. Apply wall paint

DIFFERENT FORM OF WINDOW-OPENING

DETAILS

Frost resistance of salt

- rammed earth
- adhesive
- mixed material
- plywood
- wall coating

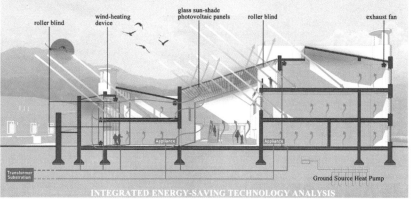

- roller blind
- wind-heating device
- glass sun-shade photovoltaic panels
- roller blind
- exhaust fan
- Transformer Substation
- Ground Source Heat Pump

INTEGRATED ENERGY-SAVING TECHNOLOGY ANALYSIS

SPECIALTY SALE | STALL | RESTAURANT

CLIMBING PATH

SECTION PERSPECTIVE RENDERING

综合奖·二等奖
General Prize Awarded · Second Prize

注 册 号：100479
Registration No：100479

项目名称：光的致意（实地建设项目）
Greeting of Light
(Field Construction Project)

作　者：李文强、张一鸣、杨敏敏、
谭景柏、张入川
Authors：Li Wenqiang, Zhang Yiming,
Yang Minmin, Tan Jingbai,
Zhang Ruchuan

参赛单位：西北工业大学
Participating Unit：Northwestern Polytechnical
University

指导教师：刘　煜
Instructor：Liu Yu

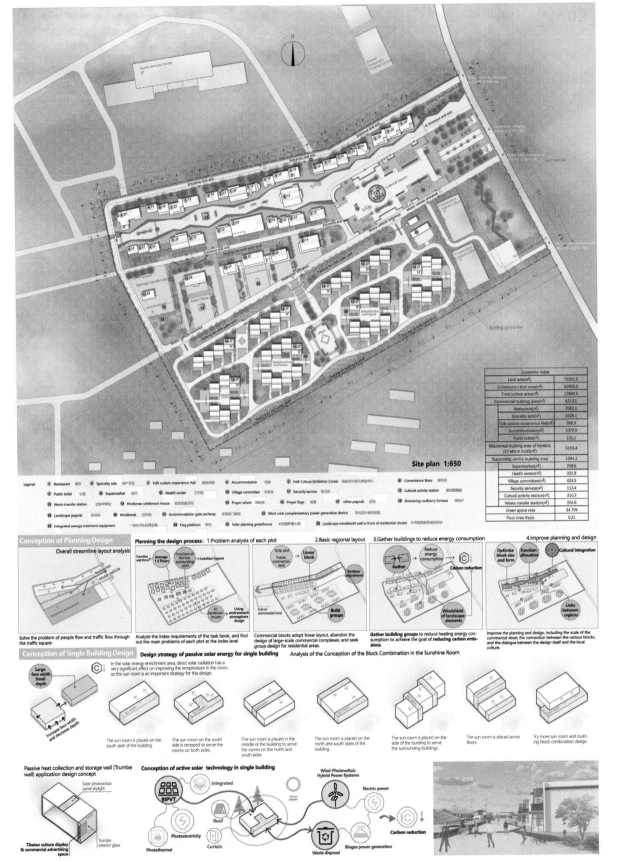

专家点评：

总体规划布局和功能流线组织合理，布局整体富有变化；在建筑空间和建筑造型方面，具有较好的地域特色；考虑了被动式阳光房设计，应用了BIPV系统、风光互补发电系统等现代节能低碳技术，实现了建筑采暖与蓄热；作品具有可实施性，技术应用具有经济性和普适性；居住部分组团设计既有逻辑性又丰富了建筑形象；设计成果表达充分、完整，设计者具有较好的专业素质；太阳能板需详细计算后增加用量。

The overall planning and layout are varied with well-designed circulation. In terms of building space and architectural image, it has some regional features. The design of the residence complexes is logical and enriches the architectural image. The passive solar house, BIPV system, solar-wind hybrid power system, and other active energy-saving and low-carbon technologies are applied to achieve building heating and heat storage. The work is implementable and the technology application is economical and universal. The architecture design is fully expressed and complete. The number of photovoltaic panels should increase after detailed calculations.

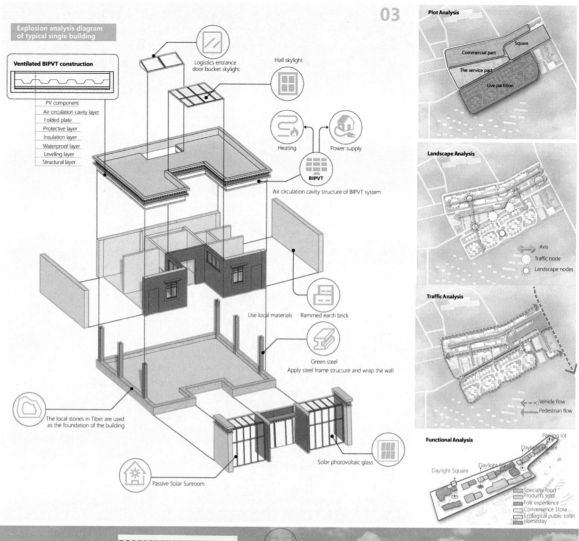

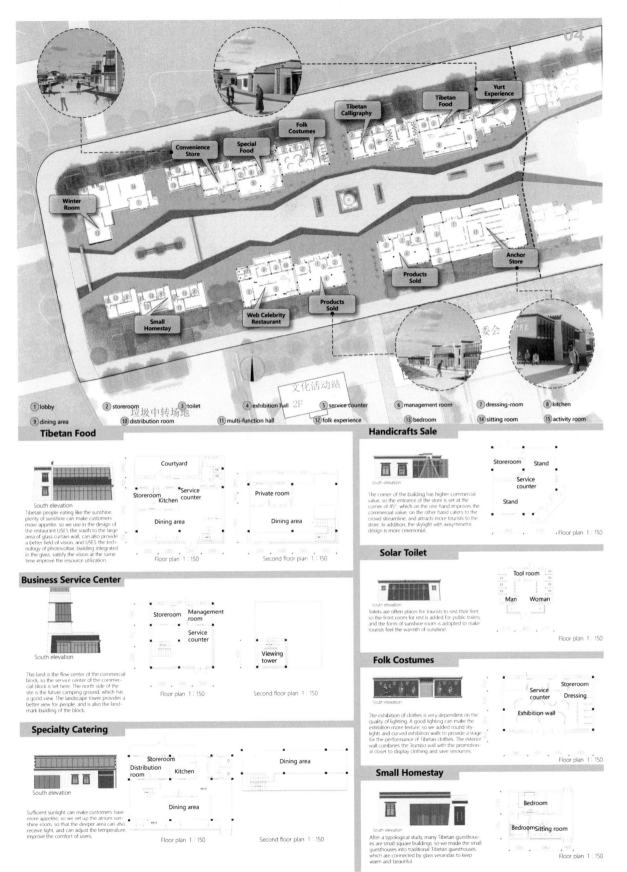

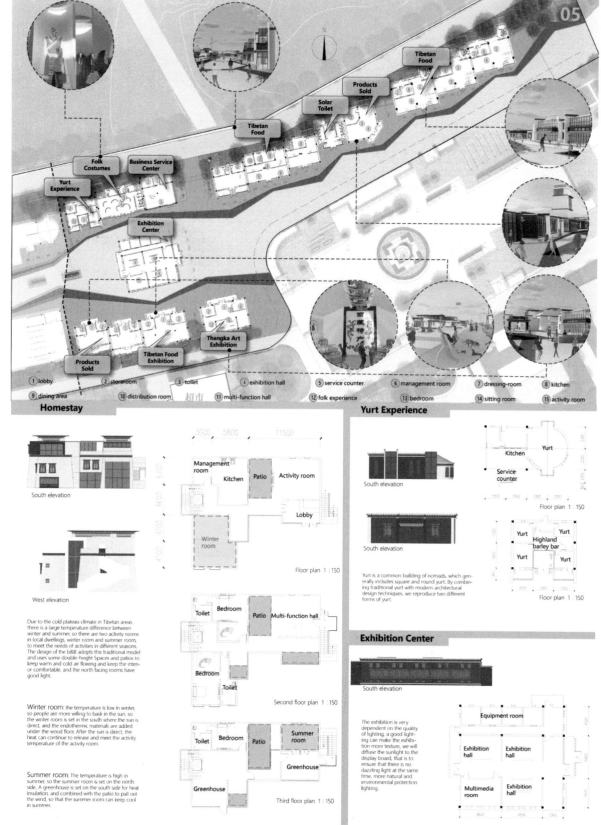

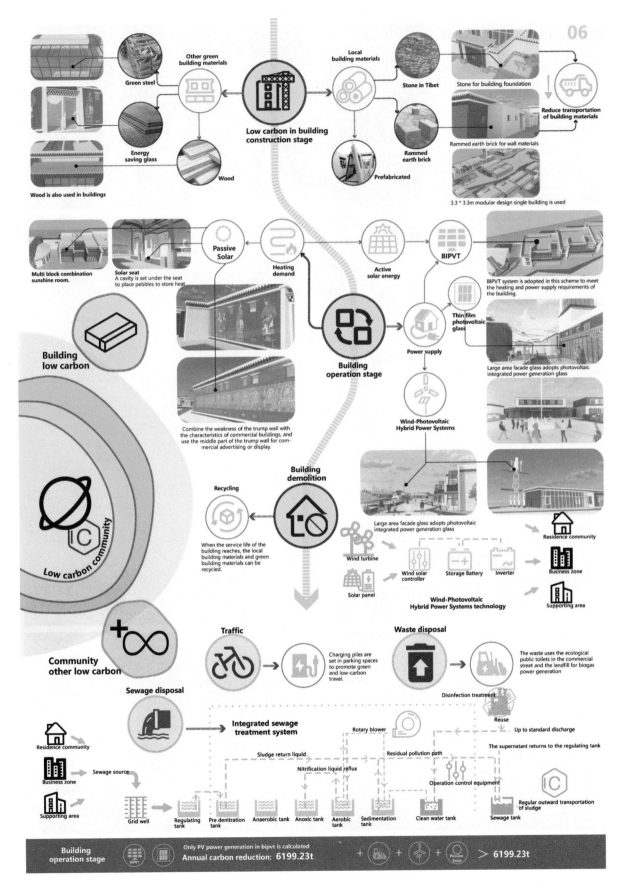

综合奖·二等奖
General Prize Awarded · Second Prize

注 册 号：100332
Registration No：100332

项目名称：与海共舞（概念设计项目）
Dancing with Sea
(Concept Design Project)

作　　者：王 岳、边新元
Authors：Wang Yue, Bian Xinyuan

参赛单位：华北理工大学
Participating Unit：North China University of Science and Technology

指导教师：孙嘉男
Instructor：Sun Jianan

Dancing with Sea

A carbon neutral floating community experiment for low-income people living by the sea

Location: Jakarta
Area: 3000
Type: Individual Work
Planning years : 2021 — 2044

Climate warming is a prominent climate problem in the 21st century. More than 90% of the current climate warming is caused by human activities. In particular, urban life consumes a large amount of fossil energy, emits about 75% of the global greenhouse gases, and produces 80% of the global pollution. Frequent climate disasters threaten the normal production and life of urban residents.

Located in Jakarta, the capital and largest city of Indonesia, coastal residents are suffering from floods and rising sea levels caused by global warming and climate change, and are falling at a rate of about 25 centimeters per year. In addition to climate change, overexploitation of groundwater is the main reason for the accelerated subsidence. Their meager income makes it impossible for them to escape or change. The project conducted an experiment on how to build a community living at sea from a positive coping perspective.

Background

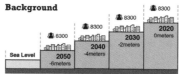

With a population of more than 10 million as of 2021, Jakarta, Indonesia's capital, is the melting point of the country's ethnic cultures and draws migrants from all over the country in search of work, yet its habitable northern plains have been eroding by the sea since 1977, declining at a rate of about 25 centimeters per year.

Environment

Due to the short flow of most of Jakarta's rivers, a large amount of meteoric water, which is difficult to store, is quickly lost to the ocean. At the same time, environmental protection measures and awareness are not in place, and many surface water sources are polluted. As a result, Jakarta's surface water resources are insufficient, so it has to turn its attention to groundwater, which aggravates land subsidence and causes the displacement of tens of thousands of people.

Conception

Development

1. Combine floating module
2. Install bamboo frame
3. Determine entrance
4. Insert traffic core
5. Set aside green space

Energy System & Interior Design

1. One Module
2. Two Modules
3. Half Module
4. One and A Half Modules

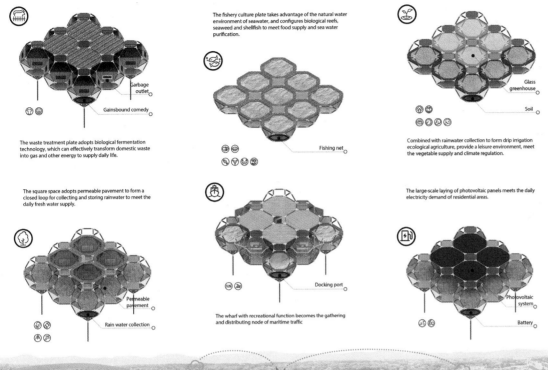

专家点评：

项目设计思路巧妙，针对可持续未来的思考，比较全面。需要完善的内容主要集中在两个方面：第一，涉及聚落层级的内容，其复杂度似乎本方案并没有考虑到，仅仅是简化的模型示意系统；第二，以竹材为主的结构体系，其在海边的耐久性以及竹材的承载力是否能适配多个系统所需要的结构荷载，存在较大的疑问。

The design of the project is clever and comprehensive in terms of a sustainable future. There are two main areas for improvement. First, the complexities of the project have not been adequately considered, as this work is only a simplified model of the settlement. Second, the durability of the bamboo-based structural system at the seaside and whether the bearing capacity of the bamboo can be adapted to the structural loads required by multiple systems are in doubt.

综合奖·二等奖
General Prize Awarded · Second Prize

注 册 号：100551
Registration No：100551

项目名称：能源控制实验站
（概念设计项目）
Energy Control Experimental Station (Concept Design Project)

作　　者：黄佳怡、倪子璇
Authors：Huang Jiayi, Ni Zixuan

参赛单位：浙江工业大学
Participating Unit：Zhejiang University of Technology

指导教师：仲利强
Instructor：Zhong Liqiang

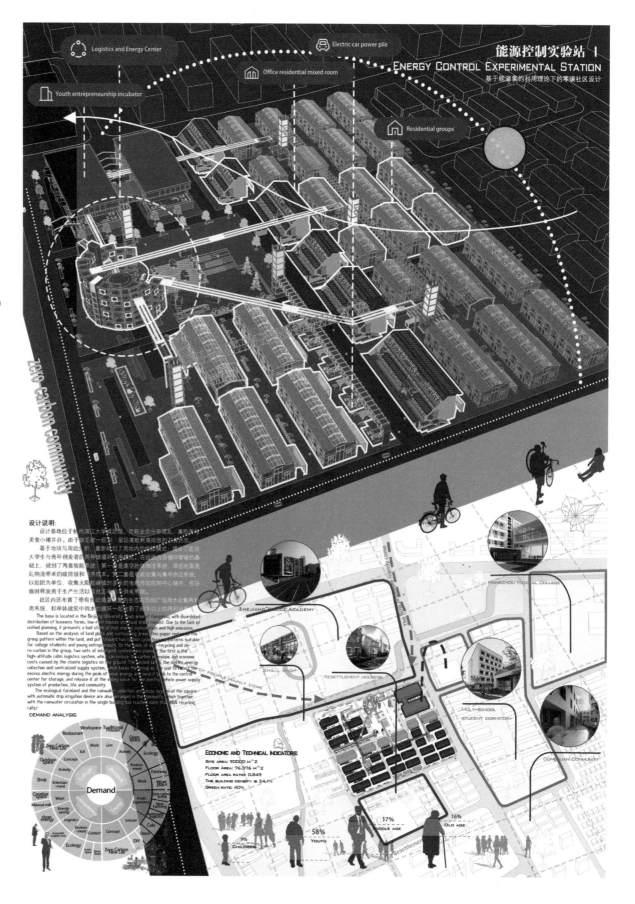

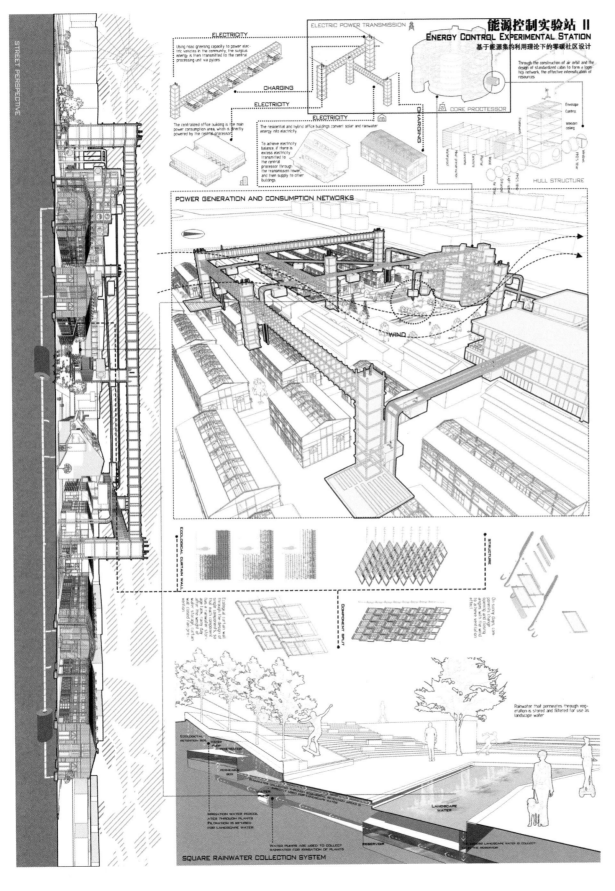

专家点评：

设计者构思了一个核心处理器，周边分布着零碳住宅、零碳办公等不同功能模块。核心处理器设计了广场雨水系统，用来对雨水进行收集和景观灌溉。住宅和办公模块的功能研究与使用模式研究及展示，均比较薄弱。能源系统效率应该不如理想状态效率高，而花费了大量心思设计的核心处理器的作用和价值同样没有充分展示。总体感觉是一个细节丰富的设计方案，但整体概念需要进一步提高和梳理设计。

The designer has set up a core processor at the community center, surrounded by different functional modules such as the zero-carbon residential and zero-carbon office. A square rainwater system is designed next to the core processor for rainwater collection and landscape irrigation. The research of the functions of the residential and office areas as well as the research and display of their modes are generally innovative. However, the role and value of the core processor are also not fully demonstrated, so the overall concept needs to be further improved.

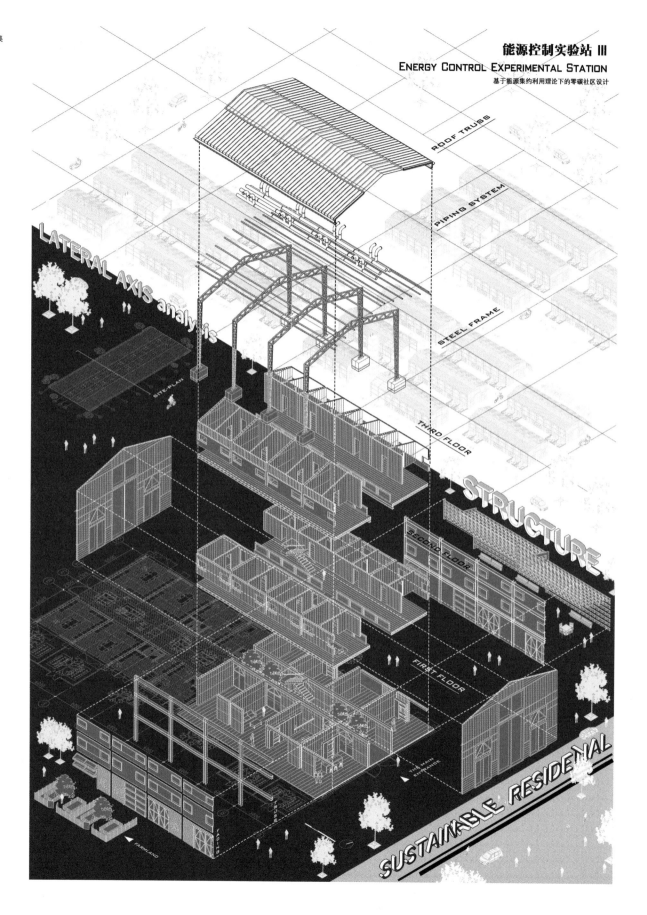

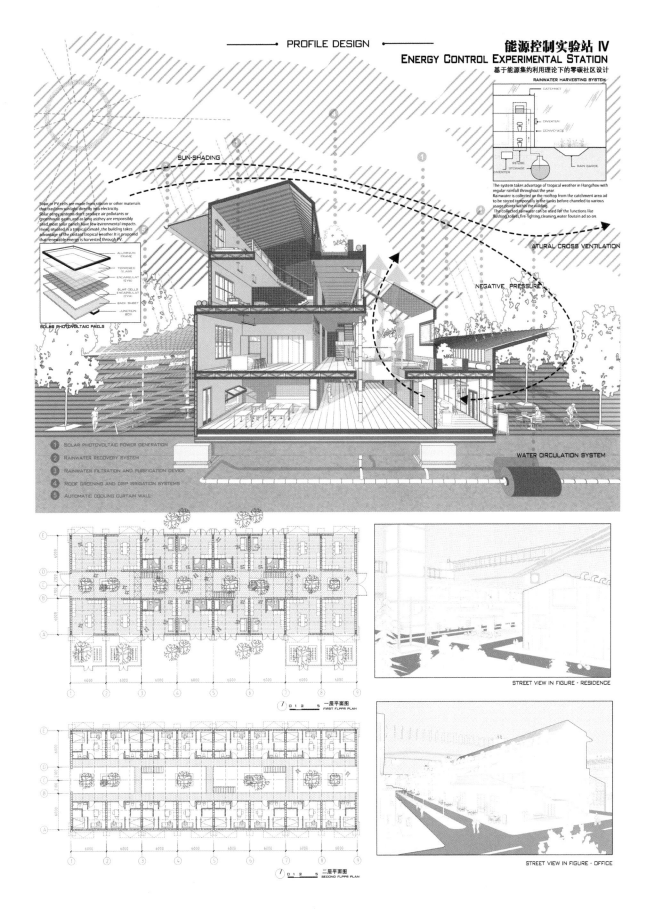

综合奖·三等奖
General Prize Awarded · Third Prize

注　册　号：100320
Registration No：100320

项目名称：光·窖（实地建设项目）
　　　　　The Cellar of Light
　　　　　(Field Construction Project)

作　　者：丁阳权、冯磊磊、龙泓昊、
　　　　　冉大林
Authors：Ding Yangquan, Feng Leilei,
　　　　　Long Honghao, Ran Dalin

参赛单位：重庆大学
Participating Unit：Chongqing University

指导教师：周铁军、张海滨
Instructors：Zhou Tiejun, Zhang Haibin

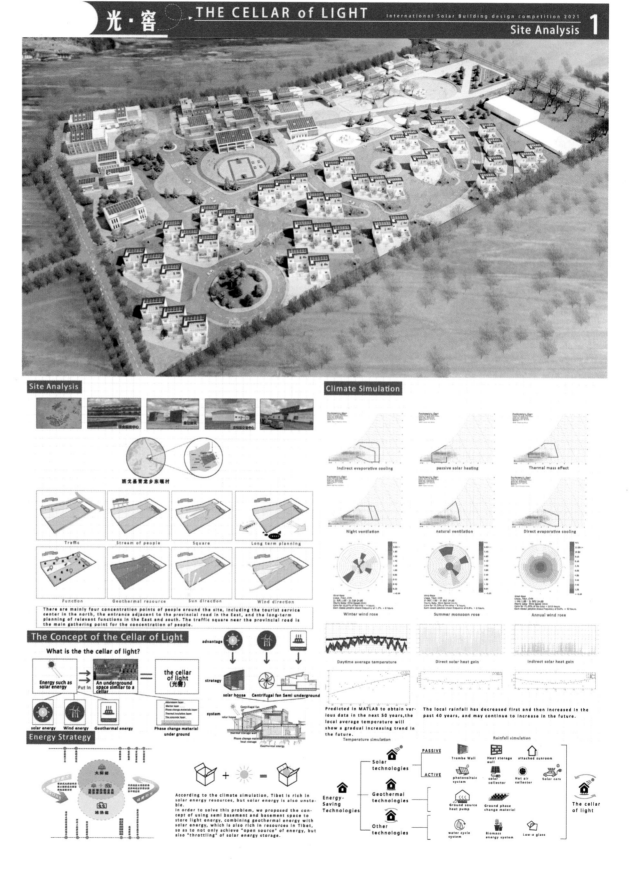

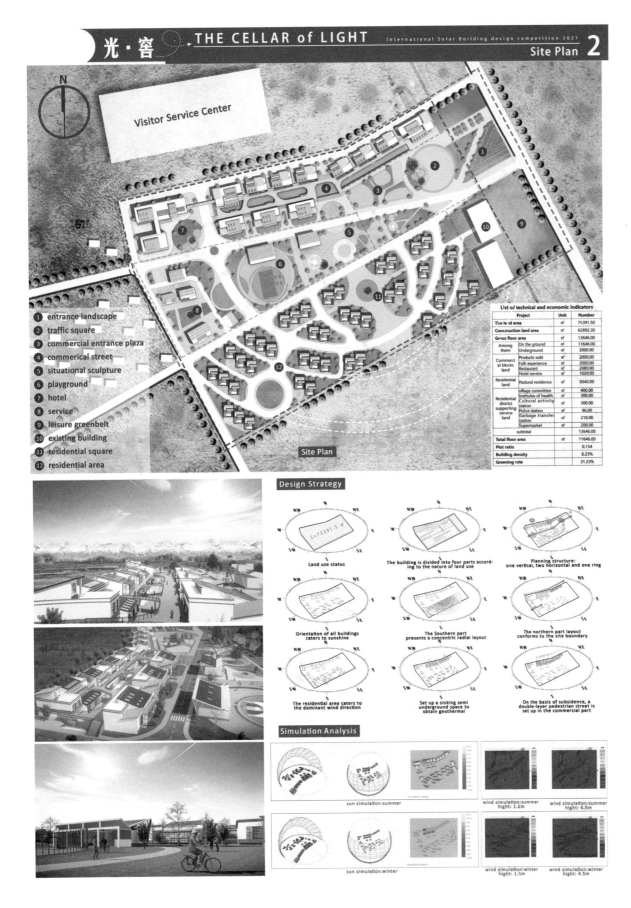

专家点评：

作品规划分区明确，住宅组团有趣味性，商业区布局规整，流线合理。建筑竖向空间布局丰富，对太阳能主、被动技术的应用较为合理，采用半地下式建筑，有效减少了建筑的体型系数，降低了建筑能耗，在被动技术应用上同时利用地下蓄能与被动集热的结合，为室内提供热量。

The work is clearly planned and zoned, with an interesting residential complex design, a neat layout of commercial areas, and a reasonable circulation. The vertical spatial layout of the building is well-designed. The use of active and passive solar energy is more reasonable. The semi-underground construction effectively reduces the shape coefficient of the building and lowers the building's energy consumption. In addition to the passive technologies, the combination of underground energy storage and passive heat collection is also applied to provide heat indoors.

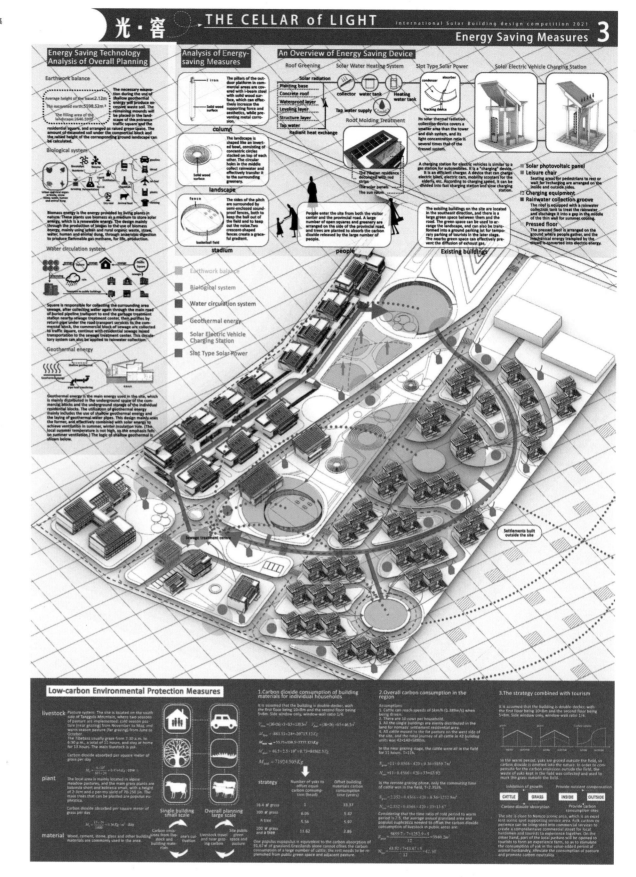

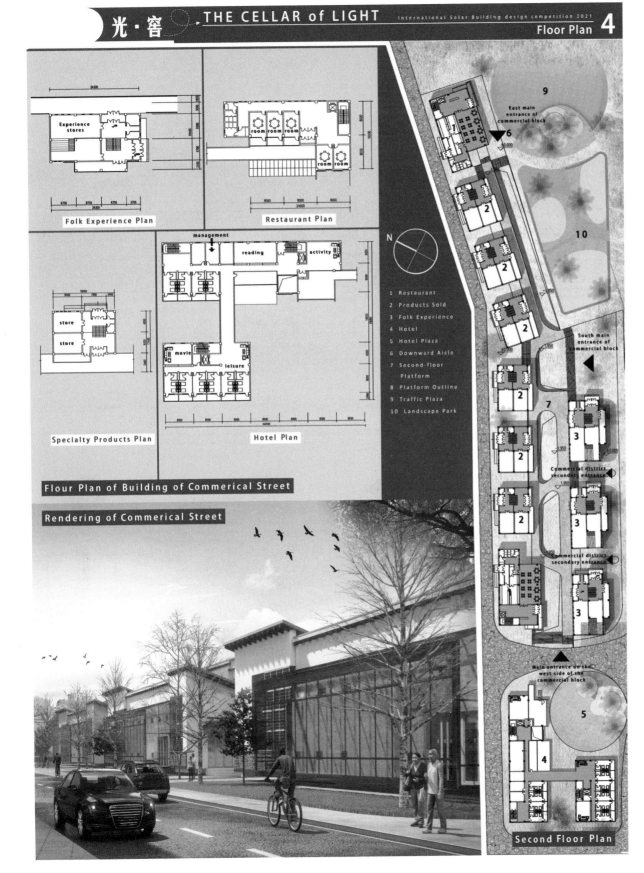

光·窘 — THE CELLAR of LIGHT
Technology Application 5

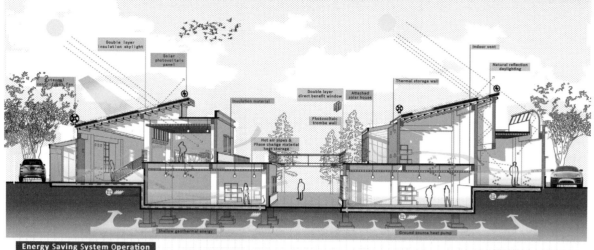

Energy Saving System Operation

Daytime operation in winter

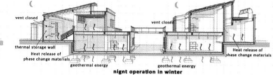
night operation in winter

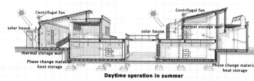
Daytime operation in summer

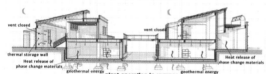
night operation in summer

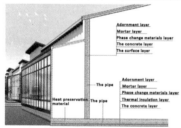
Solar house Structural map

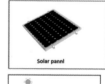
Solar panel

Skylight daylighting

The solar house with sloping roof
The solar house with sloping roofs conducive to receiving more sunlight

Photovoltaic panels Structural map

Photovoltaic panels trumbo wall

Thermal storage wall
The thermal storage wall serves two purposes. 1. Absorb solar energy directly and releases heat at night. 2. Heat the passing cold air

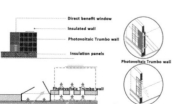

Hot air heat storage system (热风蓄热系统)
When the indoor temperature is detected, the air supply port is closed, and the heat energy in the air is stored in the phase change material under the floor through the pipe

Roof daylighting
Direct light is obtained through mirrors in the room

Ground source heat pump

THE CELLAR of LIGHT
Energy Saving Simulation

Energy Saving Calculation

Energy saving model

Carbon footprint model

On the one hand, building energy consumption index can be measured by building annual power consumption. On the other hand, building energy comes from two aspects: solar energy and geothermal energy. Through calculation, the community energy acquisition exceeds the consumption, which indicates that the community has achieved the energy-saving goal.

In order to simulate the carbon footprint of community families, a family of four is set. Families use 120 square meters of housing. They use electricity from photovoltaic power generation throughout the year, and use walking, community bus and electric vehicle for travel. In addition, they eat organic food produced and sold by the community. As a result, the household's carbon consumption will be close to the global average and much lower than the Chinese average.

Building Simulation

R1-1441	R2-1447
R1-1444	R2-1450
	R2-1453

Shop 1F model — Shop 2F model

TABLE 4 — Settings

	Month	Date
Start date of heating season	11	15
End date of heating season	3	15
Start date of air conditioning season	6	1
Air conditioning end date	8	30

In DeST software, the building energy-saving simulation calculation is carried out. Taking the selling store in the commercial street as an example, the upper and lower models of the store are established respectively, and the rooms are numbered. Finally, the calculation results of each room and the whole building can be output.

TABLE 1 — Power consumption

建筑类型 Building type	数量 Amount	建筑用电指标 (w/m²) Power Consumption Index	建筑年耗电量 (kw·h) Power Consumption
住宅 Residence	42		331128
商店 Shop	6	40	404400
体验店 Experience store	3	40	312000
民宿 Inn	1	50	160600
餐饮 Restaurant	2	30	157680
配套服务建筑 Supporting service building	6	30	92163
总计 Total	60		1457971

TABLE 2 — Solar panel gain

建筑类型 Building type	数量 Amount	发电面积 (m²) Electricity Generating Area	年光伏发电收益 (kw·h) Solar Panel Gain
住宅 Residence	42	701.4	226800
商店 Shop	6	422.4	136739
体验店 Experience store	3	426.6	137982
民宿 Inn	1	362.7	117196
餐饮 Restaurant	2	295.2	95372
配套服务建筑 Supporting service building	6	853.6	275778
总计 Total	60	3062.1	989767

TABLE 3 — Geothermal gain

建筑类型 Building type	地热得热面积 (m²) Geothermal Area	节约燃煤量 (t) Saving Coal consumption	折算省电量(kw·h) Converted power saving	系统运行耗电量(kw·h) System Operation Consumption	年净省电量 (kw·h) Net Energy Saving
住宅 Residence	1512	52.7	428520	51734	376786
商业街 commercial street	1305	45.5	369853	44651	325202
总计 Total	2817	98.2	798373	96385	701988

Table 1 shows the annual power consumption statistics of all buildings, with a total of 1457971kw·H.
Table 2 shows the annual heat gain statistics of photovoltaic power generation in the community, with a total of 989767kw·H.
Table 3 shows the statistics of energy saving converted by community access to geothermal energy, with a total of 701988kw·H.
It can be seen that the community can still have a surplus of 233784kw·h per year after deducting the annual power consumption of buildings through solar and geothermal energy compensation of 1457971kw·H.

TABLE 5 — Building load statistics

项目名称 Project name	总建筑空调面积 Total building air conditioning area	全年最大热负荷 Annual maximum heat load	全年最大冷负荷 Annual maximum cooling load	全年最大加湿量 Annual maximum humidification capacity	全年累计热负荷 Annual cumulative heat load	全年累计冷负荷 Annual cumulative cooling load	全年累计加湿量 Annual cumulative humidification	全年最大热负荷指标 Annual maximum heat load index	全年最大冷负荷指标 Annual maximum cooling load index	全年累计热负荷指标 Annual cumulative heat load index	全年累计冷负荷指标 Annual cumulative cooling load index	加湿季节累计 Annual cumulative humidification index	空调采暖季节 Air conditioning heating season	空调供冷季节 Air conditioning seasonal cooling load index
单位 Unit	m²	kW	kW	kg/s	kWh	kWh	kg	W/m²	W/m²	kWh/m²	kWh/m²	kg/m²	W/m²	W/m²
数值 Value	412.4	66.6846018	17.46676706	72.98227912	64403.62144	11737.87033	49743.81934	106.6811641	42.291541	174.3400365	107.4035957	20.295501	120.2300489	25.36476049	5.09270611

TABLE 6 — Room heat load statistics

房间名称 Room name	房间功能 Room function	房间面积 Room area	全年最大时刻 Maximum time of the year	全年最大热负荷(W) Annual maximum	冷房负荷(W) room annual	加湿量(kg/h) Humidification	新风量(m³/h)(负荷) Indoor sensible	室内显热负荷(kw) Sensible heat load	室内潜热负荷(kw) Indoor calorific	空调累计加湿量(m³) Accumulated m³	热负荷指标(W/m²) (W/m²)	层数 Number of layers
R1-1441		123.6	6247	129.4876409	16.00467242	17.51668167	7.621417522	8.383235505	0	25.87097761	15.27212374	1
R1-1444	一般商店类	61.8	6367	166.4663463	10.2876202	9.890398028	6.049653053	4.237967014	0	73.63940194	43.47072133	1
R2-1447	一般商店类	72	79	147.0141812	10.58502105	5.976381669	6.057012558	4.528008461	0	86.51964539	51.07180956	1
R2-1450	一般商店类	84	6247	104.9421464	6.815140301	11.90452062	3.117782593	5.687637655	0	18.57485855	14.07467699	1
R2-1453	一般商店类 (空共区)	72	55	186.7415022	13.64530816	5.607281685	8.608300209	4.837078631	0	126.9245824	74.32572749	1

TABLE 7 — Room cooling load statistics

房间名称 Room name	房间功能 Room function	房间面积 Room area	全年最大时刻 Maximum time of the year	全年最大冷负荷(W) Annual maximum	冷房负荷(W) room annual	除湿量(kW) Dehumidification load	新风量(m³/h)(负荷) Indoor sensible	室内显热负荷(kw) Sensible heat load	室内潜热负荷(kw) Indoor calorific	空调累计除湿量(m³) Accumulated m³	冷负荷指标(W/m²) (W/m²)	层数 Number of layers
R1-1441	一般商店类	123.6	7219	86.79389545	10.72648453		-13.22659486	2.599870205	9.462519646	14.070443	11.04430376	1
R1-1444	一般商店类	61.8	4002	15.20138589	0.939445648		-1.122497679	0.183052078	2.767750025	0.280146914	0.312517201	1
R2-1447	一般商店类	72	4003	35.85685628	2.581693652		-2.952588558	0.370809809	4.714963436	3.489436518	2.73896116	1
R2-1450	一般商店类	84	7719	79.65544056	6.891057007		-8.457959175	1.766062089	6.430839585	9.152865646	7.384360051	1
R2-1453	一般商店类 (空共区)	72	5034	10.32191182	0.743177651	-0.743177651	-0.631426632	0.937804264	3.543436766	0.0300043941	0.102519652	1

| 胡杨 Populus euphratica | 云杉 Spruce | 高寒草甸 Alpine meadow | 西藏落叶松 Tibetan larch | 大果圆柏 Juniperus tibetica | 圆柏灌丛 Juniper shrub |

South elevation of catering service 1:200 — South elevation of Specialty sales 1:200 — Folk Experience Hall 1:200

1-1 Section of commercial street 1:300

综合奖·三等奖
General Prize Awarded · Third Prize

注 册 号：100376
Registration No：100376

项目名称：格桑社区（实地建设项目）
Gesang Community
(Field Construction Project)

作　者：杨新哲、姚其郁
Authors：Yang Xinzhe，Yao Qiyu

参赛单位：河南工业大学
Participating Unit：Henan University of Technology

指导教师：张　华
Instructor：Zhang Hua

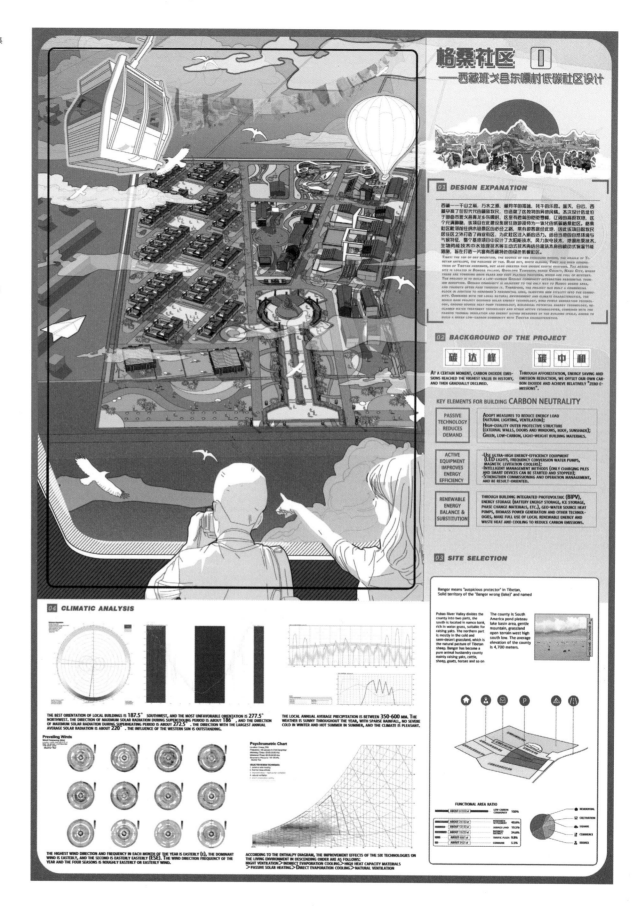

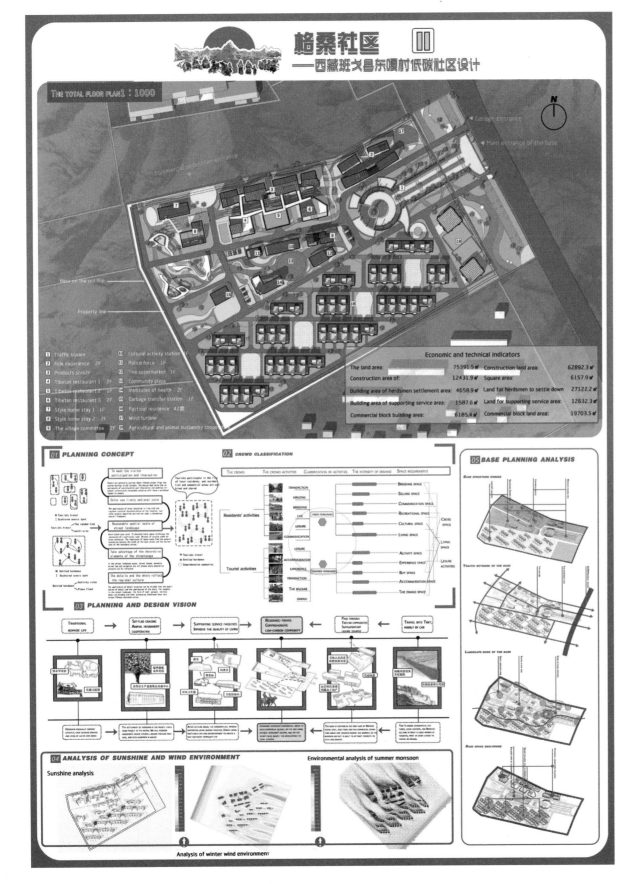

专家点评：

作品表现形式新颖，建筑组群规划合理，平面功能布局和朝向适宜高原气候环境，在被动技术应用上采用高性能围护结构，并注重被动集热与主动太阳能系统结合，主动太阳能系统选择合理、适宜。采用的部分生态节能技术缺乏地域适应性。

The work is creative in its expression. The planning of the building complexes is reasonable, with the functional layout and orientation suitable for the plateau climate. As for the passive technology, a high-performance envelope enclosure is applied. Besides, the designer focused on the combination of passive heat collection and active solar energy systems, with a reasonable and appropriate selection of active solar energy systems. However, some of the ecological energy-saving technologies lack geographical adaptability.

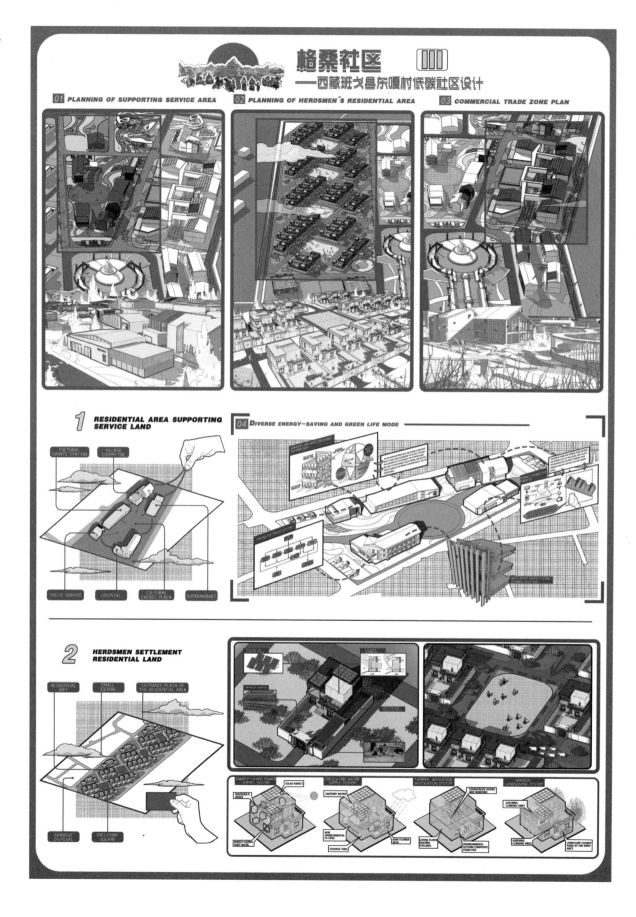

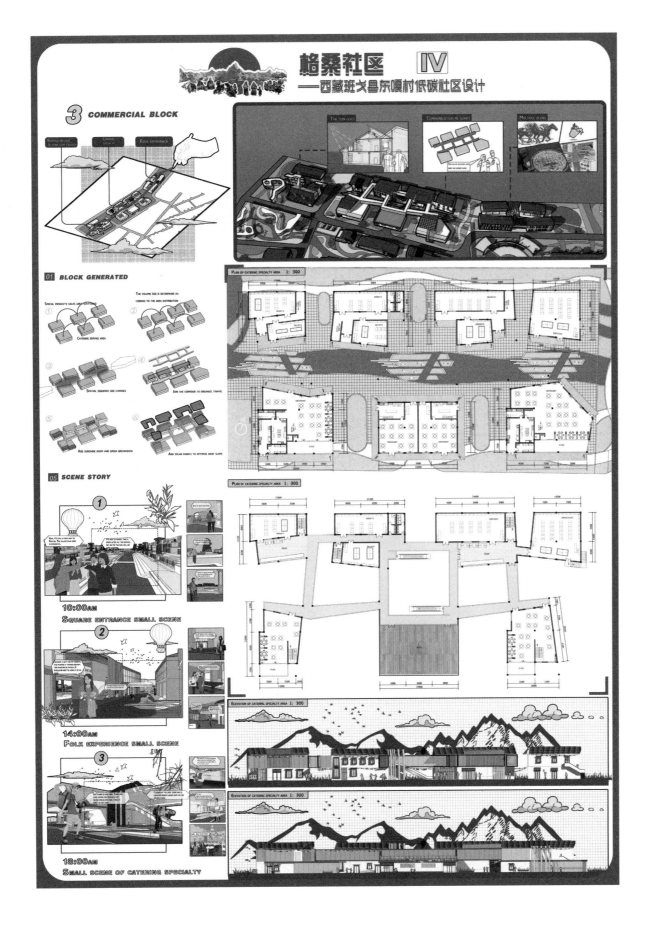

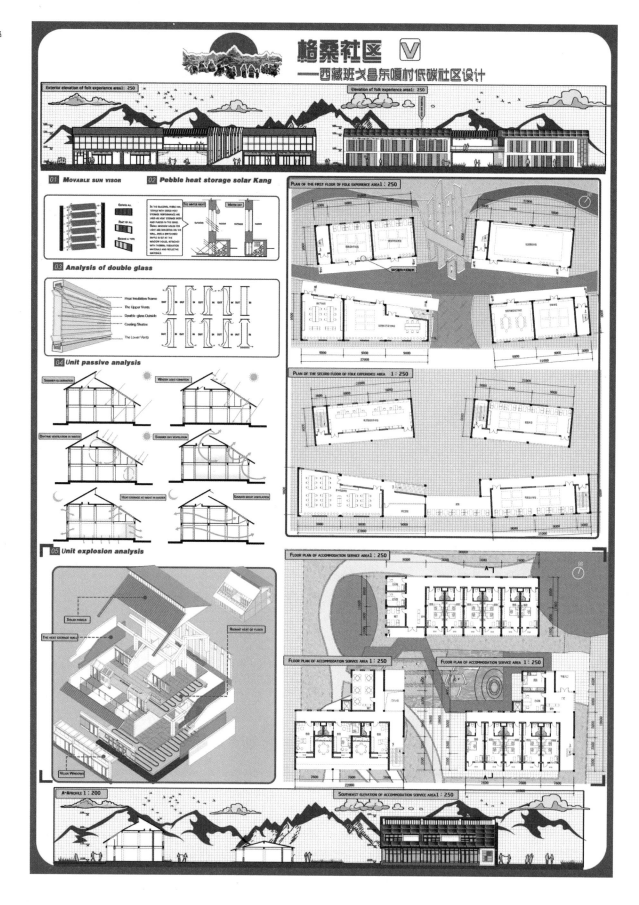

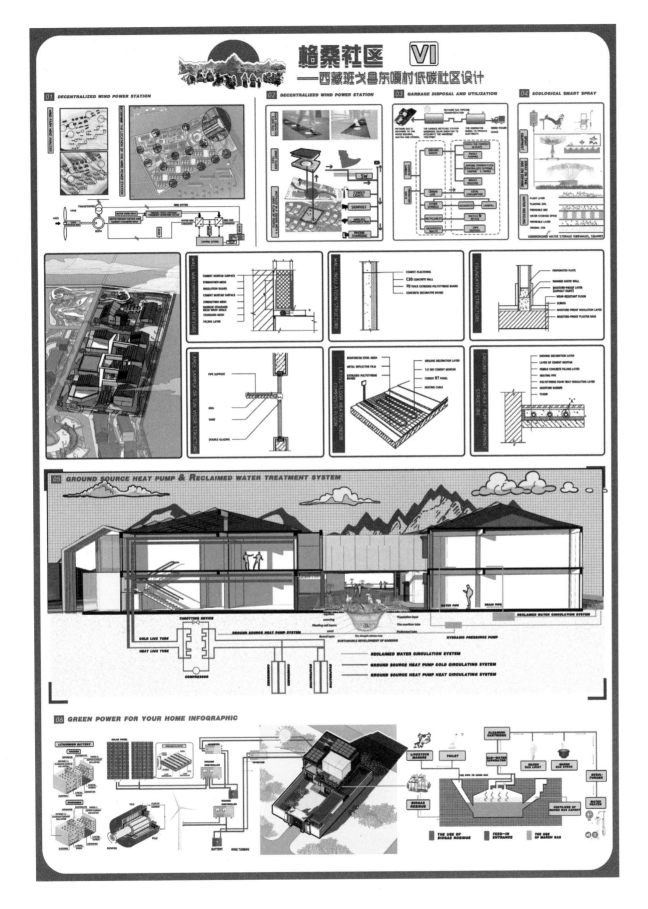

综合奖·三等奖
General Prize Awarded · Third Prize

注册号：100424
Registration No：100424

项目名称：故土·亲尘（实地建设项目）
The Gift of the Land
(Field Construction Project)

作　者：严凡辉、高　兴、唐　超、邓　棋
Authors：Yan Fanhui, Gao Xing, Tang Chao, Deng Qi

参赛单位：河南工业大学
Participating Unit：Henan University of Technology

指导教师：刘　强
Instructor：Liu Qiang

THE GIFT OF THE LAND
故土·亲尘 01

该方案以西藏地区传统的碉楼和土石砌筑手法为切入点，本着遵从本土性、地域性和文化性的原则，将材料设计融入建筑设计之中。建筑以现代语言转换传统原与碉楼，吸收其建造逻辑，并体现出墙厚、收分、立面小窗、体量敦实等藏区特点。夯土材料的使用使建筑与环境有机地融为一体，也为其生命周期的回收处理增加可行性。绿建方面，亲尘关注太阳能、被动式技术，设计手法以吸能吸热和围护人的行为为主，并以"烟囱效应"为母题设计建筑形象，将技术手段和视觉形象融合，以取得更为和谐的整体效果。

The plan takes the traditional Tibetan towers and earth-stone masonry techniques as the starting point, and integrates the material design into the architectural design in accordance with the principles of locality, regionality and culture. The building uses modern language to translate traditional houses and towers, absorbs its construction logic, and reflects the characteristics of thick walls, divisions, small windows on the facade, and solid and heavy volume. The use of rammed earth materials integrates the building and the environment organically, and also increases the feasibility of recycling throughout the life cycle. Green construction focuses on passive solar energy technology. The design techniques focus on absorbing and retaining the sun's light and heat. The building usage is designed with the "chimney effect" as the motif, and the technical means and visual image are integrated to achieve a more harmonious overall effect.

BASIC DATA ANALYSIS OF CLIMATE
BEST ORIENTATION

TEMPERATURE AND RADIATION

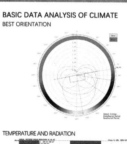

ANNUAL DISTRIBUTION ANALYSIS

| AVE TEM | MAX TEM | MIN TEM |
| RELATIVE HUMIDITY | DIRECT SOLAR RADIATION | AMBIENT DIFFUSE |

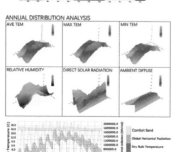

| WIND SPEED | VISIBILITY | PRECIPITATIO |

THE THREE REALMS

Desire Realm	The realm with appetites and carnal desires
Form Realm	The gradually spiritualized realm
Formless Realm	The highest pure realm

| Formless Realm | Form Realm | Desire Realm |
| accommodation | culture | restaurant |

Combined with the concept of "three realms" in Buddhism, the commercial section is divided into three sections on the north-south axis, corresponding to a group of functions. The terrain gradually rises from east to west, towards the western paradise.

REINCARNATION

reincarnation are closely related to the culture of the Tibetan people.

The image of mandala is selected, abstracted and translated to make it an important clue.

PLANNING IDEAS

① Two "spiritual axes" are inserted into the site to establish the basic order.
② Divide the axis into three parts according to the "three realms".
③ Place three "Gates" along the path.
④ Identify 5 nodes and set them up as mandala square.
⑤ Delineate the location of residential areas and supporting facilities.
⑥ Set up Prayer flags and shakans in the central square.

ANNUAL CLIMATE COMFORT

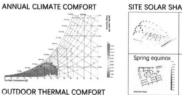

OUTDOOR THERMAL COMFORT
sun and wind

sun and no wind

no sun and wind

no sun and no wind

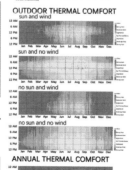

ANNUAL THERMAL COMFORT

SITE SOLAR SHADOW

| Spring equinox | Summer solstice |
| | Winter solstice |

TECHNOLOGY STRATEGY

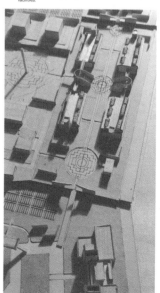

THE GIFT OF THE LAND

故土·斋尘 02

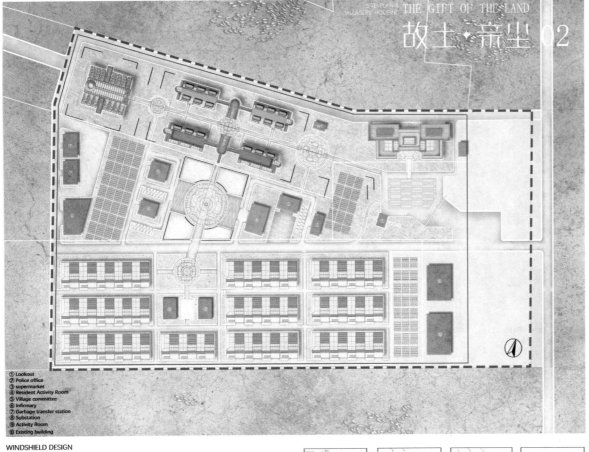

① Lookout
② Police office
③ Supermarket
④ Resident Activity Room
⑤ Village committee
⑥ Infirmary
⑦ Garbage transfer station
⑧ Substation
⑨ Activity Room
⑩ Existing building

WINDSHIELD DESIGN

The residential area adopts a semi-underground design, using insulation materials to keep the frozen soil isolated. The building is composed of three to five groups, and the traditional open courtyard is replaced by a greenhouse courtyard embedded in the ground. The solar radiation is used to obtain heat, while reducing heat loss and reducing the shape factor of the building.

The original wind speed of the site has an impact on people's activities	Increase the wall perpendicular to the wind direction to slow down the wind speed	The wall is bent to form more space to enhance the wind-shield effect	Adjust the wall according to the visual needs, with both landscape and practicality

LIVING UNIT ORIENTATION DESIGN

◀ Schematic diagram of the distribution of solar radiation when the house faces the southeast

◀ Schematic diagram of the distribution of solar radiation when the house faces due south

PLANE DIVISION OF RESIDENTIAL UNITS

Main heating　Secondary heating　Non-heated space　outdoor

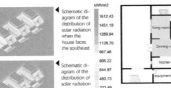

▲ south-to-east orientation to increase the sunshine duration on the wall on the side of the greenhouse courtyard. This design increases the energy utilization rate by 8% to 10%

ECONOMIC INDEX

Thermal zone	Severe cold-A
Body shape coefficient limit	≤0.5
Single-family building area	120m²
Building volume	496m³
Building surface area	167m²
Building shape factor	0.34

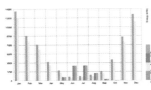

Greenhouse courtyard　Greenhouse courtyard　Slope 40 degree sunroof　insulation wall

concrete base
Concrete retaining wall 400-600mm
Insulation layer 200mm
Surface soil 0-30mm
Frozen soil layer 2500mm-3300mm
Permafrost boundary

专家点评：

总体规划布局合理，功能流线组织合理；朝向布局适宜高原气候环境，注重单体建筑在被动式太阳能集热上的最佳效果，尤其在主动式太阳能系统应用上，光热与光电以及其他能源形式的结合上具有特色。公共建筑部分的建筑空间和建筑造型方面，既地域特色浓郁又具有现代感。在公共建筑部分充分考虑了被动式节能，将传统建造方式与现代节能低碳技术相结合，实现建筑采暖与蓄热。作品具有可实施性，技术应用具有经济性和普适性。居住建筑部分与公共建筑部分在总体空间设计上缺少有机联系；居住建筑部分在形象设计和技术应用上与公共建筑部分不匹配。

The overall planning is clear. Its circulation is reasonable. The buildings' orientation is suitable for the plateau climate. The designer focused on the best effect of passive solar energy collection in single buildings. Particularly, in the application of active solar energy, the work creatively combines solar thermal and photoelectricity, as well as other forms of energy. Public buildings have abundant regional features and a modern style in terms of architectural space and form. Passive energy efficiency has been fully considered in public buildings, combining traditional construction methods with modern energy-saving and low-carbon technologies to achieve building heating and heat storage. The work is implementable, with economical and universal technologies. The residential and public sub-areas lack a connection in the overall spatial design.

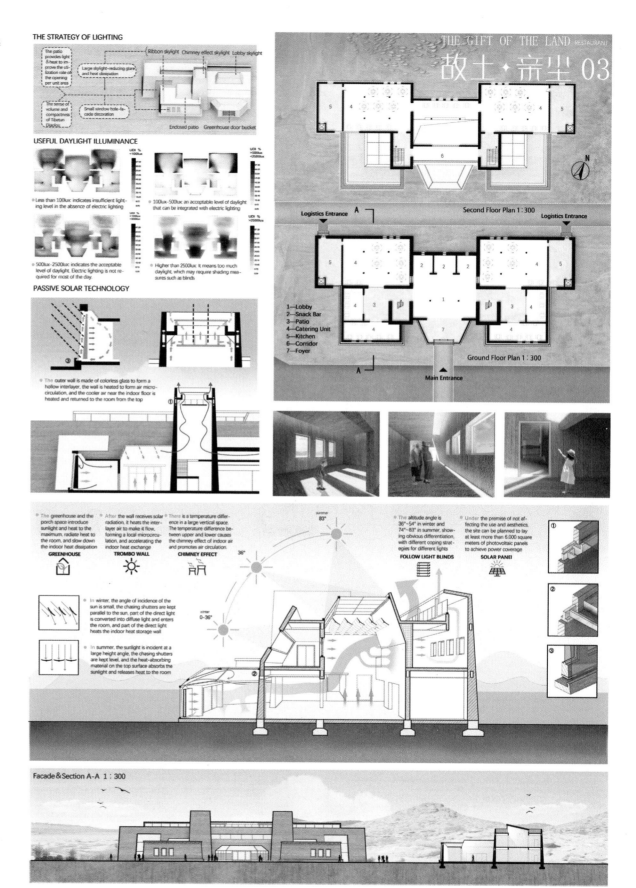

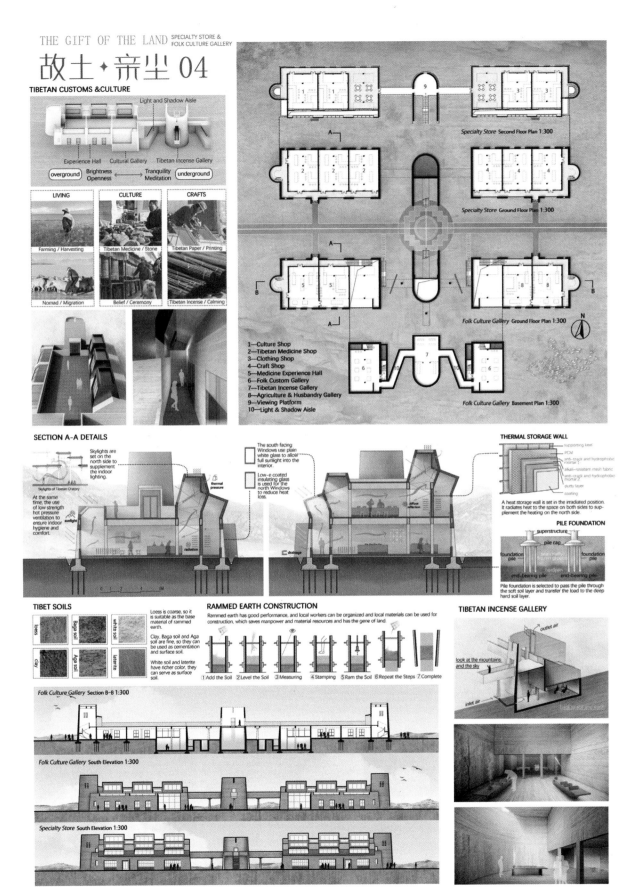

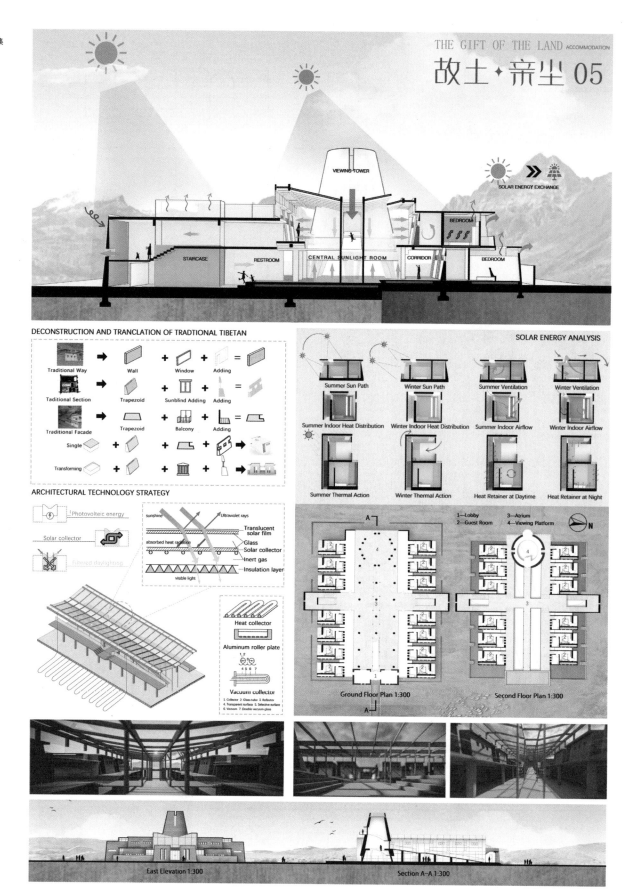

HEAT, ELECTRICITY AND RAINWATER SYSTEM

According to site conditions, the ground source heat pump and solar collector plate are used to provide users with ground heating and hot water, and the solar array and translucent solar film are used to collect electric energy.

THE GIFT OF THE LAND
故土·亲尘 06

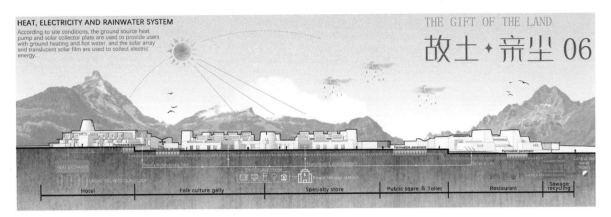

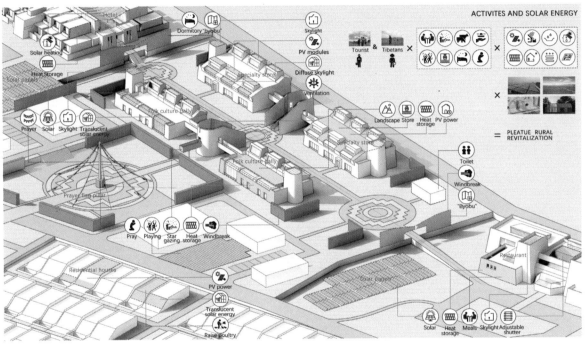

CENTRAL CULTURE SQUARE

These square, angular and strip-shaped flags are orderly fixed on the door Jingfan pillar and fengma banner have a long history in Tibet and are the beautiful spiritual sustenance of the Tibetan people. Therefore, prayer flags and Wind Horse flags are set in the central activity square to make it a cultural center of the whole community.

综合奖·三等奖
General Prize Awarded · Third Prize

注 册 号：100485
Registration No：100485

项目名称：向阳（实地建设项目）
Towards to Sun
(Field Construction Project)

作 者：田彬、吴凡、张栩峰
Authors：Tian Bin, Wu Fan, Zhang Xufeng

参赛单位：重庆大学
Participating Unit：Chongqing University

指导教师：周铁军、张海滨
Instructors：Zhou Tiejun, Zhang Haibin

西藏班戈县青龙乡东嘎村低碳社区 02 向阳

Low Carbon Community of Dongga Village, Tibet

2021 台达杯国际太阳能建筑设计竞赛获奖作品集

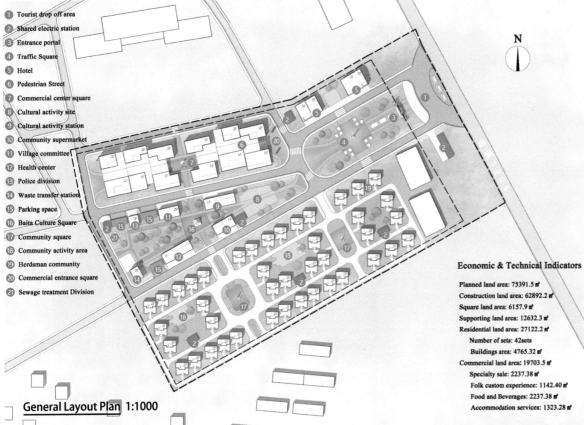

1. Tourist drop off area
2. Shared electric station
3. Entrance portal
4. Traffic Square
5. Hotel
6. Pedestrian Street
7. Commercial center square
8. Cultural activity site
9. Cultural activity station
10. Community supermarket
11. Village committee
12. Health center
13. Police division
14. Waste transfer station
15. Parking space
16. Baita Culture Square
17. Community square
18. Community activity area
19. Herdsman community
20. Commercial entrance square
21. Sewage treatment Division

General Layout Plan 1:1000

Economic & Technical Indicators
Planned land area: 75391.5 ㎡
Construction land area: 62892.2 ㎡
Square land area: 6157.9 ㎡
Supporting land area: 12632.3 ㎡
Residential land area: 27122.2 ㎡
Number of sets: 42sets
Buildings area: 4765.32 ㎡
Commercial land area: 19703.5 ㎡
Specialty sale: 2237.38 ㎡
Folk custom experience: 1142.40 ㎡
Food and Beverages: 2237.38 ㎡
Accommodation services: 1323.28 ㎡

专家点评：

建筑规划与平面布置适宜高原气候环境，设计充分考虑了单体建筑在被动集热、建筑室内通风循环以及与主动太阳能系统的结合，在被动、主动与通风控制上与室内热环境综合考量，富有创新特色。商业部分考虑了内街的处理，适合当地的气候。居住部分建筑朝向较好，但建筑形象略显平淡，缺少美感和特色。

The architecture planning and floor plan adapt to the plateau climate. Passive heat collection of the single building, indoor ventilation of the building, and the active solar system are combined. The integrated consideration of passive and active solar energy, ventilation control, indoor thermal environment, and other elements is really innovative. The design of the commercial area takes into account the inner street and makes it suitable for the local climate. The residential area, however, is not so attractive in the architectural image and lacks aesthetics and features.

■ Region Feature

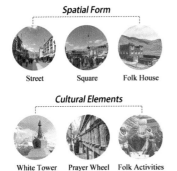

Spatial Form: Street, Square, Folk House

Cultural Elements: White Tower, Prayer Wheel, Folk Activities

■ Climate Simulation

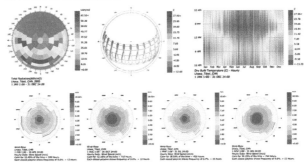

According to the climate simulation, Bango county has strong annual sunshine radiation and rich solar energy resources. The winter is cold and the temperature difference between day and night is large. The main wind direction is southwest.

■ Upper Planning

According to the superior planning, there will be three axes around the site. The direction of people flow on the site can be divided into three main directions as shown in the figure.

■ Passive Strategy Simulation

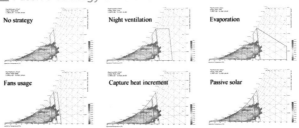

No strategy | Night ventilation | Evaporation
Fans usage | Capture heat increment | Passive solar

By simulating the human comfort area under five passive strategies, it can be concluded that the suitable passive strategies for this area are: capture heat increment and passive solar energy utilization.

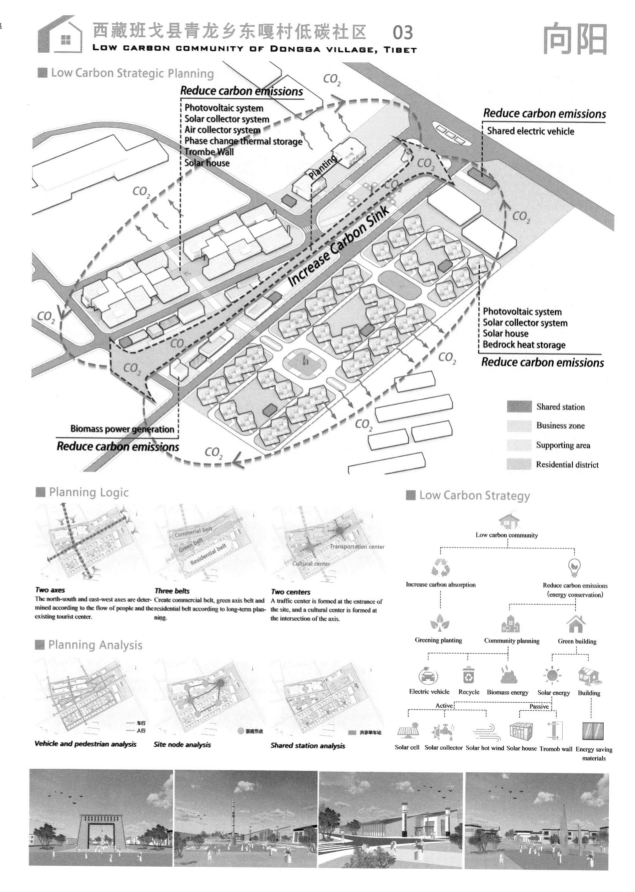

西藏班戈县青龙乡东嘎村低碳社区 04
Low Carbon Community of Dongga Village, Tibet

向阳

2021 台达杯国际太阳能建筑设计竞赛获奖作品集

1. Hotel
2. Restaurant
3. Specialty sale
4. Folk custom experience
5. Entrance square
6. Central square

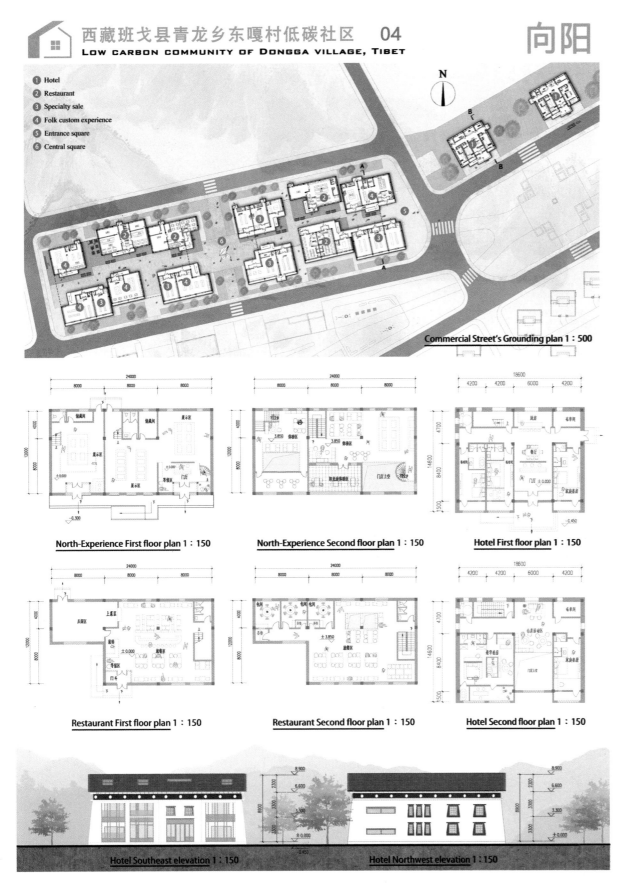

西藏班戈县青龙乡东嘎村低碳社区 05 向阳

Low Carbon Community of Dongga Village, Tibet

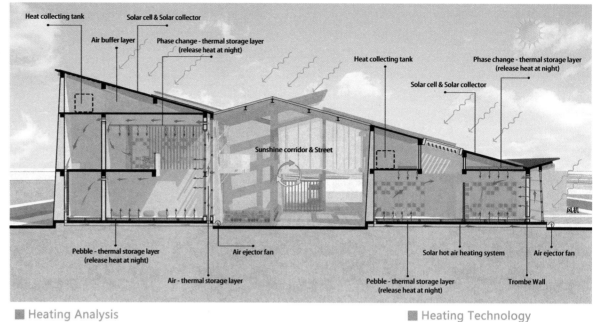

■ Heating Analysis　　　　　　　　　　　　　　　■ Heating Technology

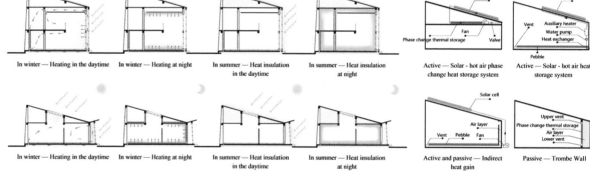

■ Unit Structural & Material

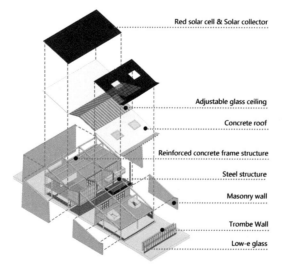

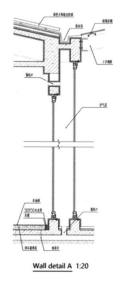

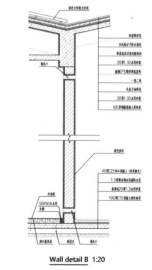

Wall detail A 1:20　　　Wall detail B 1:20

西藏班戈县青龙乡东嘎村低碳社区 06 向阳
Low carbon community of Dongga village, Tibet

■ Carbon Computing

■ Energy Computing

Building type	Building area (m²)	Roof PV area (m²)	PV power generation efficiency (KW)	Total PV power generation (KW*h)	Building thermal energy consumption (KW*h/m²*a)	Building lighting energy consumption (KW*h/m²*a)	Energy consumption of building equipment (KW*h/m²*a)	Annual energy consumption (KW*h)	Net electric energy output (KW*h)
Residential buildings	4654	1316	65.78	136055	7.34	4.50	13.50	117922	18133
Supporting service buildings	1572	875	43.75	90490	15.50	18.0	19.5	75456	15034
Commercial buildings	6150	3096	154.8	320179	18.45	27.5	30.5	470167	-149988

Annual total building energy consumption (KW*h)	Annual total PV power generation (KW*h)	Annual total net energy consumption (KW*h)
663545	546724	116821

Season setting	Month	Day
Start date of heating season	11	15
End date of heating season	3	15
Start date of air conditioning season	6	1
End date of air conditioning season	8	31

■ Layout Simulation

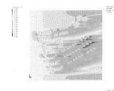

Wind speed simulation—Layout optimization

Solar radiation simulation

Commercial Buildings

Project Statistics	Units	Statistics
Total building air conditioning area	m²	451.80
Project load statistics		
Annual maximum heating load	kW	23.16
Annual maximum cooling load	kW	0.00
Annual maximum humidification capacity	kg/h	3.51
Annual cumulative heating load	kW·h	8335.70
Annual cumulative cooling load	kW·h	0.00
Annual cumulative humidification capacity	kg	1093.38
Project load area index		
Annual maximum heating load index	W/m²	37.24
Annual maximum cooling load index	W/m²	0.00
Annual maximum humidification index	g/h/m²	5.02
Annual cumulative heating load index	kW·h/m²	18.45
Annual cumulative cooling load index	kW·h/m²	0.00
Annual cumulative humidification index	kg/m²	2.42
Seasonal load index of the project		
Heating load index in heating season	W/m²	5.62
Cooling load index in air conditioning season	W/m²	0

Residential Buildings

Project Statistics	Units	Statistics
Total building air conditioning area	m²	478.50
Project load statistics		
Annual maximum heating load	kW	3.77
Annual maximum cooling load	kW	9.43
Annual maximum humidification capacity	kg/h	18.86
Annual cumulative heating load	kW·h	1655.61
Annual cumulative cooling load	kW·h	1952.28
Annual cumulative humidification capacity	kg	6804.27
Project load area index		
Annual maximum heating load index	W/m²	8.51
Annual maximum cooling load index	W/m²	24.13
Annual maximum humidification index	g/h/m²	37.34
Annual cumulative heating load index	kW·h/m²	3.46
Annual cumulative cooling load index	kW·h/m²	4.08
Annual cumulative humidification index	kg/m²	14.22
Seasonal load index of the project		
Heating load index in heating season	W/m²	0.94
Cooling load index in air conditioning season	W/m²	0

Hotel Buildings

Project Statistics	Units	Statistics
Total building air conditioning area	m²	110.80
Project load statistics		
Annual maximum heating load	kW	3.75
Annual maximum cooling load	kW	0.00
Annual maximum humidification capacity	kg/h	1.27
Annual cumulative heating load	kW·h	1162.55
Annual cumulative cooling load	kW·h	0.00
Annual cumulative humidification capacity	kg	273.62
Project load area index		
Annual maximum heating load index	W/m²	23.14
Annual maximum cooling load index	W/m²	0.00
Annual maximum humidification index	g/h/m²	7.95
Annual cumulative heating load index	kW·h/m²	7.34
Annual cumulative cooling load index	kW·h/m²	0.00
Annual cumulative humidification index	kg/m²	1.73
Seasonal load index of the project		
Heating load index in heating season	W/m²	1.58
Cooling load index in air conditioning season	W/m²	0

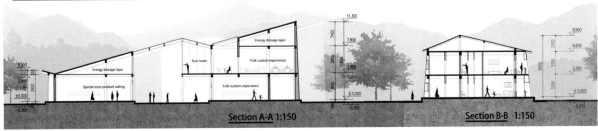

Section A-A 1:150   Section B-B 1:150

综合奖·三等奖
General Prize Awarded · Third Prize

注册号：100296
Registration No: 100296

项目名称：疆域零城（概念设计项目）
　　　　　Uyghur-Zero
　　　　　(Concept Design Project)

作　　者：徐明昊、李帅杰
Authors: Xu Minghao, Li Shuaijie

参赛单位：东南大学
Participating Unit: Southeast University

指导教师：杨　靖、李　丹
Instructors: Yang Jing, Li Dan

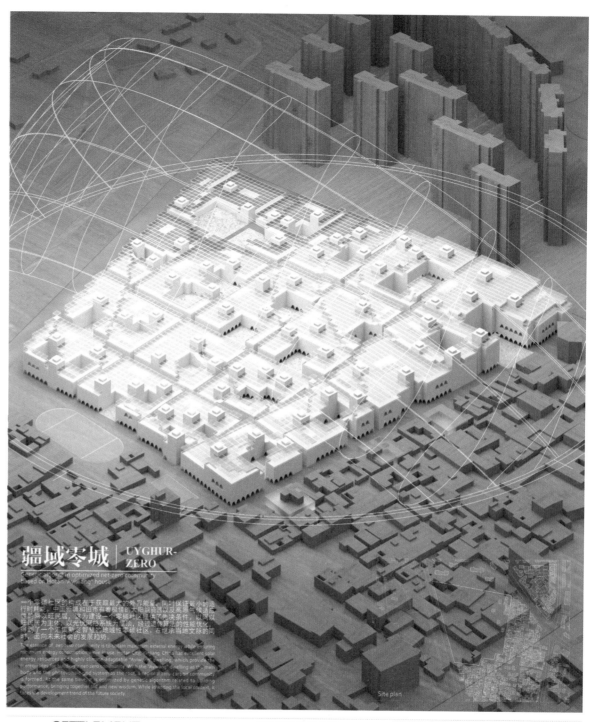

疆域零城 | UYGHUR-ZERO

一个零碳社区的构成在于获取最大的外界能量，同时保证最小的运行时耗能。中国新疆和田市有着极佳的太阳能资源以及高原气候适应性的阿以旺民居。文为建设一个零碳社区提供了先决条件。以可以旺民居为主体，以光伏网络系统为辅助，经过遗传算法的性能优化，得到了一个汇集新老智慧的地域性零碳社区。在继承当地文脉的同时，面向未来社会的发展趋势。

The essence of net-zero community is to sustain maximum external energy while ensuring minimum energy consumption in use. Hotan City, Xinjiang, China has excellent solar energy resources and a highly climate-adaptable "Ayiwang" dwelling, which provides the prerequisites for building a net-zero community. With the "Ayiwang" dwelling as the main body and the photovoltaic grid system as the auxiliary, a net-zero carbon community is formed. At the same time, it is optimized by genetic algorithm related to building performance, bringing together old and new wisdom. While inheriting the local context, it faces the development trend of the future society.

Site plan

SETTLEMENT

Local buildings need to avoid the negative effects of high temperature and sunshine, but it also provides conditions for obtaining sufficient solar energy.

PROTOTYPE

"Ayiwang" has a courtyard in the center and a roof higher than the surrounding rooms on the top of the courtyard, and side windows under the courtyard roof for lighting and ventilation.

PHOTOVOLTAIC

Photovoltaic power plants can absorb solar energy and convert it into electrical energy. Its efficiency is related to factors such as shape, orientation and angle.

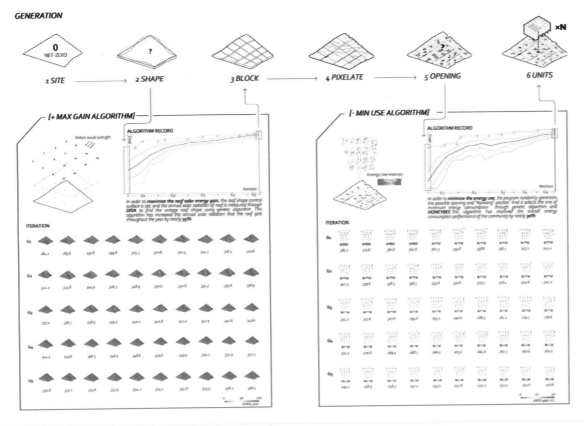

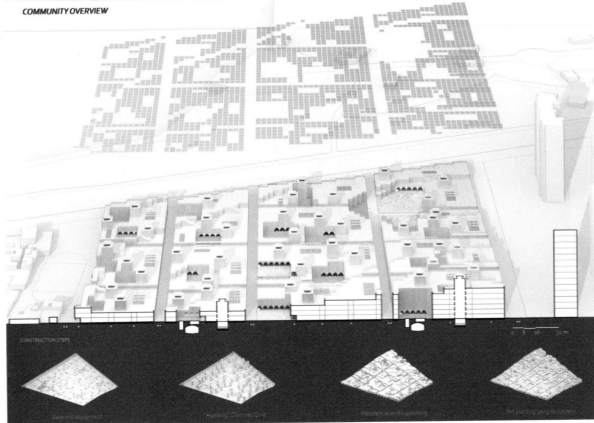

专家点评：

作品在充分吸纳传统阿以旺民居以庭院为中心、自遮阳与垂直通风并重的气候适应性建筑特点的基础上，将光伏发电遮阳屋顶覆盖在由住宅和社区公共服务设施组成的混合社区及其街道之上，形成了天然采光、街道自遮阳、天井被动通风、屋面休闲种植于一体的新型阿以旺居民社区。密集型集聚体量应较符合于高温避热（阳光地区），体量除通井元素外细节考虑尚不足，减排效益有论证空间。多层公共配套设施空间进深较大，需要加强采光通风的研究。

The work fully incorporates the climate-adapted architectural features of traditional residence Aywang. Such residence, centered on courtyards, is equipped with local shading and vertical ventilation. Considering these features, the designer placed photovoltaic shaded roofs over a mixed community of houses and public facilities and their streets, creating a new type of Aywang residential community with natural light, local shading streets, the passive patio ventilation, and recreational roof planting. However, the densely clustered volumes better suit areas with high temperatures. The volume, except for the ventilating shaft, has room for improvement in details. The depth of the multi-story public amenity requires enhanced lighting and ventilation.

NET-ZERO SYSTEM

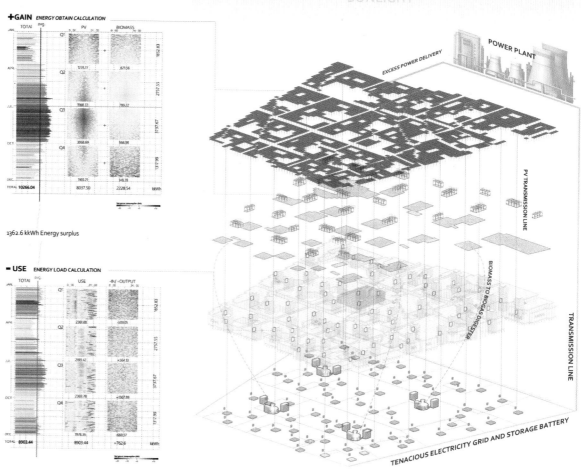

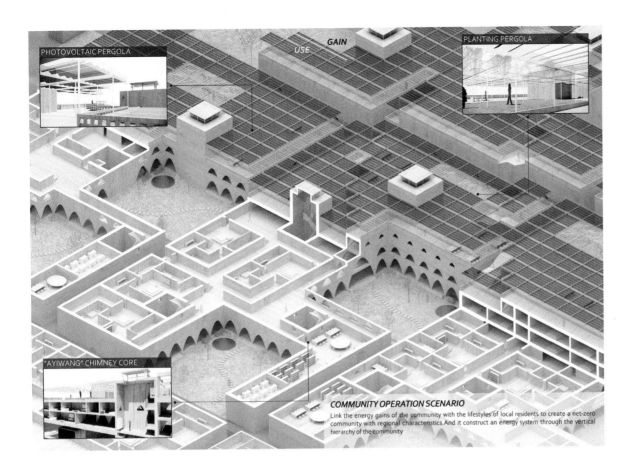

PHOTOVOLTAIC PERGOLA

GAIN
USE

PLANTING PERGOLA

"AYIWANG" CHIMNEY CORE

COMMUNITY OPERATION SCENARIO
Link the energy gains of the community with the lifestyles of local residents to create a net-zero community with regional characteristics. And it construct an energy system through the vertical hierarchy of the community

POWER TEST MODEL

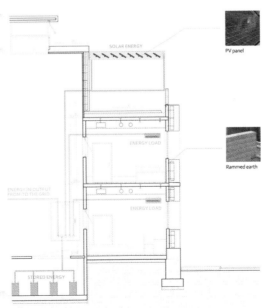

PV panel

Rammed earth

Annual power test

Spring equinox   Summer solstice   Autumnal equinox   Winter solstice

Power test table (Voltage)

	7:00	8:00	9:00	10:00	11:00	12:00	13:00	14:00	15:00	16:00	17:00	18:00
Spring equinox	03.32 V		08.44 V	10.36 V	12.32 V	13.89 V	14.45 V	11.78 V				03.87 V
Summer solstice			08.68 V	11.97 V	13.25 V	16.77 V	17.35 V	18.55 V	11.32 V	10.39 V		
Autumnal equinox	03.37 V			07.77 V	09.53 V	11.49 V	10.55 V	08.11 V				
Winter solstice	02.38 V	04.49 V	05.24 V		11.89 V	10.53 V	06.32 V			04.81 V	02.11 V	

综合奖·三等奖
General Prize Awarded · Third Prize

注 册 号：100355
Registration No：100355

项目名称：巨构社区：律之海
（概念设计项目）
Mega Community:
Sea of Melody
(Concept Design Project)

作 者：阮立凡、郭艳春、何 强、
颜端帅
Authors：Ruan Lifan, Guo Yanchun,
He Qiang, Yan Duanshuai

参赛单位：联创新锐设计顾问（武汉）
有限公司——高扬工作室
Participating Unit：United Design Group
Co. Led, Wuhan

指导教师：高 杨、刘晓峰
Instructors：Gao Yang, Liu Xiaofeng

巨构社区：律之海
Mega Community: Sea of Melody

DESIGN INTRODUCTION

We set the site for "Zero Carbon Community" design competition in a dense residential area in Jiang'an District, Wuhan, Hubei. Hubei Wuhan has always been known as the "province of a thousand lakes and the city of hundreds of lakes", and is well-known for its unique natural environment. This design starts from a new community concept and focuses on a new community model, trying to break the traditional residential space and activate the vitality of the community. As is shown in the picture on the right, the traditional community type has clear hierarchy, simplification and closedness. If we want to reach a location, we must walk out of the community, wasting long and boring traffic time. As the various resources of the community are structured in a tree-like structure with distinct levels, we will walk back and forth in the "two points and one line" mode. Comparing with the traditional type, the super mode we create will integrating all community resources via a simple and efficient transportation system, and achieve the purpose of green travel, low carbon and environmental protection.

This design covers an area of 90,000 square meters, with a construction area of about 240,000 square meters, which can accommodate Over 300 families. At the same time, this design also uses solar photovoltaic power generation, wind power generation, roof planting, water circulation system, passive energy saving and other technologies to improve the environment of the building, achieve energy self-sufficiency, and achieve the goal of low carbon or even zero carbon. In addition, the design also considers a large area of green roof and a walking system with rich layers, which will greatly return the precarious green leisure space that was originally affected by urbanization to the citizens. Affected by the natural environment of the water system in Wuhan, we are trying to plant natural wetlands in the community environment, accelerate the natural circulation of the community system, and promote the harmonious coexistence of man and nature. So far, this design has optimized the living mode of the residential area in terms of housing, transportation, activities, etc., and created a positive, low-carbon and sustainable community space.

Model Selection

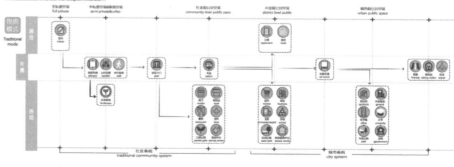

The design uses the Y-shape as the basic type for combination. The first starting point is "the ecological settlement where people can interact with each other". The Y-shape can be understood as a "branch" shape representing the environmental foundation of natural ecology, and it can also be understood as the Chinese character "people". The gathering of people can form settlements. 《Han Feizi·Five beetles》 says: The wooden room is a nest to avoid group harm; the so-called nesting is our most primitive form of community. In addition, this form is very adaptable. In the actual thermal environment, no matter how the form is placed, it can be guaranteed that there are two branches that can face the south and get better sunlight. Therefore, there are two branches suitable for residence, and the remaining branch can be used for public services.

Traditional mode VS Super mode

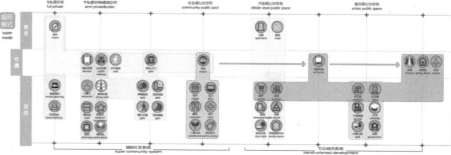

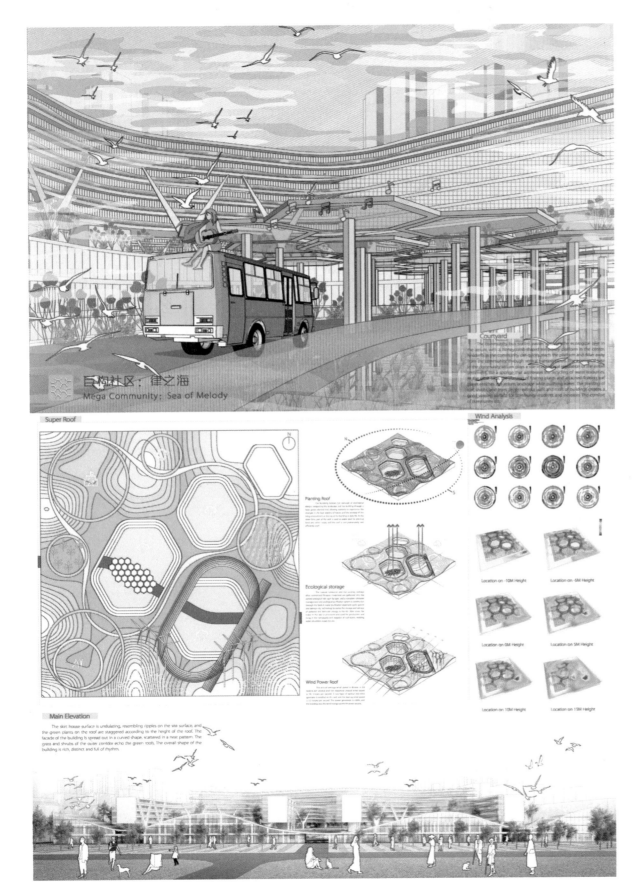

专家点评：

作品规划与建筑特色突出，从新型社区的理念出发，试图采用巨构的模式打破传统的居住空间，充分利用简单、高效的交通体系整合所有的社区资源。同时，在低碳设计方面还使用了太阳能光伏发电、风力发电、屋顶种植、水循环系统、被动式节能等技术改善建筑的环境，实现能源的自给自足，达到零碳的目的。但巨构形式对零碳诉求形成基本挑战，需在构造方式与建筑材料上补充说明来增加合理性。

The planning and architectural features are unique. Guided by the idea of a new community, the designer tried to innovate the traditional residential space with the mode of mega-structure. The design makes good use of the simple but efficient transportation system to integrate all community resources. Besides, as for low-carbon design, the designer also uses technologies such as solar photovoltaic, wind power, roof planting, water circulation systems, and passive energy conservation, to achieve energy self-reliance and zero-carbon emission. However, the mega-structure would inevitably pose challenges to the appeal of zero-carbon emission. As a result, more illustrations of structure and materials should be added to demonstrate the rationality.

巨构社区：律之海
Mega Community: Sea of Melody

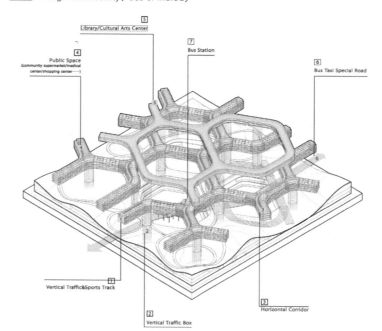

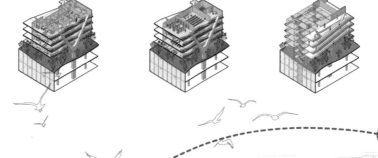

Transport Analysis

Community traffic consists of a pedestrian system and a car system. The pedestrian system is composed of horizontal traffic and vertical traffic. A bus lane crosses the border and runs through the entire plot. The horizontal corridor connects each level, and the same level can be communicated with each other through the horizontal corridor. The vertical traffic consists of stairwells, interlayer runways and vertical traffic boxes. The stairwells and interlayer runways connect the levels above the super roof, and the vertical traffic boxes connect the ground and underground floors of the entire site. All levels and districts of the entire community. All rooms are accessible to each other, and the passage is convenient, and they are connected to each other.

a. Closed housing system, high capacity and low density, the space in the residential area is relatively dull, and the utilization rate of the landscape is low.

a. Semi-open roof space, integrated use, more efficient, both residents and other citizens can use.

b. The management level of the residential area is distinct, and you must pass through the gate to enter and exit the community and reach the public part.

b. The transportation connection part is intensive and efficient, the public and the residential area are mixed with each other, and the transportation core is connected.

c. Due to closed management, the gate will become a more congested area, and the traffic distance from the check-in area will be longer.

c. The bus station is connected with the vertical transportation, which greatly shortens the traffic time and improves the efficiency.

Section Analysis

The community is composed of living space, public space and transportation space. The residential space is modular, diverse in form, and flexible in layout to meet the living needs of various family populations. The public space is composed of community supermarkets, shopping malls, medical centers, libraries, cultural and art centers, etc., which are distributed on all floors, serving various areas and meeting daily needs. The whole community integrates solar photovoltaic power generation, wind power generation, roof planting, water circulation system, passive energy saving and other technologies to achieve energy self-sufficiency and achieve the goal of low carbon or even zero carbon.

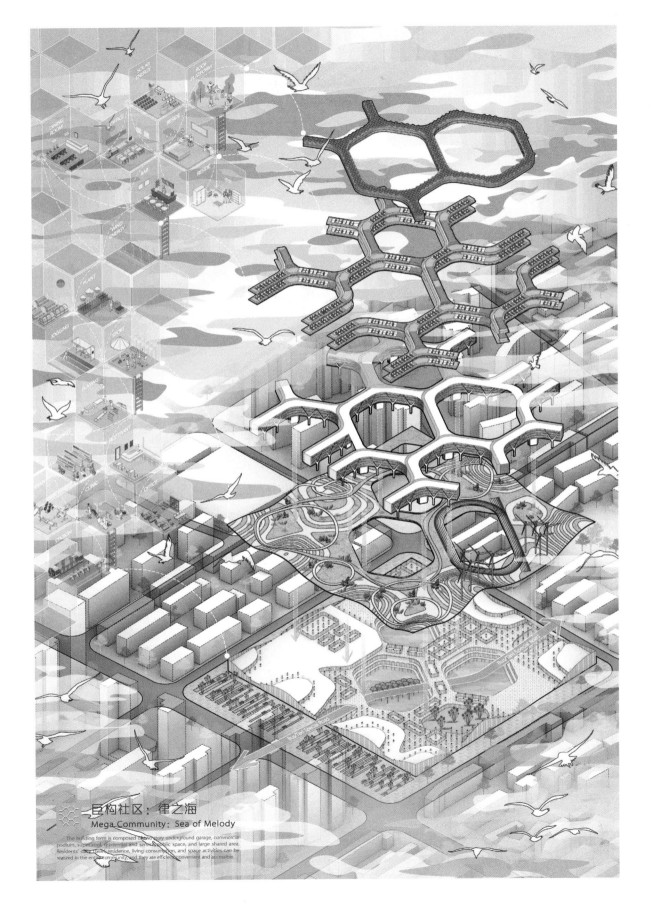

综合奖・三等奖
General Prize Awarded · Third Prize

注 册 号：100361
Registration No：100361

项目名称：生生不息（概念设计项目）
Loop (Concept Design Project)

作　　者：潘翼舒、华　颖、张彦彤、
　　　　　赵宇杰、陆文凯、李志伟
Authors：Pan Yishu, Hua Ying,
　　　　　Zhang Yantong, Zhao Yujie,
　　　　　Lu Wenkai, Li Zhiwei

参赛单位：浙江大学
Participating Unit：Zhejiang University

指导教师：罗晓予、高　峻
Instructors：Luo Xiaoyu, Gao Jun

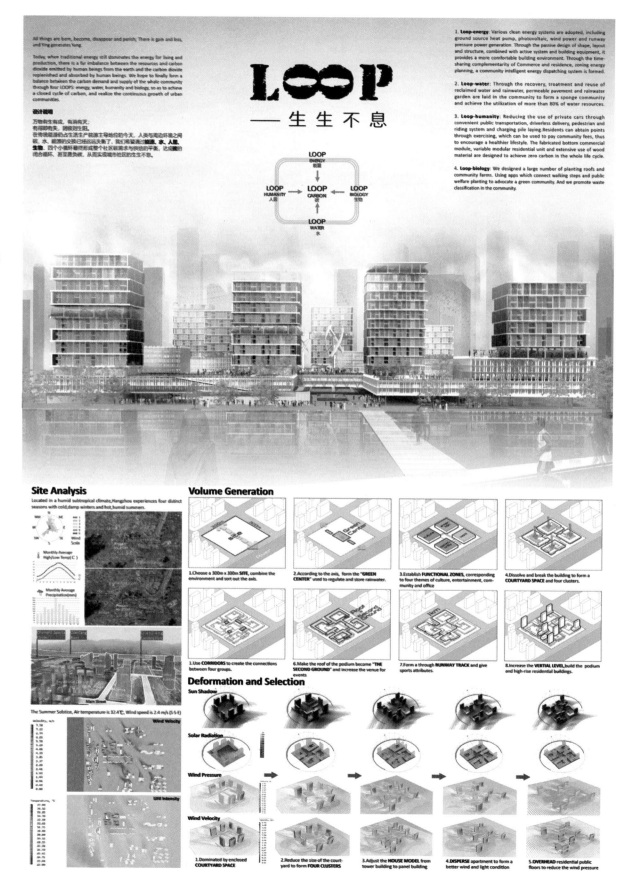

2F Plan 1:1200

1F Plan 1:1200

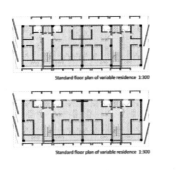

Standard floor plan of variable residence 1:300

Standard floor plan of variable residence 1:300

First floor plan of shared space 1:300

Second floor plan of shared space 1:300

Residence Plan 1:300

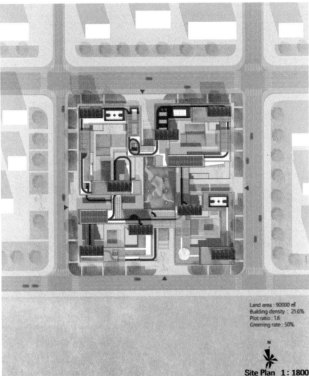

Land area : 90000 ㎡
Building density : 21.6%
Plot ratio : 1.6
Greening rate : 50%

Site Plan 1:1800

- SOLAR PANEL
- INTELLIGENT LIGHTING
- PV FACADE
- WIND POWER
- ROOF PLANTING
- WATER RECYCLING
- BIKE PARKING
- WASTE RECYCLING

专家点评：

作品试图从能源、水、人居和生物四个层面的循环达成低碳社区的目标，实现人与自然的和谐共生与生生不息。因此，优化低层社区共享空间和高层住宅布局成为零碳社区设计的关键环节，以期达到高效便捷、健康低碳、自我循环的人居环境；同时充分利用建筑屋面、墙面、平台安装新能源设施和绿化种植，实现遮阳、通风、发电的同时，促进人与自然的互动、人与人的交往。提出"行为"维度作为零碳的路径之一，十分可取。量化尝试较好，值得肯定。

This work attempts to build a zero-carbon community and promote the coexistence of man and nature in four aspects: energy, water, habitats, and creatures. Optimizing the shared space of the low-rise residence and the layout of the high-rise residence is regarded as the key part of the zero-carbon community design. In this way, the human settlement can be efficient, convenient, healthy, low-carbon, and self-cycling. In addition, new energy facilities fixed on roofs, walls, and platforms can achieve the targets of shading, ventilating, and power generation. Those facilities can also enhance the interaction between man and nature and people-to-people exchanges. The dimension of "behavior" in this work, which is set as one of the ways of achieving zero-carbon emission, is a good quantitative try.

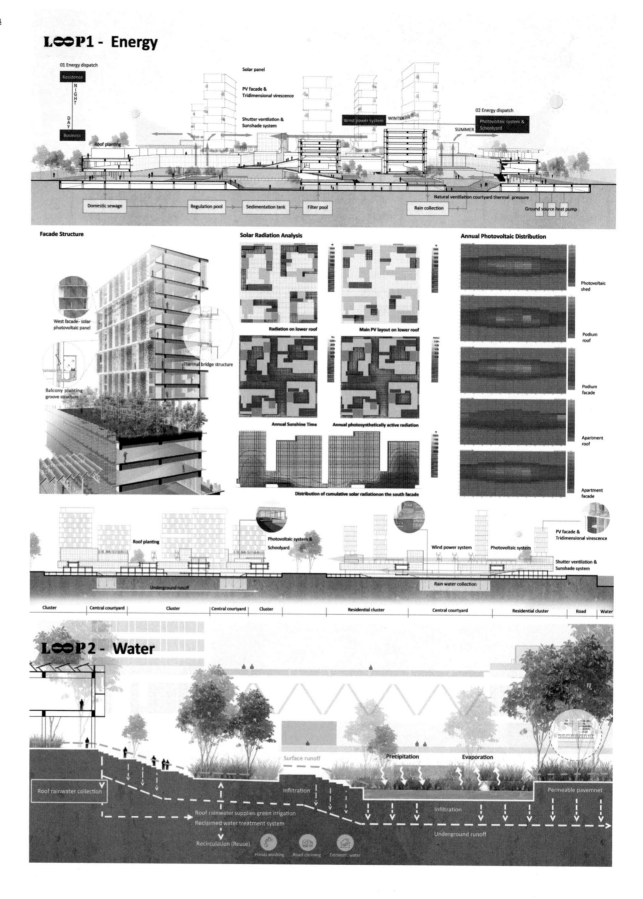

LOOP3 - Humanity

EASY ACCESS TO **PUBLIC TRANSPORTATION** | **ELECTRIC VEHICLE AND BIKES** | **DRIVERLESS DELIVERY** | **RUNWAY POWER GENERATION** | **EXERCISE** TO GET *E-POINT* | *E-POINT* TO PAY OFF **LIVING EXPENCES**

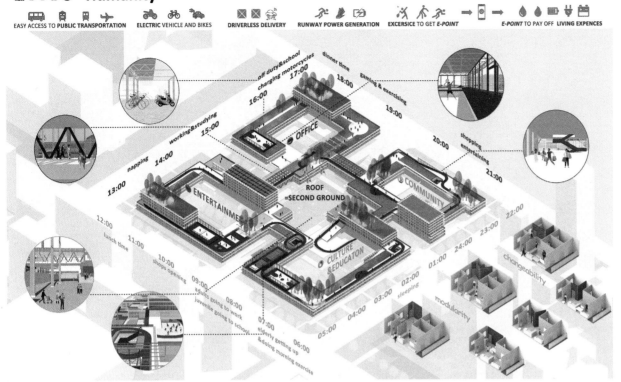

LOOP4 - Biology

REduce. REuse. REcycle. REpeat.

Smart community · Roof planting | Waste classification · Composting regeneration

Seasonal crop planting: Spring·carrot, gourd, Summer·pumpkin, Autumn·medlar, cauliflower, Winter·tomato

Harvest · Watering · Gardening · Watering

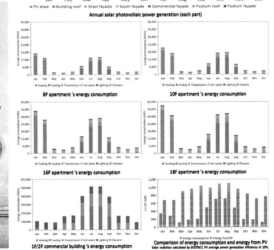

Energy Consumption Analysis

Annual solar photovoltaic power generation (each part)

8F apartment's energy consumption | 10F apartment's energy consumption

16F apartment's energy consumption | 18F apartment's energy consumption

1F/2F commercial building's energy consumption | Comparison of energy consumption and energy from PV

综合奖·三等奖
General Prize Awarded · Third Prize

注 册 号：100546
Registration No：100546

项目名称：衣聚·宜居（概念设计项目）
Clothing in Living ·
Living in Comfort
(Concept Design Project)

作 者：潘烨成、潘珏宜、郑宇豪、朱倩婷
Authors：Pan Yecheng, Pan Jueyi, Zheng Yuhao, Zhu Qianting

参赛单位：广州大学
Participating Unit：Guangzhou University

指导教师：席明波、万丰登
Instructors：Xi Mingbo, Wan Fengdeng

衣聚宜居 Zero Carbon Community Concept Design
零碳社区概念设计

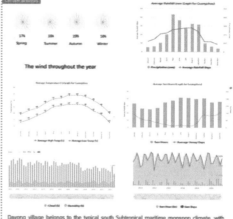

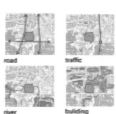

Design description

本次设计强调产业和居住的平衡，旨在改变广东制衣村作为旧的制造业代表，以压缩居住和透支环境的方式换取生产利益的行为。方案在整体布局上，通过装配式功能模块和网架结构还原场地肌理，降低建造能耗；重新梳理道路系统，提倡公共交通和BRT轨道交通；模块化的地面网格结合雨水花园系统，与河岸形成微气候景观，并收集处理雨水和污水。在生产方面，丰富单一加工产业链，结合规划，融入会展旅游业，使其具有独特的行业竞争力；提倡废料回收和余热能源利用，使产业零消耗、全产出；网架与墙面附着的橡胶静电除尘装置在减振降噪的同时吸附悬浮粒子。在生活方面，对不同人群的需求分区布置，减少他们作息上的相互影响；模块化住宅方便便捷的同时提供了垂直多元的绿化空间；结合道路的振动发电装置和大片铺设的太阳能板将提供生活和生产所需的全部能源。构想方案作为环境友好的零碳社区，将成为海珠区服装制造业对外输出的体验点，联动海珠湿地共同构成该区的风貌展示带。

The design emphasizes the balance between industry and residence, aiming to change the behavior of Guangdong garment village, as the representative of the old manufacturing industry, to compress the living and overdraft environment in exchange for production benefits.
In terms of overall layout, prefabricated functional blocks and grid structures are used to restore the site texture and reduce construction energy consumption. Reorganizing the road system to promote mass transit and BRT rail transit; Modular ground grids combine with a rain garden system to create a microclimate landscape with the riverbank and to collect and treat rain pollution. In terms of production, it enriches the single processing industry chain, combines planning, and integrates exhibition tourism to make it have unique industry competitiveness. Promote waste recycling and waste heat energy utilization, so that the industry zero consumption, full output; The rubber electrostatic dust removal device attached to the network frame and the wall absorbs suspended particles while absorbing shock and noise. In terms of life, the needs of different groups of people are partitioned to reduce their mutual influence on work and rest; Modular housing is convenient and provides vertical and diversified green space; Combined with the road's vibratory power generation units and extensive paved solar panels will provide all the energy needed for living and production.
As an environmentally friendly zero-carbon community, the proposal will become an experience point for the export of the garment manufacturing industry in Haizhu District, and jointly form the landscape display belt of the area by linking the Haizhu wetland.

SITE PLAN

Site area	90000㎡	Building height	21m
Area of structure	62300㎡	Plot ratio	0.7
Building density	25000㎡	Total Households	1250
Green space rate	35%	Resident population	2500

Population analysis

Type of crowd: migrant workers
Age composition: 16~45
Living environment: dormitory
Behavior:

Type of crowd: shop owners
Age composition: 30~65
Living environment: residential
Behavior:

Type of crowd: Fashion designers
Age composition: 22~40
Living environment: apartments, residential
Behavior:

Population type: general residents
Age composition: 6~70
Living environment: residential
Behavior:

Active timeline

衣聚宜居 Zero Carbon Community Concept Design
零碳社区概念设计

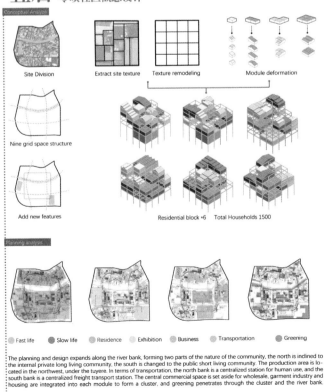

Conceptual Analyze: Site Division · Extract site texture · Texture remodeling · Module deformation · Nine grid space structure · Add new features

Residential block *6   Total Households 1500

Planning analysis

● Fast life ● Slow life ● Residence ● Exhibition ● Business ● Transportation ● Greening

The planning and design expands along the river bank, forming two parts of the nature of the community, the north is inclined to the internal private long living community, the south is changed to the public short living community. The production area is located in the northwest, under the tuyere. In terms of transportation, the north bank is a centralized station for human use, and the south bank is a centralized freight transport station. The central commercial space is set aside for wholesale, garment industry and housing are integrated into each module to form a cluster, and greening penetrates through the cluster and the river bank.

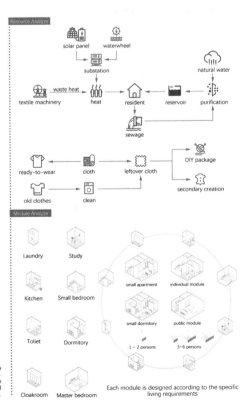

Resource Analyze: solar panel, waterwheel, substation, textile machinery, waste heat, heat, resident, reservoir, natural water, purification, sewage

ready-to-wear → cloth → leftover cloth → DIY package
old clothes → clean → secondary creation

Module Analyze: Laundry, Study, Kitchen, Small bedroom, Toilet, Dormitory, Cloakroom, Master bedroom — small apartment, individual module, small dormitory, public module — 1~2 persons, 3~6 persons

Each module is designed according to the specific living requirements

专家点评：

作品以混合社区为研究对象，以建筑工业化为手段，基于标准功能模块和通用结构框架，在满足生产、生活、商业和运输等复合功能的基础上达成零碳目标。通过架空层实现垂直方向的功能复合，通过新型交通设施和架空环路实现不同功能组团的联络和生产生活交通，为实现生产生活复合型社区的零碳目标提供了系统的解决方案。建筑形式与效果过于单一，设计手法稍显保守传统，仍需突破。

This work set the mixed community as the subject. Based on the standard function module and the general structure framework and through architectural industrialization, the design realizes compound functions, including production, living, commerce, and transportation, and then achieves the zero-carbon goal. The vertical functional complex is achieved through the empty space. The new transport facilities and overhead loops enable the connection between different functions and the connection between production and living, providing a systematic solution to achieve the zero-carbon goal of the complex community in daily life and work. The architectural form and effect are too homogeneous. The design method is slightly conservative and traditional and needs to be innovated.

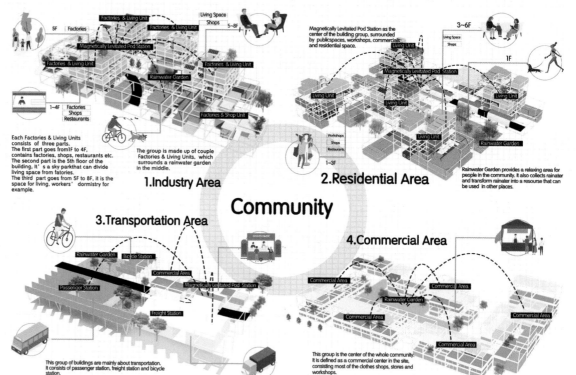

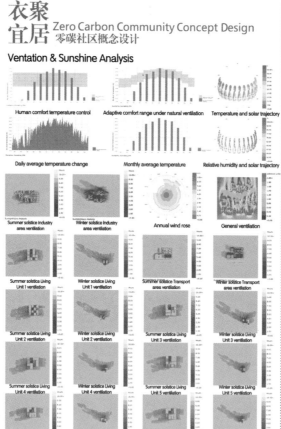

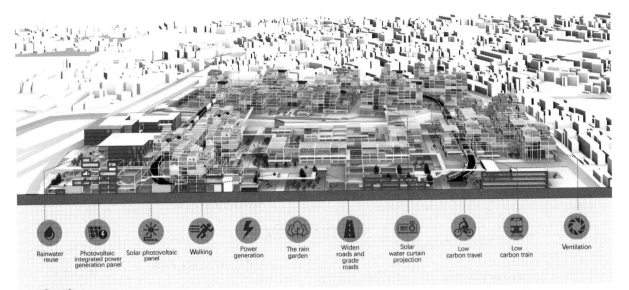

| Rainwater reuse | Photovoltaic integrated power generation panel | Solar photovoltaic panel | Walking | Power generation | The rain garden | Widen roads and grade roads | Solar water curtain projection | Low carbon travel | Low carbon train | Ventilation |

衣聚宜居 Zero Carbon Community Concept Design
零碳社区概念设计

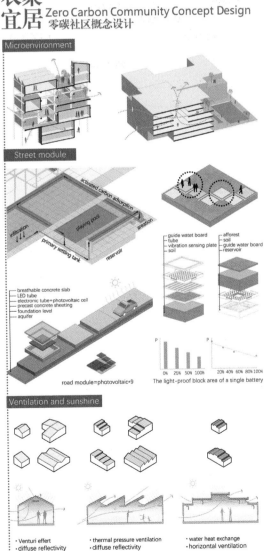

Microenvironment
Street module
- breathable concrete slab
- LED tube
- electronic tube+photovoltaic cell
- precast concrete sheeting
- foundation level
- aquifer

road module=photovoltaic*9   The light-proof block area of a single battery

Ventilation and sunshine
- Venturi effert
- diffuse reflectivity
- solar panels

- thermal pressure ventilation
- diffuse reflectivity
- solar panels

- water heat exchange
- horizontal ventilation

Module and energy saving

Transmittance curve at noon on the 21st of each month

Comparison of WBGT between shaded area and illuminated area

The use of louvres on summer solstice

The use of louvres on winter Solstice

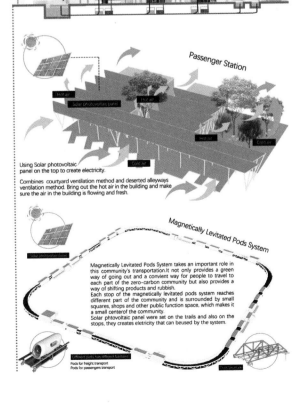

Passenger Station
Using Solar photovoltaic panel on the top to create electricity.
Combines courtyard ventilation method and deserted alleyways ventilation method. Bring out the hot air in the building and make sure the air in the building is flowing and fresh.

Magnetically Levitated Pods System
Magnetically Levitated Pods System takes an important role in this community's transportation. It not only provides a green way of going out and a convient way for people to travel to each part of the zero-carbon community but also provides a way of shifting products and rubbish.
Each stop of the magnetically levitated pods system reaches different part of the community and is surrounded by small squares, shops and other public function space, which makes it a small centerof the community.
Solar phtovoltaic panel were set on the trails and also on the stops, they creates electricity that can beused by the system.

Pods for freight transport
Pods for passengers transport

实地建设项目
Field Construction Project

综合奖·优秀奖
General Prize Awarded·
Honorable Mention Prize

注 册 号：100015
Registration No：100015

项目名称：暖巢
　　　　　Warm Nest

作　者：武玉艳、高新萍、杨　晴、
　　　　巩博源、张　莹
Authors：Wu Yuyan, Gao Xinping,
　　　　Yang Qing, Gong Boyuan,
　　　　Zhang Ying

参赛单位：西安建筑科技大学
Participating Unit：Xi'An University
　　　　　　　　　 of Architecture and
　　　　　　　　　 Technology

指导教师：陈景衡
Instructor：Chen Jingheng

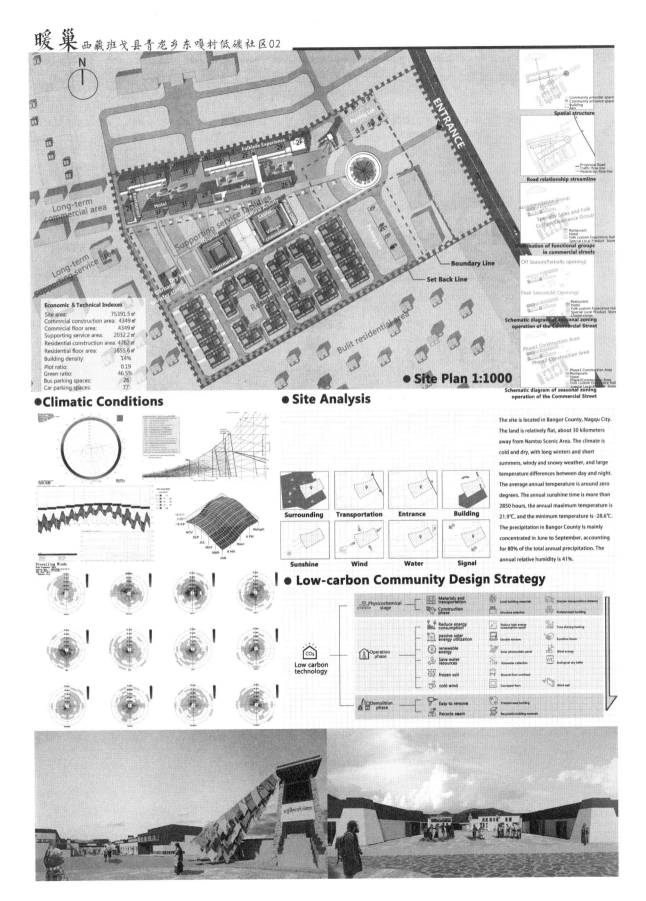

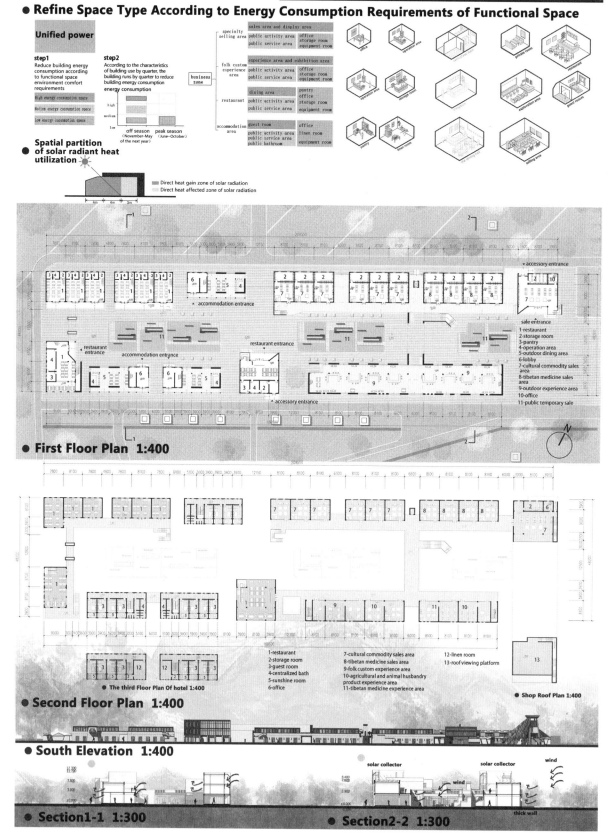

暖巢 西藏班戈县青龙乡东嘎村低碳社区 04

● Solar radiation heat

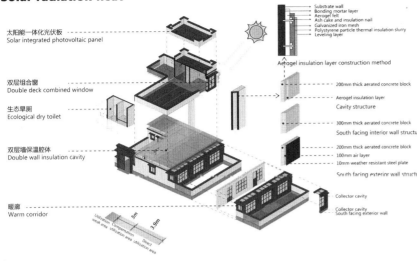

● Model Photos

● Residential Indoor Temperature ● Wind Simulation

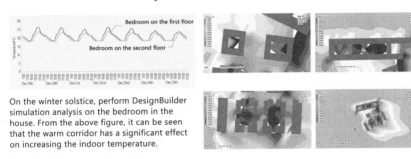

On the winter solstice, perform DesignBuilder simulation analysis on the bedroom in the house. From the above figure, it can be seen that the warm corridor has a significant effect on increasing the indoor temperature.

● Partial Perspective View

寨心·光·聚 01
The Heart of the Village · Light · Gather

综合奖·优秀奖
General Prize Awarded ·
Honorable Mention Prize

注 册 号：100054
Registration No：100054

项目名称：寨心·光·聚
The Heart of the Village ·
Light · Gather

作　　者：郭嘉钰、张玉琪、王松瑞
Authors：Guo Jiayu, Zhang Yuqi, Wang Songrui

参赛单位：北京建筑大学
Participating Unit：Beijing University of Civil Engineering and Architecture

指导教师：俞天琦、丁光辉
Instructors：Yu Tianqi, Ding Guanghui

设计说明

西藏那戈县青龙乡东嘎村低碳社区的规划设计以低碳与"寨心光聚"两个方面入手。基于西藏独特的自然条件，对当地等环境、降水量等分析。采用太阳能光伏板、锯齿形聚热蓄热墙等技术，建造低碳社区。并结合当地独有藏区文化与传统，将"坛城、寨心、经轮"等概念相融合。并结合太阳能技术，打造出"一核、两轴、多节点"的社区规划。

Design description

The planning and design of the low-carbon community in Dongga Village, Qinglong Township, Bangor County, Tibet started from two aspects: low-carbon and "zhaixin-guangju". Based on Tibet's unique natural conditions, analyze the local environment and precipitation, and use technologies such as solar photovoltaic panels and zigzag heat collection and storage walls to build low-carbon communities.
Combining the unique local Tibetan culture and traditions, combining the concepts of "mandala, fast heart, prayer flags", and solar energy technology to create a community plan of "one core, two axes, and multiple nodes".

Climate analysis

It can be seen from the monthly climate data of Qinglong Township:
1. Because it is located in the severely cold area of the Qinghai-Tibet Plateau, Qinglong Township has a low temperature throughout the year. The average temperature in June is only 16 ℃, while the average temperature of the coldest month reaches minus 12 ℃. It can be seen that heat preservation and antifreeze should be put in the building design. Consider in the first place;
2. Because it is located in the high-altitude area of the Qinghai-Tibet Plateau, Qinglong Township has a relatively high UV index. The UV index is the highest in June of the year. The UV index is 4, which does not reach the risk index.
3. From the snowfall statistics table, it can be seen that there is more snowfall in Qinglong Township, and there has been an increasing trend of snowfall in recent years
4. From the wind speed statistics chart, it can be seen that the wind speed in this area is relatively high, and wind protection should be considered in the architectural design to prevent the invasion of cold wind.

① Mandala thought employed in Tibetan areas

The idea of mandala has a subtle influence on the development of temple buildings and settlements in Tibetan areas.
The image of mandala applied to architecture and settlement space becomes a kind of metaphor, which is embodied in all levels of material and spirit.

— Samye Temple Restoration Picture
— Secret Mandala Painting Frame
— Satellite picture of Samye Monastery
— Samye Temple Floor Plan (Examples of Mandala Thoughts)
— Mandala in Thangka (Ideal planning form)

The image of Mount Xume is a concentric, hierarchical figure, with square, circle, square and circle, lotus shape and other forms.

② Prayer Flags(Fengqi, Long Da)

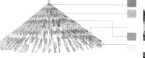

The wind horse flag symbolizes the harmony of the sky, the earth, people, and animals; the herdsmen tie the wind horse flag to the tents they have just set up, in order to win the blessing of migrating through the water and grass; the pilgrims trek through the desert to carry the eye-catching The Fengma Bag is to pray for a safe journey; people around the riverside and lakes planted Fengma Bags, showing their awe of the water god; the people living in the forests and ridges hanging the Fengma Bag as an offering to the mountain god.

③ Tibetan architectural element extraction

— Tibetan windows
— Tibetan roof
— Street space
Labrang Monastery
Translation of Tibetan architectural
Labrang Monastery

Line chart of monthly average temperature (highest/lowest)

Line graph of annual average temperature (highest/lowest)

Monthly average sunshine hours analysis chart

Annual average cloud cover and humidity analysis chart

Yearly wind speed analysis chart

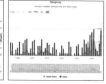
Analysis chart of annual average snowfall

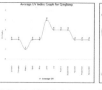
Monthly average UV index line chart

Visibility analysis chart

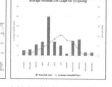
Analysis chart of monthly average snowfall

	FESTIVAL	CELEBRATION	BELIEF	DIET
Convention				
Culture				
Crowd	elder	Child	Government	tourist	young people
Activity				

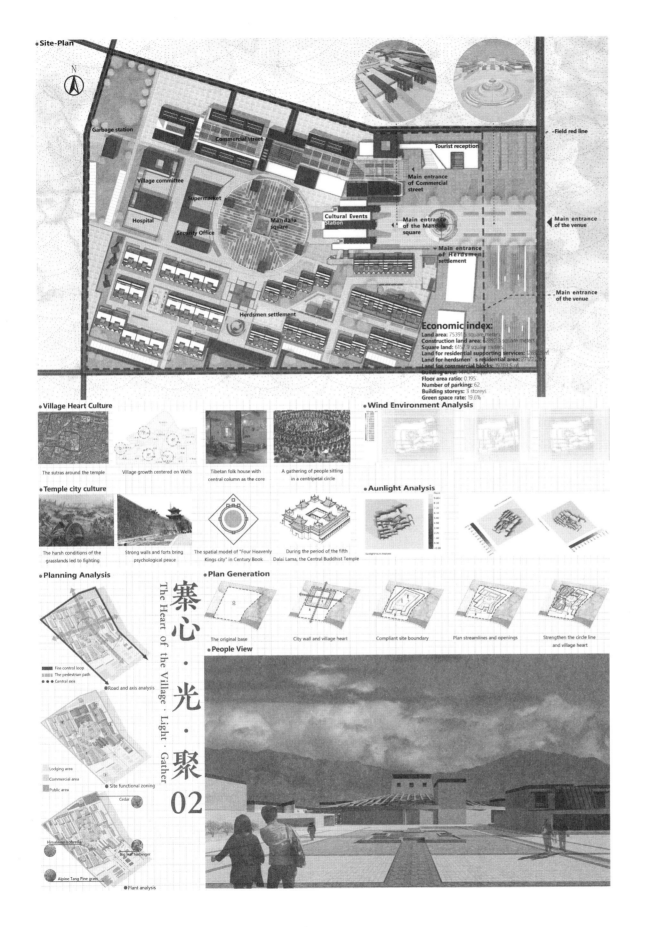

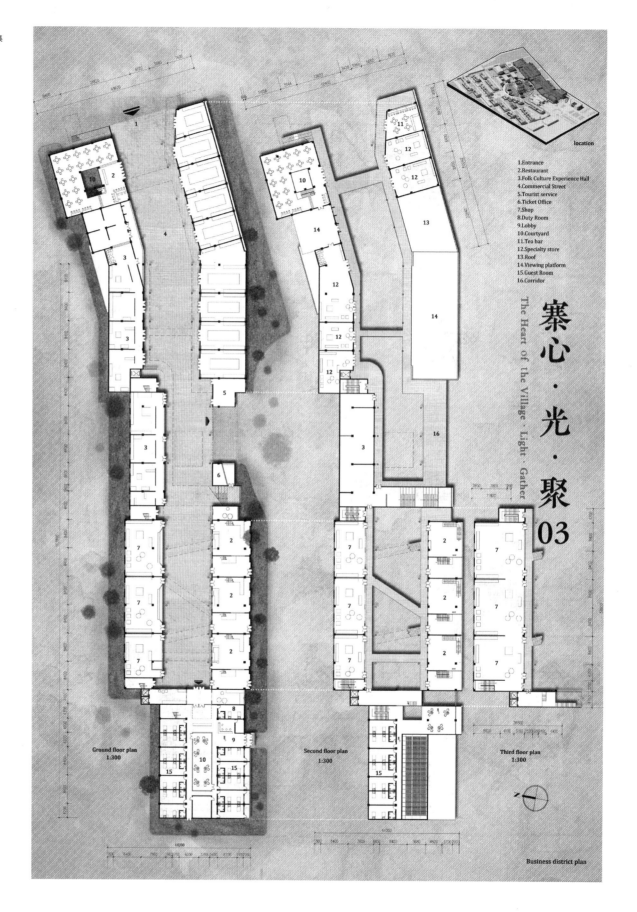

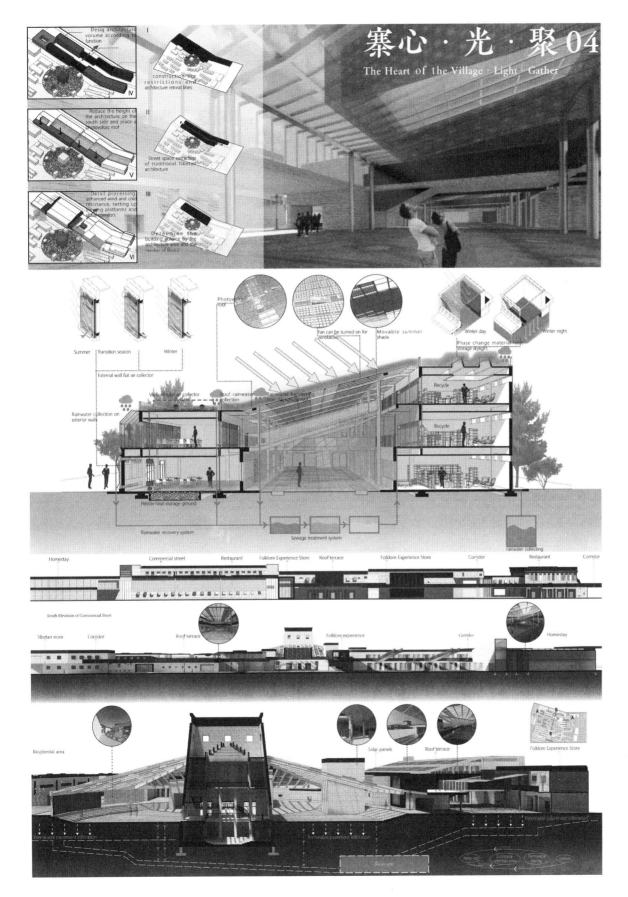

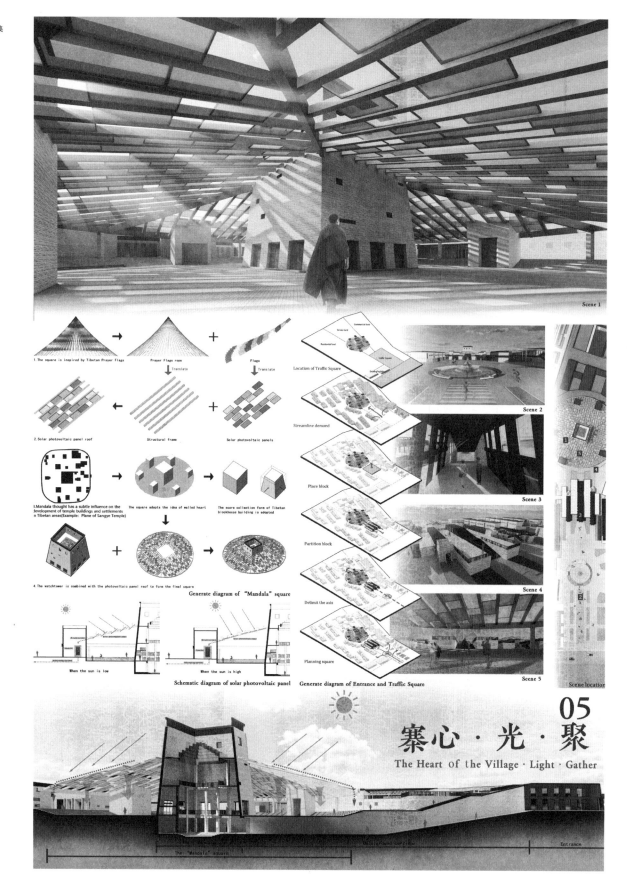

综合奖·优秀奖
General Prize Awarded · Honorable Mention Prize

注　册　号：100069
Registration No：100069

项目名称：伴日——低碳社区综合体设计
Accompanying the Sun

作　　者：马远林、阎丽伊、王慎薇、
　　　　　刘　佳、张偌涵
Authors：Ma Yuanlin, Yan Liyi,
　　　　　Wang Shenwei, Liu Jia,
　　　　　Zhang Ruohan

参赛单位：石家庄铁道大学
Participating Unit：Shijiazhuang Tiedao
　　　　　University

指导教师：高力强
Instructor：Gao Liqiang

低碳社区综合体设计二

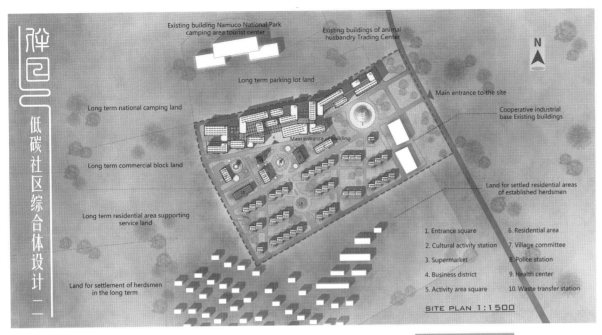

SITE PLAN 1:1500

1. Entrance square
2. Cultural activity station
3. Supermarket
4. Business district
5. Activity area square
6. Residential area
7. Village committee
8. Police station
9. Health center
10. Waste transfer station

LOCATION ANALYSIS

Item Position in Baingoin Qinglong Township Donggacun, Hokutan 90.801°, Longitude 31.091° Baingoin Tree Measures Scenic Area Approx. Provincial Highway S206 Group Raku Line Yu Base East side north-south direction. Baingoin Qinglong Township Donggacun, Baingoin Highland Baingoin Semi-Drought Seasonal Zone. Climate cold, air sparse, four seasons unclear, winter long summer short, heavy snowfall, annual temperature difference phase opposite large to daily difference.

ANALYSIS OF SURROUNDING ENVIRONMENT

Item ground comparison flat. Site East Taiwan S206 Provincial Highway (Team Raku Line) close Street Processed Established Livestock and Livestock Trade Center Bunch Wakosakusha Industrial Base Bunch, North Park Period Parking Lot and National Park Park, Park, West Park, Park, Park, Park, Namtso, Namtso, Namtso, Namtso, Namtso, Namtso Land for nomadic settlements. Items: Gunsan Gunsan Environment, Yushan Valley Area, Summer Winds, Southeastern Winds, Winter Winds, Northwestern Winds.

1. Existing buildings
2. Surrounding roads
3. Surrounding mountains
4. Noise
5. Winter wind direction
6. Summer wind direction
7. Sunshine direction

CLIMATE ANALYSIS

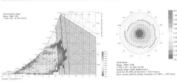

OPTIMUM TEMPERATURE HUMIDITY   ALL YEAR WIND ROSE

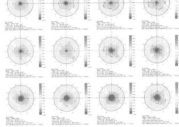

MONTHLY WIND ROSE

WIND AND SUNSHINE ANALYSIS

SUNSHINE   SUMMER WIND   WINTER WIND

This design introduces the concept of wind environment into the design process of building layout form, improves the wind environment around the building, and is conducive to improving the health and comfort of the community.

Based on the climate characteristics of Tibet Naqu region, the wind environment in summer and winter is calculated by using Phoenix software. Through the analysis of simulation results, the numerical range of wind speed in summer is 2.18-3.3. The overall wind speed is comfortable and suitable for outdoor people to rest and walk; The wind speed range in winter is 3.5-6, and the wind speed range of the two-story building in the east of the community is 3-4, which can effectively improve the problem of excessive wind speed in winter.

ANALYSIS OF GREEN BUILDING TECHNOLOGY

After analysis, what kind of green building technology can be used in our buildings?......

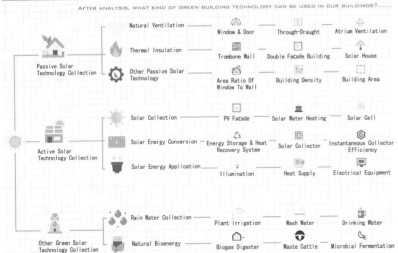

- Passive Solar Technology Collection
  - Natural Ventilation: Window & Door, Through-Draught, Atrium Ventilation
  - Thermal Insulation: Trombone Wall, Double Facade Building, Solar House
  - Other Passive Solar Technology: Area Ratio Of Window To Wall, Building Density, Building Area
- Active Solar Technology Collection
  - Solar Collection: PV Facade, Solar Water Heating, Solar Cell
  - Solar Energy Conversion: Energy Storage & Heat Recovery System, Solar Collector, Instantaneous Collector Efficiency
  - Solar Energy Application: Illumination, Heat Supply, Electrical Equipment
- Other Green Solar Technology Collection
  - Rain Water Collection: Plant Irrigation, Wash Water, Drinking Water
  - Natural Bioenergy: Biogas Digester, Waste Cattle, Microbial Fermentation

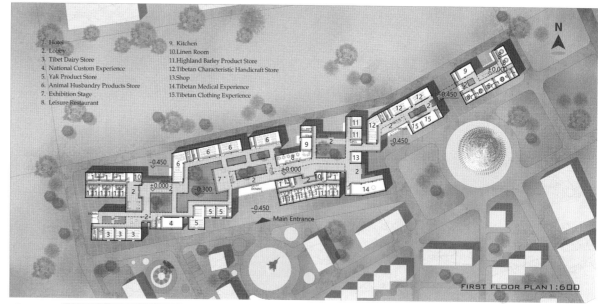

禅包 低碳社区综合体设计（二）

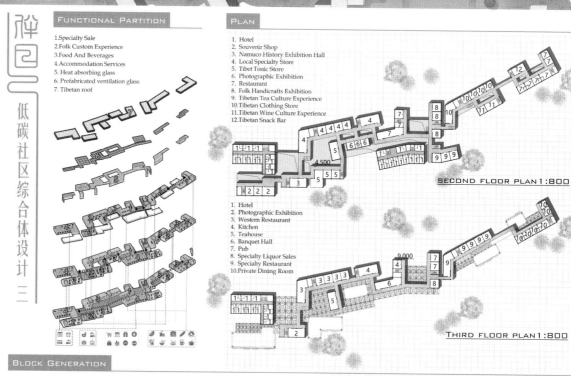

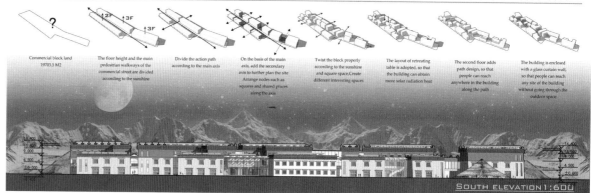

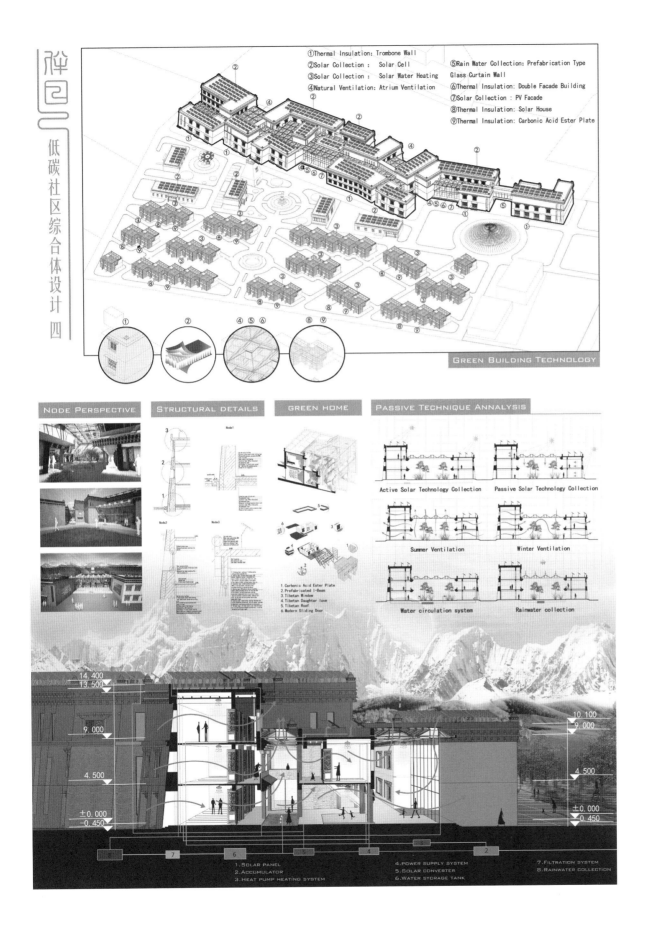

综合奖·优秀奖
General Prize Awarded · Honorable Mention Prize

注　册　号：100072
Registration No: 100072

项目名称：引璨流光
　　　　　Welcome, Miss Sunshine

作　　者：赵时、张玉琪、郭嘉钰、
　　　　　周文宇
Authors: Zhao Shi, Zhang Yuqi,
　　　　　Guo Jiayu, Zhou Wenyu

参赛单位：天津大学、河北工程大学、
　　　　　哈尔滨工业大学
Participating Unit: Tianjin University,
　　　　　Hebei University of
　　　　　Engineering, Harbin
　　　　　Institute of Technology

指导教师：刘家伟华、张伟亮、侯万钧
Instructors: Liu Jiaweihua, Zhang
　　　　　Weiliang, Hou Wanjun

引璨流光 Welcome, Miss Sunshine 1

基于以过渡空间代替加厚维护结构的整体思路，建设包围核心空间的阳光房；在部分围护结构外设置智能活动表皮，表皮耦合太阳高度进行变化，将阳光折射入室，引璨流光。

The overall idea of this project is replacing thickened maintenance structure with transition space. Use the intelligent kinetic facade which can coupled with the solar elevation to create sunshine rooms in both north and south.

Therefore, we can say: "Welcome, Miss Sunshine."

■ Site Bitmap

班戈县青龙乡东嘎村 90.801° N, 东经 31.091° E, 距纳木措景区约 30 公里, 距青龙乡 16 公里, 距班戈县 82 公里。省道 S206 班洛线于基地东侧南北向穿过。

Dongga village, Qinglong Township, bange County, 90.801° N, 31.091° E, is about 30km away from Namucuo scenic spot, 16km away from Qinglong Township and 82km away from bange county. Provincial Highway S206 ban-luo line passes through the east of the base in a north-south direction.

■ Site Photos

■ Frozen Soil Analysis

Seasonal frozen soil occurs in Namtso area from October to May of the next year. The depth of the frozen soil is about 1m. Therefore, relevant structural measures need to be taken in this design.

■ Techno-economic Index

Total Floor Area: 12520 m²
Building Area of Shopping Street: 5980 m²
Shopping Street Occupancy Rate: 52.3 %
Building Area of Supporting Facilities: 1500 m²
Construction Area of Herdsmen's Residence: 5040 m²

■ Design Note

在青龙乡巍峨的群山脚下，建设一片属于高原的建筑。小碉房、窄长窗、煨桑塔、藏红墙，逛街愉快吃得香，商业活泼，建筑稳重。
在商业街中理清客、商、后勤的流线关系，慎重处理了商业片区与居住片区的分隔和联系。

Designing native architectures under the towering mountains of Qinglong Village. Small blockhouse, narrow and long windows, simmering mulberry tower and Tibetan red wall. Convenient shopping conditions, delicious local food, rich commercial types and architectural types with plateau characteristics.

In the commercial street, the streamline relationship of customers, businesses and logistics have been clarified clearly. And the separation and connection between commercial area and residential area were carefully handled.

■ Extraction of Architectural Elements

Architecture　Border Marble Wall　Rammed Earth Wall　White Walls Combined With Murals　Outdoor Walkways

■ Climate Analysis

■ Wind Rose - Summer　■ Wind Rose - Winter　■ Best Orientation　■ Passive Technology

Prevailing Winds　Prevailing Winds　Optimum Orientation

■ Solar Radiation　■ Psychrometric Chart

Psychrometric Chart　Psychrometric Chart

■ Building Usage

■ Movement of Building Occupants

■ Building Visitors Flowrate

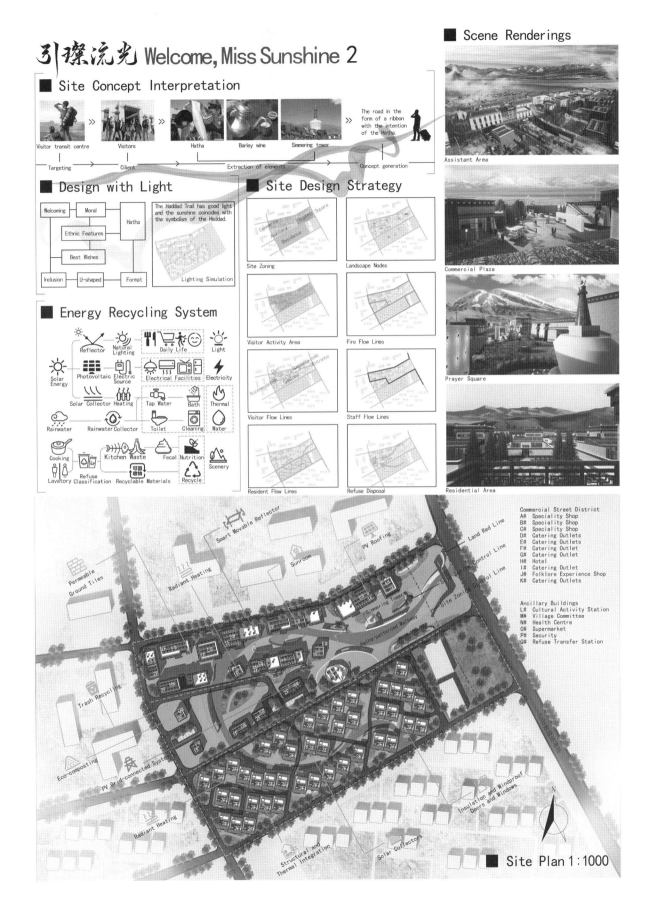

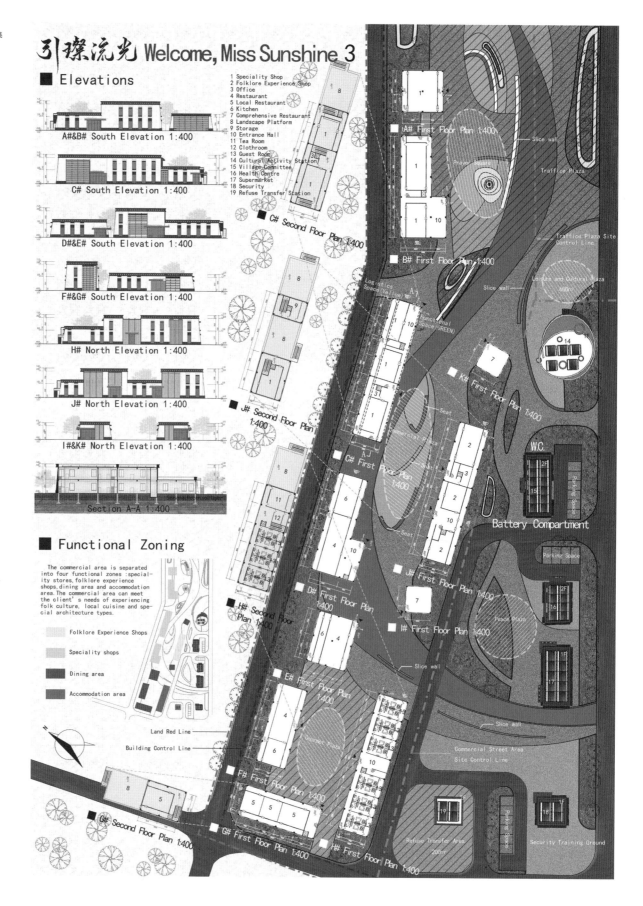

引璨流光 Welcome, Miss Sunshine 4

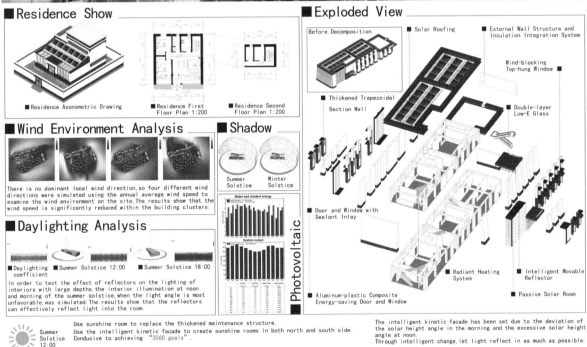

综合奖·优秀奖
General Prize Awarded · Honorable Mention Prize

注 册 号：100100
Registration No：100100

项目名称：念青
　　　　　Together to an Ecological Habitat

作　　者：李　婧、李玉盟、齐　琪、蒋东霖、李敬良、桑弘翼
Authors：Li Jing, Li Yumeng, Qi Qi, Jiang Donglin, Li Jingliang, Sang Hongyi

参赛单位：中国矿业大学（北京）
Participating Unit：China University of Mining & Technology, Beijing

指导教师：李晓丹
Instructor：Li Xiaodan

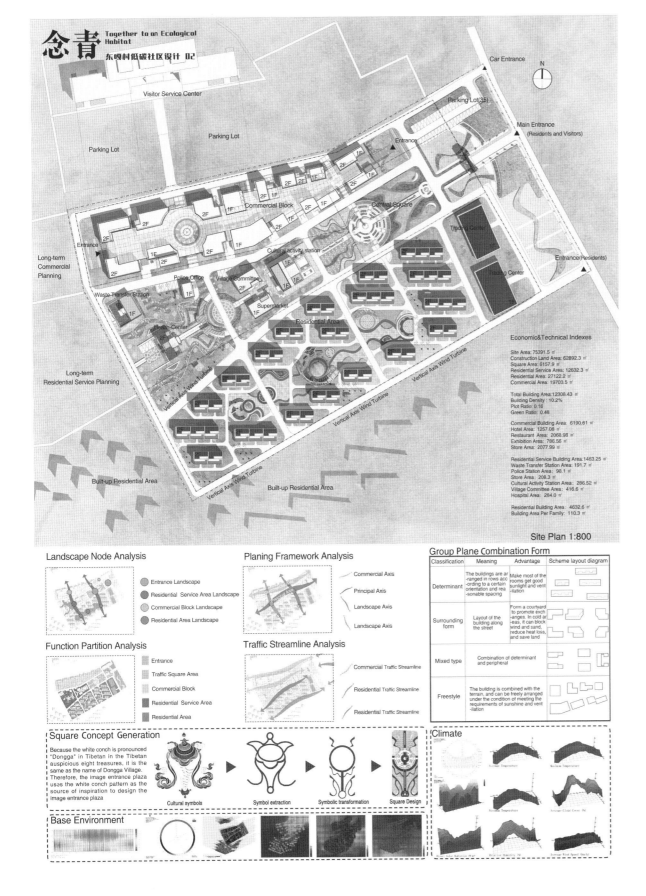

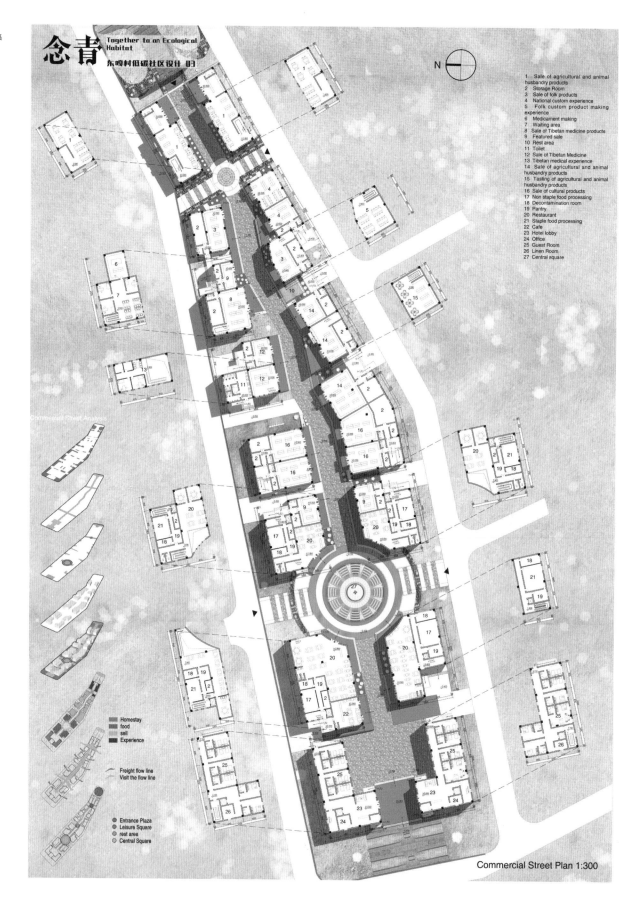

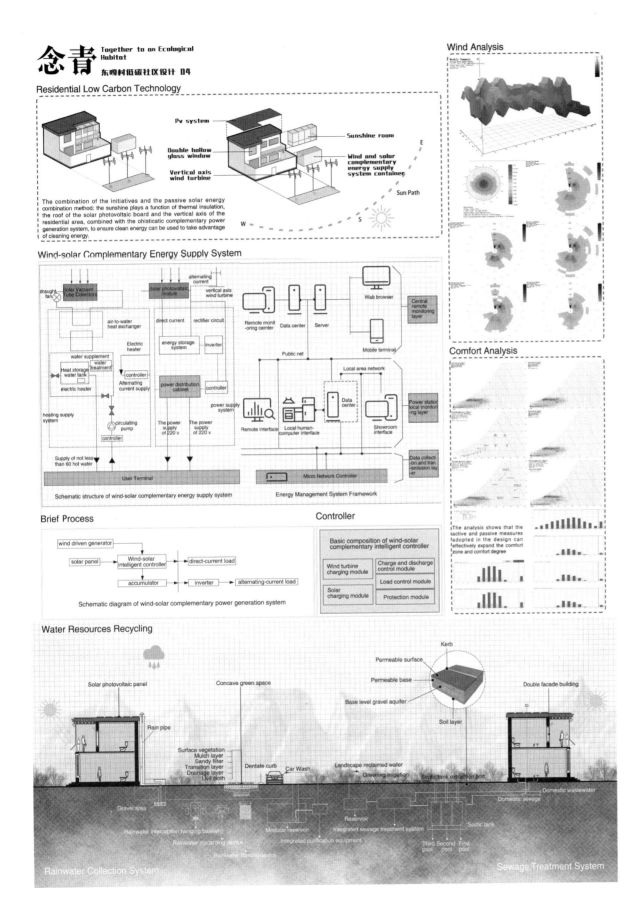

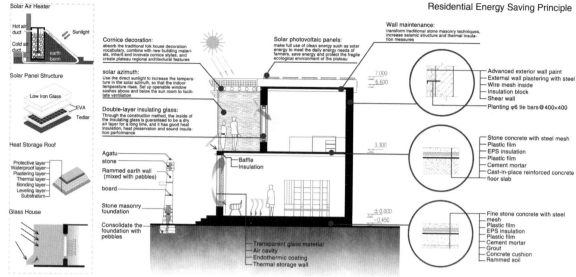
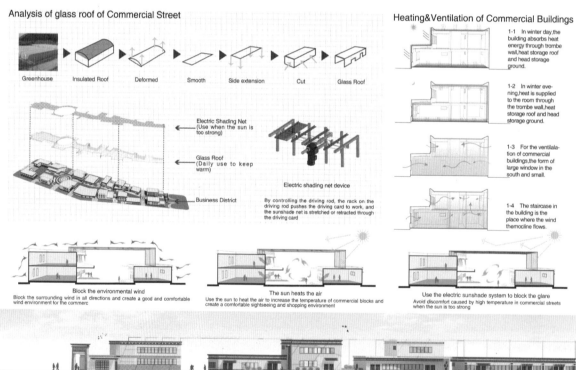

念青 Together to an Ecological Habitat
东嘎村低碳社区设计 06

2021 台达杯国际太阳能建筑设计竞赛获奖作品集

1-1 Section 1:300

Node Perspective

Landscape Node

Symbolic Evolution

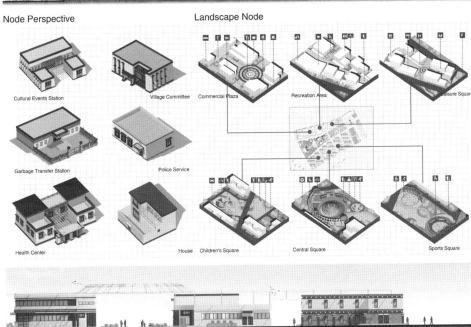

- Ethnic Window
- Decorative Pattern
- Landscape Component
- Prayer Flags
- Health Center Gate
- Image Plaza Entrance Door
- Garbage Transfer Station Gate

North Elevation-2 1:250

South Elevation-2 1:250

综合奖·优秀奖
General Prize Awarded · Honorable Mention Prize

注　册　号：100118
Registration No：100118

项目名称：向光
　　　　　Embracing Sunshine

作　　者：张嘉倩、曾雅清、林凯思
Authors：Zhang Jiaqian, Zeng Yaqing, Lin Kaisi

参赛单位：福州大学
Participating Unit：Fuzhou University

指导教师：王　炜、赵丽珍
Instructors：Wang Wei, Zhao Lizhen

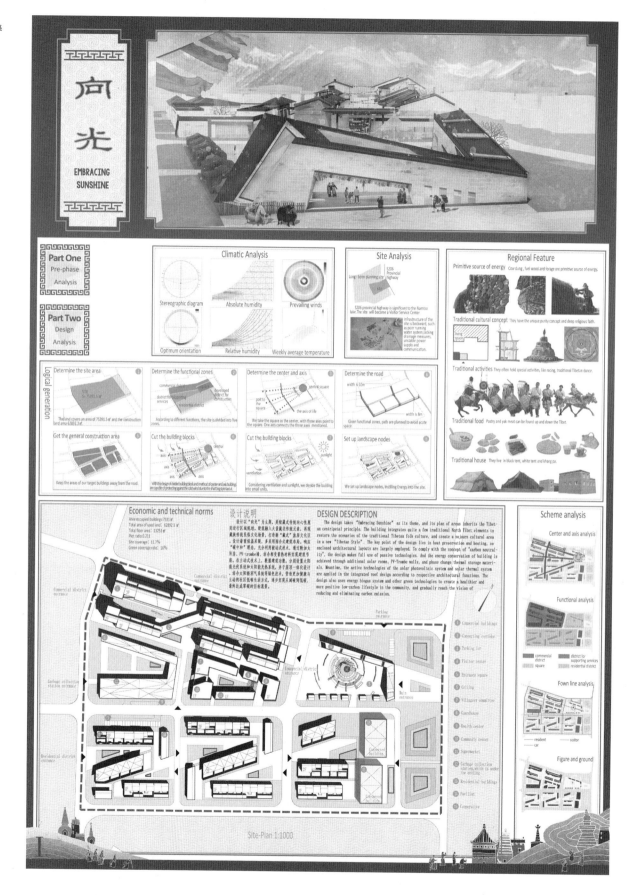

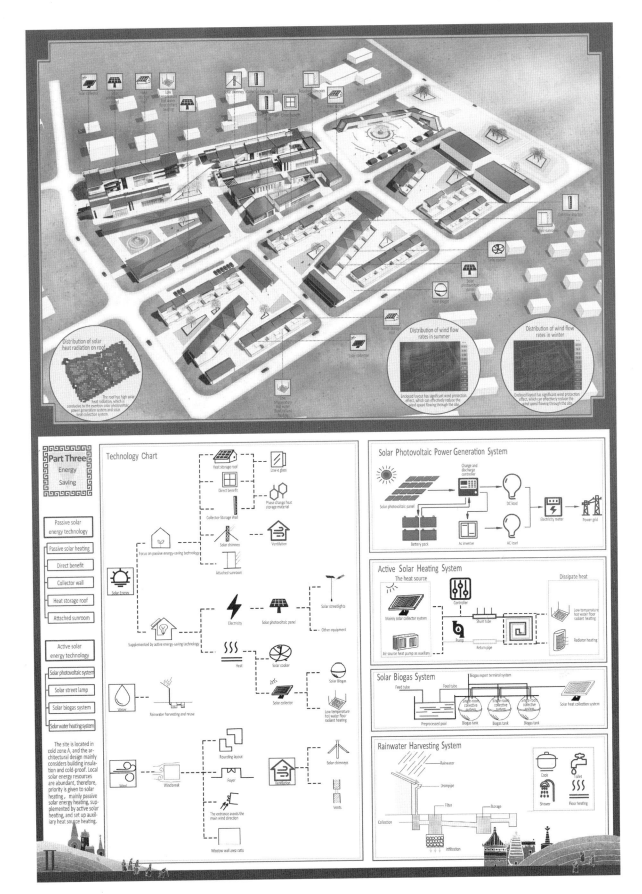

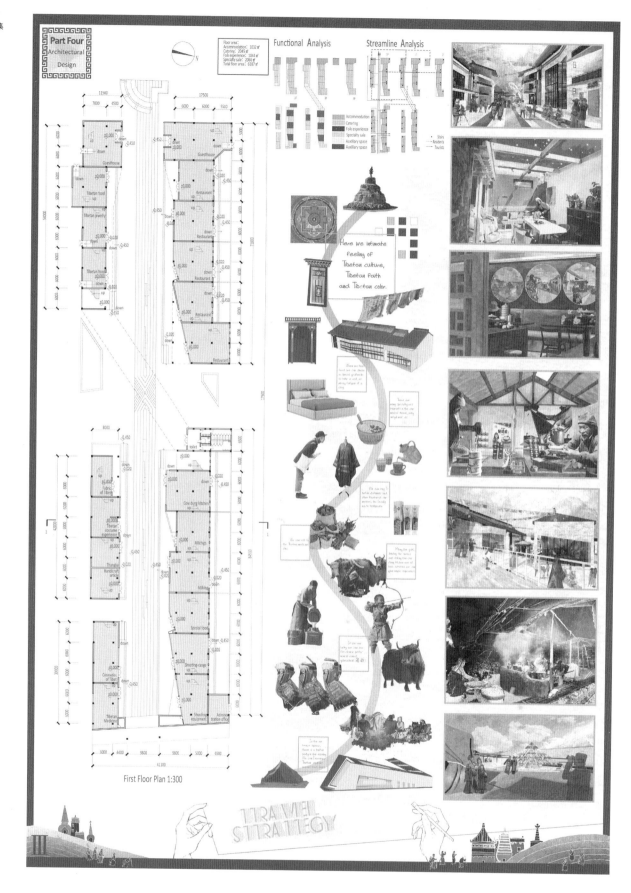

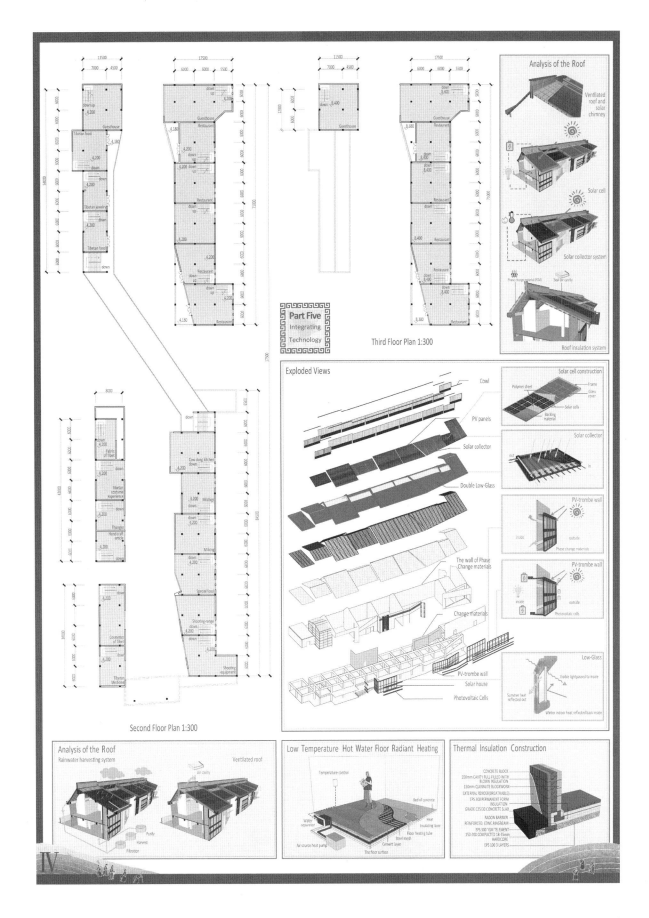

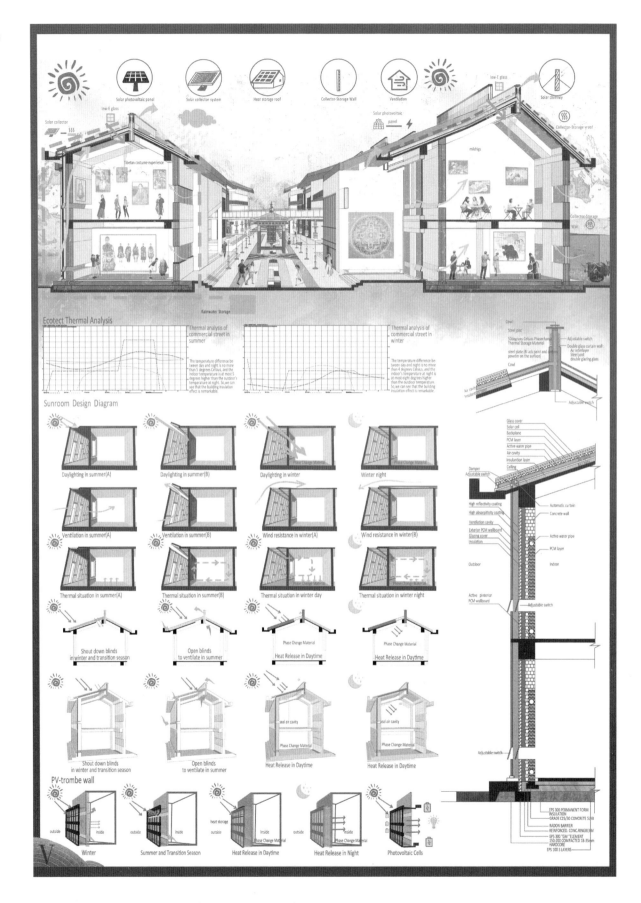

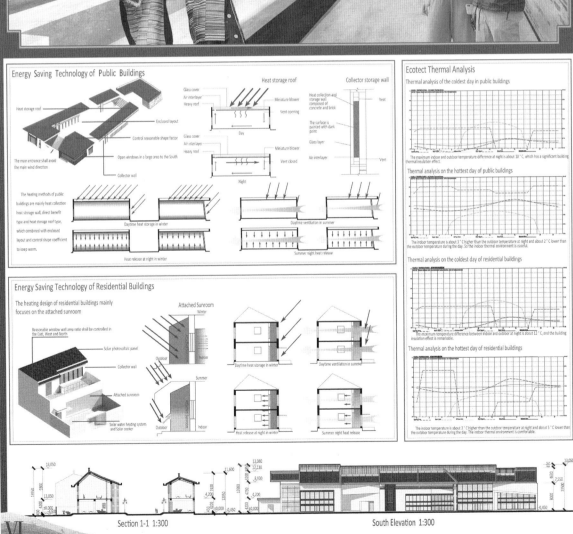

综合奖·优秀奖
General Prize Awarded · Honorable Mention Prize

注 册 号：100129
Registration No：100129

项目名称：藏进太阳里
　　　　　Hide in the Sun

作　　者：郝旌潮、胡梦娇
Authors：Hao Jingchao, Hu Mengjiao

参赛单位：河南工业大学
Participating Unit：Henan University of Technology

指导教师：马　静
Instructor：Ma Jing

藏进太阳里 HIDE IN THE SUN

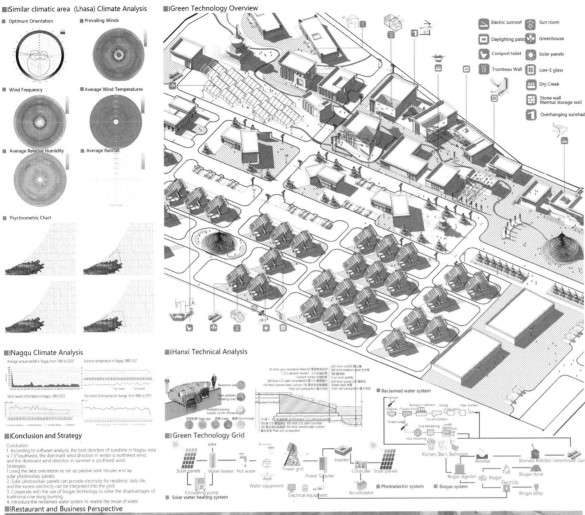

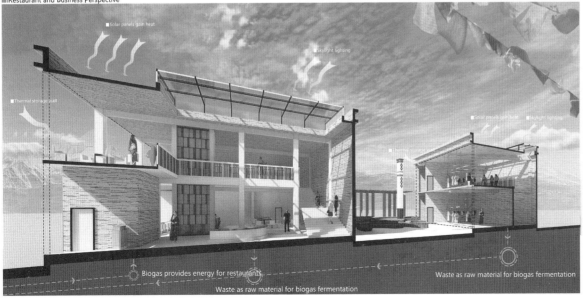

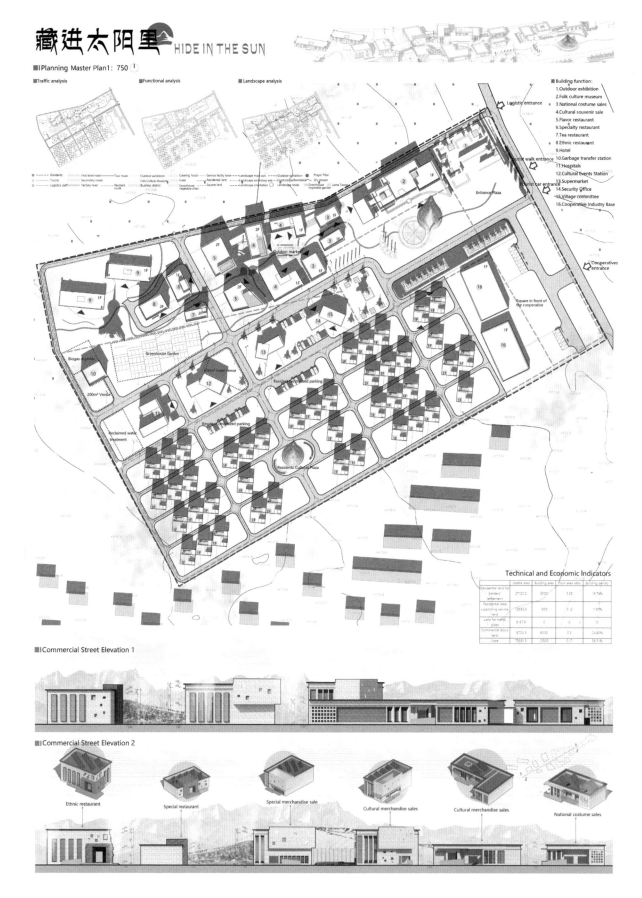

藏进太阳里 HIDE IN THE SUN

Folklore Experience and Special Product Sales Area Plan 1:300

Folk Culture Experience Hall
Cultural product sales

1. 储藏间 Storage
2. 采光天井 Light patio
3. 办公室 Office
4. 会议室 Conference Room
5. 展厅 The exhibition hall
6. 藏族服饰售卖 Tibetan clothing sales
7. 更衣间 Changing room
8. 文化纪念品售卖 Cultural souvenir sales
9. 卫生间 Bathroom
10. 藏医体验 Tibetan medicine experience
11. 看台 The stands
12. 广场 The square
13. 玛尼堆 Mani pile
14. 瞭望台 The observatory
15. 民俗集市 Folk market
16. 旱溪 The dry creek

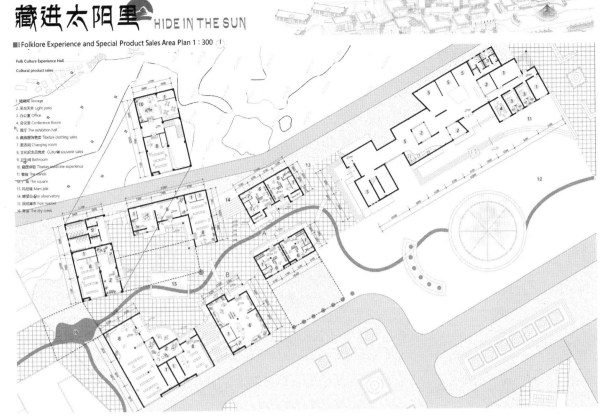

Planning Logic Generation

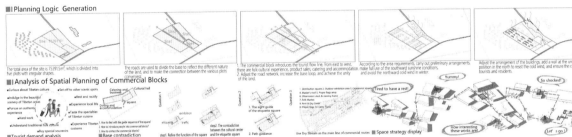

Analysis of Spatial Planning of Commercial Blocks

街区效果图

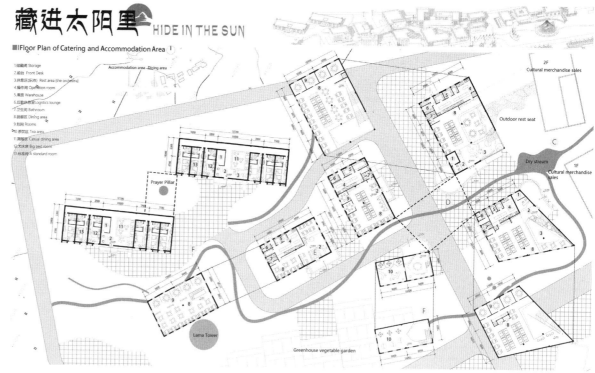
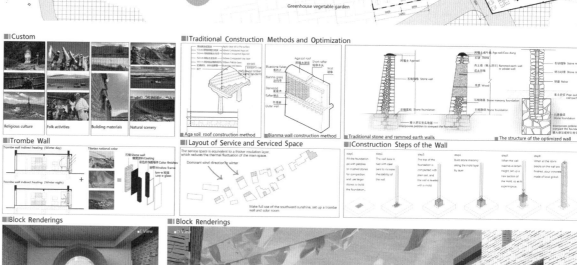

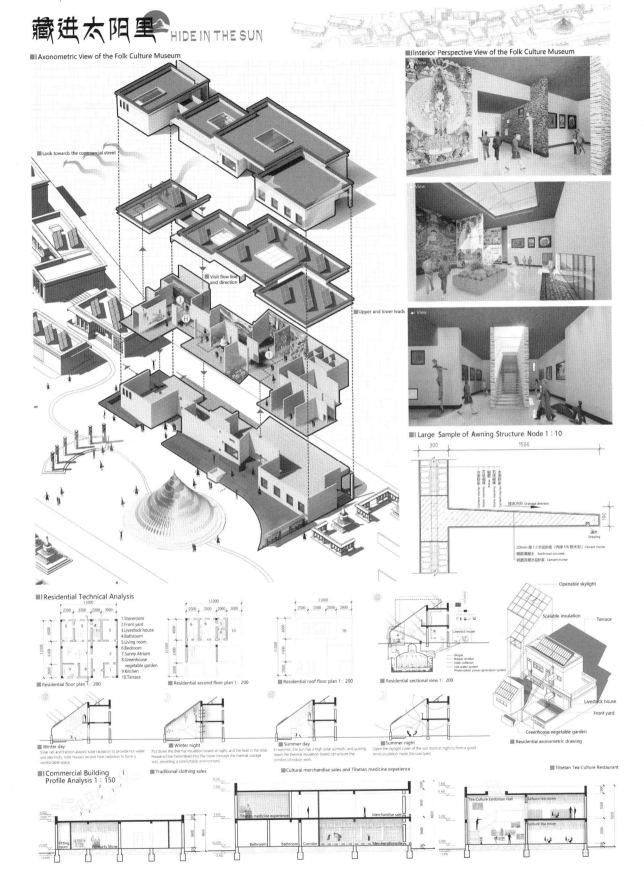

综合奖·优秀奖
General Prize Awarded·
Honorable Mention Prize

注 册 号：100140
Registration No：100140
项目名称：沐光
　　　　　Bathe in Light
作　 者：韩楚玉、王 璐
Authors：Han Chuyu, Wang Lu
参赛单位：河南工业大学
Participating Unit：Henan University of Technology
指导教师：张 华
Instructor：Zhang Hua

沐光
西藏班戈县青龙乡东嘎村低碳社区规划
Dongga village, Qinglong Town, Bange county, Xizang province
Low Carbon Community Programme

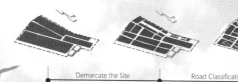

Demarcate the Site　　Road Classification

总用地面积 Land area: 75391.5m²
建设用地面积 Construction land area: 62892.3m²
广场用地面积 Square land area: 6157.9m²
牧民定居用地面积 Land for herders' settlement: 27122.2m²
牧民定居建筑面积 Building area of herdsmen settlement area: 4897.5m²
配套服务用地面积 Supporting service area land: 12632.3m²
配套服务建筑面积 Building area of supporting service area: 1874.2m²
商业区用地面积 Commercial block land: 19703.5m²
商业区建筑面积 Building area of commercial block: 6023.3m²

植被的低覆盖率、电力资源的匮乏与缺水问题是我们对西藏高原地区的一贯印象，但其背后蕴含着丰富的自然资源：风能、太阳能、生物质能等。怎样在最大程度上保持当地地域特色的前提下改善居民生活并通过合理规划推动其旅游业发展，是本次设计面临的主要问题。

本规划从低碳节能的角度出发，结合太阳能光伏发电系统、沼气系统、污水处理系统以及风能利用，以居民的生活体验感受为主要线索，调研提取了其特色民居形式来规划未来东嘎村居民。

日常生活：畜牧、种植、交流……体现在规划后的集中饲养点、种植大棚以及活动广场。在商业区的规划中，以各种功能分区组团的形式使游客的浏览体验最佳，帮助激活旅游产业。

——Low vegetation coverage, lack of power resources, and water shortages are our consistent impressions of the Tibetan Plateau, but behind this is a wealth of other natural resources: wind energy, solar energy, biomass energy, etc. How to improve the lives of residents and promote the development of their tourism industry through reasonable planning while maintaining the local regional characteristics to the greatest extent is the main problem faced by this design.
From the perspective of low carbon and energy saving, this plan combines solar photovoltaic power generation system, biogas system, sewage treatment system and wind energy utilization, taking the residents' life experience as the main clue, research and extraction of its characteristic residential form to plan the future residents of Dongga Village Daily life: animal husbandry, planting, communication... is reflected in the planned centralized breeding sites, planting greenhouses and activity squares.
In the planning of the business district, various functional zones are used to optimize the browsing experience of tourists and help activate the tourism industry.

沐光 — 西藏班戈县青龙乡东嘎村低碳社区规划

Preliminary investigation

Tibetan Architechture

Residence | **Temple** | **House ingredients**

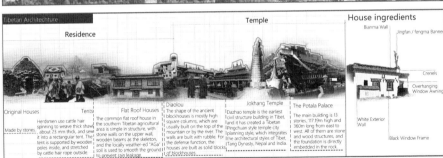

- **Original Houses** — Herdsmen use cattle hair spinning to weave thick shawl, about 23 mm thick, and sewn it into a rectangular tent. The tent is supported by wooden poles inside, and stretched by cattle hair rope outside. Made by stones.
- **Tents**
- **Flat Roof Houses** — The common flat roof house in the southern Tibetan agricultural area is simple in structure, with stone walls on the upper wall, wooden beams as the skeleton, and the locally weather-ed "AGa" soil is used to smooth the ground to prevent rain leakage.
- **Diaolou** — The shape of the ancient blockhouses is mostly high square columns, which are usually built on the top of the mountain or by the river. The walls are built with rubble. For the defense function, the houses are built as solid blocks of blockhouses.
- **Jokhang Temple** — Dazhao temple is the earliest civil structure building in Tibet, and it has created a Tibetan (Pingchuan style temple city planning style, which integrates the architectural styles of Tibet, Tang Dynasty, Nepal and India.
- **The Potala Palace** — The main building is 13 stories, 117.19m high and 360m long from east to west. All of them are stone and wood structures, and the foundation is directly embedded in the rock.

House ingredients: Bianma Wall, Jingfan / fengma Banner, Crenels, Overhanging Window Awning, White Exterior Wall, Black Window Frame

Base Analysis

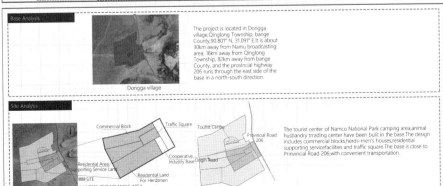

Dongga village

The project is located in Dongga village, Qinglong Township, bange County, 90.801° N, 31.091° E. It is about 30km away from Namu broadcasting area, 16km away from Qinglong Township, 82km away from bange County, and the provincial highway 206 runs through the east side of the base in a north-south direction.

Site Analysis

Commercial Block · Traffic Square · Tourist Centre · Provincial Road 206 · Residential Area · Supporting Service Land · Cooperative Industry Base · Origin Road · Residential Land For Herdsmen · SITE · LONG-TERM PLANING AREA

The tourist center of Namco National Park camping area, animal husbandry trading center have been built in the base. The design includes commercial blocks, herds-men's houses, residential supporting servicefacilities and traffic square. The base is close to Prinvincial Road 206, with convenient transportation.

Folk Culture

Derby

The annual horse race held in June of the Tibetan calendar is a grand traditional festival on the grassland in northern Tibet, also known as the "grassland Festival", which lasts for 5-15 days. Among them, Naggu Jockey Club is the largest. During the festival, herdsmen in northern Tibet, dressed in colorful festival costumes, with abundant food, set up beautiful tents and rode horses to the stadium from all directions.

Present A Hada
Xianhada is the most common etiquet-te of Ali Tibetans. They have the habit of sacrificing hada in weddings and funerals, visiting their elders, meeting Buddha statues, and seeing off on a long journey. Xianhada means purity, sincerity and loyalty.

Hua'er

"Hua'er" is also "Youth", which is a folk song. They are popular among Tibetan people because of their high pitched and bold tunes and melodious and graceful tunes. People not only sing "Hua'er" at ordinary times, but also hold large and small "Hua'er" meetings every year after spring sowing and before autumn harvest. Some of them have been handed down for more than 1000 years.

Warp Wheel

Tibetan beliefs and customs. Tibetan people generally believe in Lamaism, that is, Tibetan Buddhism. To believe in religion, one must often recite the Scriptures, but many people are illiterate, so they turn the scripture wheel and forward it once, which is equivalent to reciting it once. As a result, turning wheel has become a daily activity of many Tibetan people, and many families have hand-held wheel.

Climate simulation Analysis

Optimum Orientation
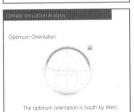

The optimum orientation is South by West.

Annual Incident Solar Radiation
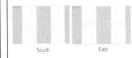

South | East

Lhasa has the highest solar radiation in the south in winter, and the highest solar radiation in the East and West in summer. The variation of solar radiation in winter and transition season is large.

Prevailing Wind Direction

The main wind direction of the base is southeast in summer and northwest in winter.

Psychrometric Chart

Judging from the current wet map and the natural natural ventilation wetting map, the more reasonable passive design that the site can adopt is passive solar heating.

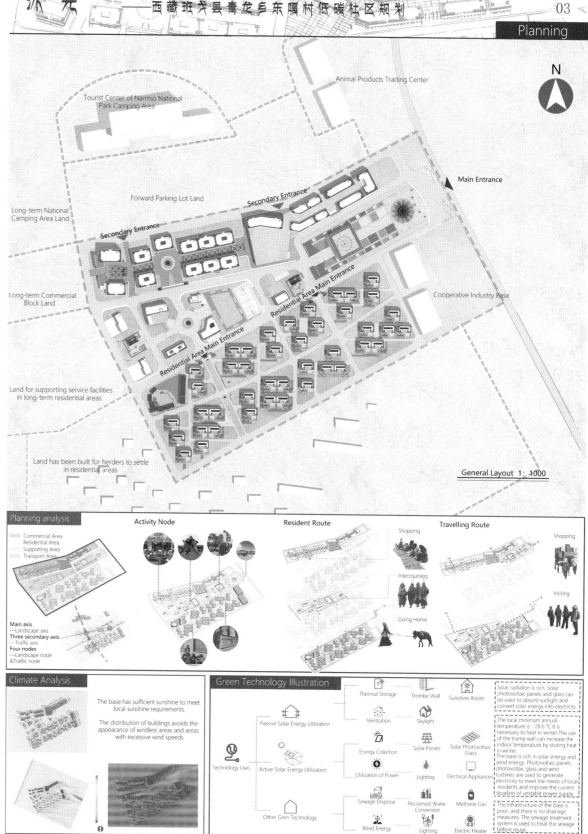

沐光——西藏班戈县青龙乡东嘎村低碳社区规划

Planning

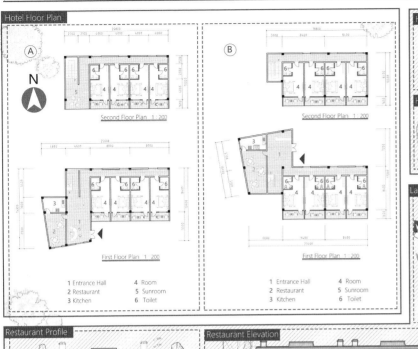

Hotel Floor Plan

Second Floor Plan 1:200
First Floor Plan 1:200

1 Entrance Hall   4 Room
2 Restaurant   5 Sunroom
3 Kitchen   6 Toilet

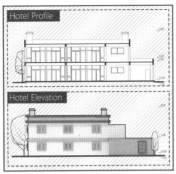

Hotel Profile

Hotel Elevation

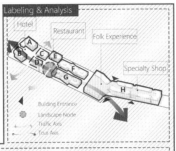

Labeling & Analysis

- Building Entrance
- Landscape Node
- Traffic Axis
- Tour Axis

Restaurant Profile

Restaurant Elevation

Restaurant Floor Plan

1 Canting
2 Sales Window
3 Kitchen
4 Pantry
5 Storeroom
6 Toilet

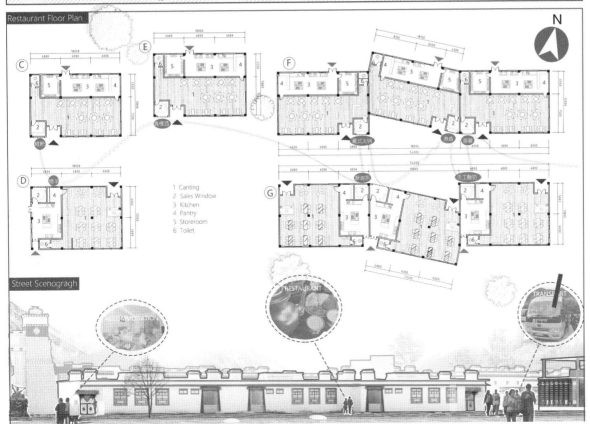

Street Scenograph

ACCOMMODATION · RESTAURANT · TRANSPORT

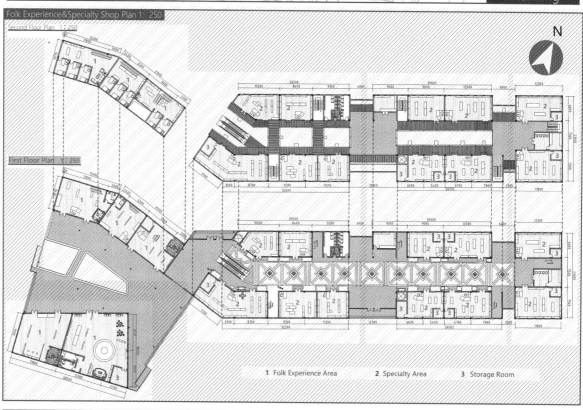

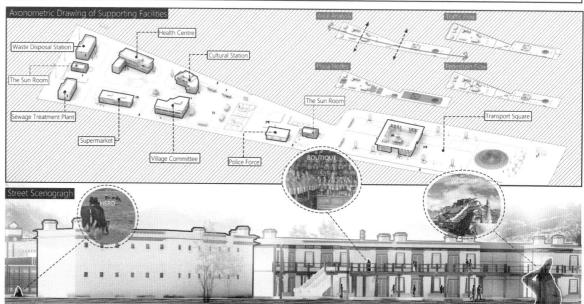

沐光 ——西藏班戈县青龙乡东嘎村低碳社区规划

Strategy

Axonometric Drawing of Residence
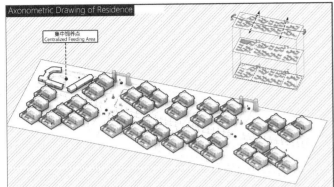

Transparent thin film photovoltaic glass
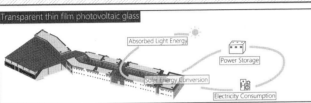

Sunshine Room
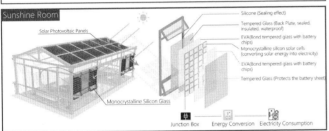

Axonometric Drawing of Residence
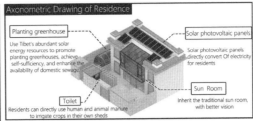

- **Planting greenhouse**: Use Tibet's abundant solar energy resources to promote planting greenhouses, achieve self-sufficiency, and enhance the availability of domestic sewage.
- **Solar photovoltaic panels**: Solar photovoltaic panels directly convert 0f electricity for residents
- **Toilet**: Residents can directly use human and animal manure to irrigate crops in their own sheds
- **Sun Room**: Inherit the traditional sun room, with better vision

Biogas system
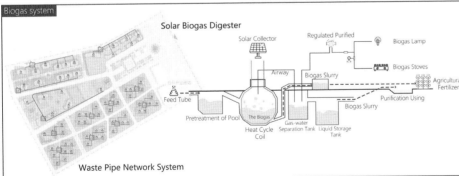

Waste Pipe Network System

Sewage treatment system
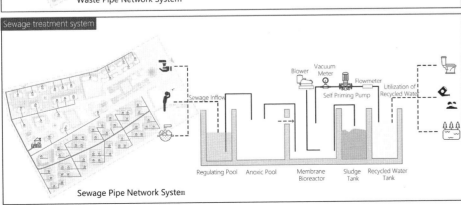

Sewage Pipe Network System

Wind Energy
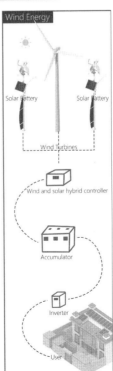

藏纹新译 | A Low-carbon Community Connected by the Dynamic Corridor

综合奖·优秀奖
**General Prize Awarded·
Honorable Mention Prize**

注　册　号：100148
Registration No：100148
项目名称：藏纹新译
　　　　　Revival of Tibetan Patterns
作　　　者：余宜冰、张颖晖、周益萍
Authors：Yu Yibing，Zhang Yinghui，
　　　　　Zhou Yiping
参赛单位：福州大学
Participating Unit：Fuzhou University
指导教师：邱文明
Instructor：Qiu Wenming

Location Analysis

Setting amidst rolling hills with a high altitude in Tibet, the location is filled with abundant solar energy, which can be made full use of in the design.

Concept Generation

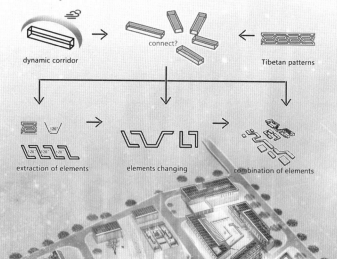

设计说明

本设计从西藏文化和气候出发，旨在设计出符合西藏地域特色的绿色舒适性社区。在文化方面，本次整体规划从藏族花纹中提取设计元素，结合当地多风气候，形成富有当地文化特色和抵御当地寒冷气候的总体布局。立面上提取西藏的经幡塔、边玛草、细长窗等元素，使立面富有地域特色，使步行于社区内的人能感受到当地的文化氛围。

在对气候回应方面，考虑到当地多风的气候特点，基于低碳交通原则，故利用"动力长廊"将室外空间室内化，打造低碳、舒适的室外活动空间。同时充分利用当地丰富的太阳辐射资源解决室内寒冷的问题。通过建筑形态控制以抵挡寒风和获得更多有利的朝向，并利用建筑材料、退台式阳光走廊、太阳能光热转换和太阳能光伏发电，为整个社区提供电力，保障建筑在冬季使用时的舒适性，同时降低社区能耗，最终达到舒适、绿色、低碳社区的设计目标。

Design Specification

Starting from Tibetan culture and climate, our objective is to design a green and comfortable community in line with Tibetan characteristic. Regarding cultural reflection, we extracts design elements from Tibetan patterns, and combine local windy climate, thus forms a general layout full of cultural characteristics and ability to resist local cold climate. On the facade, we use Tibetan Jingfan tower, bianma walls, small window and other elements to make the facade full of regional characteristics, so that people walking in the community can feel the local cultural atmosphere.

Regarding climate response, considering local windy climate and the principle of low-carbon transportation, the "dynamic corridor" is used to transform outdoor space into an indoor space to create a low-carbon and comfortable outdoor activity environment. At the same time, we make full use of local abundant solar radiation resources to solve the problem of indoor coldness. Our design withstand wind and gain more favorable aspect by shaping special modeling, and use several materials, desktop back sunshine corridor, solar-thermal conversion and solar photovoltaic for power generation, making people feel comfortable in winter, as well as lowering energy consumption in the community, finally achieving our objective of designing a comfortable, green and low-carbon community.

藏纹新译 II — A Low-carbon Community Connected by the Dynamic Corridor

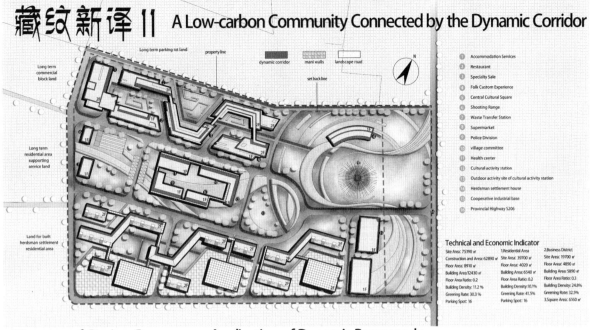

Diagram of Design Process

Application of Dynamic Promenade

1. Residential Area   2. Service Land   3. Commercial Street

- Connecting the building to form a sunshine corridor
- Outdoor Venues
- Enclose the buildings and keep out the wind
- Guide the stream of people

1. Residential Area

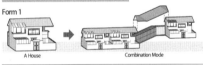
Form 1 — A House → Combination Mode

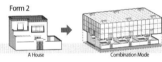
Form 2 — A House → Combination Mode

Planning Generation

Road Network Structure
Commercial Street, Square, Residential Supporting Services, Residential Area, Reserved buildings

Landscape Design
Scenery, Hard ground

Axis and Node
Landscape Axis, Landscape Node

Dynamic Corridor + Sunshine Greenhouse
Dynamic Corridor, Sunshine Greenhouse

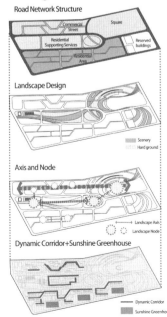

Active Solar Technology

① Solar Panel ② Trombe Wall ③ Window Storing Heat ④ Solar House ⑤ ETFE

ETFE — Pneumatic Die Cushion
Inflatable Tube, Supporting Structure
The anemometer detects the outdoor wind speed, and exchanges the air inside the air pillow with the outside world through the control box solenoid valve, so as to make the space under the film form a microclimate, increasing the thermal comfort.

Visible Light 95% / Wind 5% — Etfe inflatable membrane has good light transmittance, up to 95%.

Solar Panel
Solar Panel → Controller → Inverter → Direct Current → Used for household electricity
Charge and Discharge → Battery → Extra power for municipal facilities such as street lamps.
Working Principle Diagram of Solar Panels

Fabricated Materials, Supporting structure

Trombe Wall with the Dynamic Promenade

Summer Day — Glass Cover / Air Space / Trombe Wall

Summer Night

Winter Day

Winter Night

藏纹新译 III

A Low-carbon Community Connected by the Dynamic Corridor

2 Commercial Street

Plane Layout

Functional Distribution

Landscape Distribution

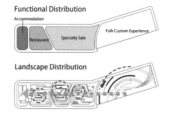

Monomer Generation

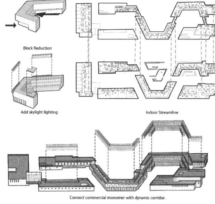

- Block Reduction
- Add skylight lighting
- Indoor Streamline
- Connect commercial monomer with dynamic corridor.

Building Materials

KMEW Photocatalyst Ceramic Plate

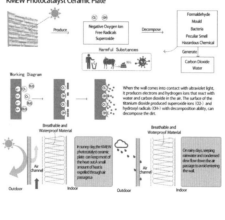

Harmful Substances → Formaldehyde, Mould, Bacteria, Peculiar Smell, Hazardous Chemical → Generate → Carbon Dioxide, Water

Produce → Negative Oxygen Ion, Free Radicals, Superoxide → Decompose

Working Diagram

When the wall comes into contact with ultraviolet light, it produces electrons and hydrogen ions that react with water and carbon dioxide in the air. The surface of the titanium dioxide produced superoxide ions (O2-) and hydroxyl radicals (OH-) with decomposition ability, can decompose the dirt.

Breathable and Waterproof Material — In sunny day, the KMEW photocatalyst ceramic plate can keep most of the heat out. A small amount of heat is expelled through air passage.

Breathable and Waterproof Material — On rainy days, seeping rainwater and condensed dew flow down the air passage to avoid entering the wall.

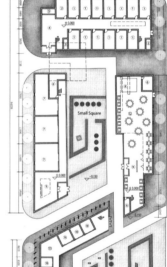

① Guest Room
② Service Room
③ Equipment Room
④ Lobby
⑤ Small Courtyard
⑥ Storeroom
⑦ Unit Restaurant
⑧ Open Specialty Kitchen
⑨ Reception
⑩ Specialty Sale
⑪ Rest Area
⑫ Restaurant
⑬ Folk Custom Experience
⑭ Logistics Room

Small Square

Outside Stair

Landscape Wall

Second Floor Plan 1:400

First Floor Plan 1:400

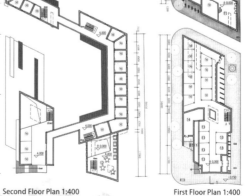

Milk Wall — The wall facade is painted with milk, honey, sugar and Tibetan medicine in proportion, so that the corrosion resistance of the wall becomes better and the wall is not easy to fall off.

Solar Photothermal Panel — Color solar photothermal panel. The color of fengma flag is extracted and designed in combination with solar thermal panels to make the building facade beautiful.

Bianma wall — The branches of bianma are peeled, dried in the sun, tied into thick and thin bricks, and finally painted with a layer of ochre red pigment on the wall. Bianma wall can not only reduce the weight of the top of the wall, but also play the decorative role of color corresponding contrast.

Double-Wall Glass — Double-Wall Glass has good air tightness and thermal insulation. It is suitable for cold areas in Tibet. It is convenient for daylighting and thermal insulation.

Solar Photovoltaic Panel — Solar photovoltaic panels are used on the top of commercial buildings and solar cells are used to directly convert solar energy into electric energy. The electricity produced meets the daily electricity demand of commercial streets.

Street Elevation 1:400

藏纹新译 IV A Low-carbon Community Connected by the Dynamic Corridor

Tibetan Campfire Party

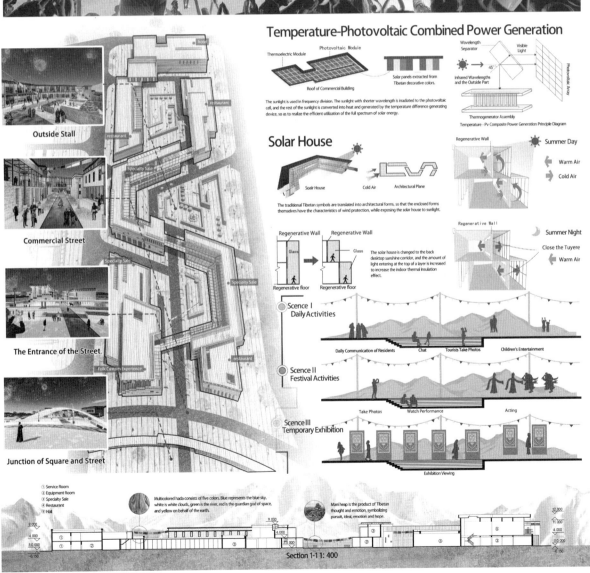

Temperature-Photovoltaic Combined Power Generation

The sunlight is used in frequency division. The sunlight with shorter wavelength is irradiated to the photovoltaic cell, and the rest of the sunlight is converted into heat and generated by the temperature difference generating device, so as to realize the efficient utilization of the full spectrum of solar energy.

Solar House

The traditional Tibetan symbols are translated into architectural forms, so that the enclosed forms themselves have the characteristics of wind protection, while exposing the solar house to sunlight.

The solar house is changed to the back desktop sunshine corridor, and the amount of light entering at the top of a layer is increased to increase the indoor thermal insulation effect.

- Scence I Daily Activities
- Scence II Festival Activities
- Scence III Temporary Exhibition

Section 1-1 1:400

综合奖·优秀奖
General Prize Awarded · Honorable Mention Prize

注 册 号：100238
Registration No：100238
项目名称：藏时风光
Tibetan Scenery
作　　者：王珂、黄志铮、张朔、朱紫悦
Authors：Wang Ke, Huang Zhizheng, Zhang Shuo, Zhu Ziyue
参赛单位：新疆大学
Participating Unit：Xinjiang University
指导教师：塞尔江·哈力克、艾斯卡尔·模拉克
Instructors：Serjian Halick, Eskar Morak

Scheme Description
The design of unloading area is inspired by the square space of traditional Tibetan dwellings. Combined with the prefabricated construction method, the dwellings and businesses are developed with 2.5m and 5m as the basic modules respectively. The site layout comprehensively considers the base boundary and sunshine direction. The residential parts are arranged in a staggered manner, which enhances the wind shelter and the site space experience on the one hand. In the commercial part, air corridors are added for connection to increase the experience of commercial space. At the same time, the three functional areas are organically connected through public space and green corridor in the base, which are independent and connected with each other. The small solar panel unit inspired by the traditional residential roof is reflected in the whole building base. The traditional style of the single part is combined with green technology, and different practical functions are combined to form modern Tibetan buildings with different charm.

设计说明
设计受到西藏传统民居方形空间启发，结合装配式建造方法，民居和商业分别以2.5米和5米为基本模数展开，场地布局综合考虑基地边界和日照朝向，民居部分采用错落式排布一方面增强避风，另一方面增强场地空间体验。商业部分增加空中廊道进行连接，增加商业空间的体验感。同时，在基地内部通过公共空间与绿色廊道将三大功能区有机联系在一起，彼此相互独立又相互联系。以传统民居屋顶为启发形成的太阳能板小单元在整个建筑基地都有体现，单体部分传统风格与绿色技术相互结合，同时结合不同的实用功能形成不同韵味的现代藏族建筑。

Location Analysis

Dongga village, Qinglong Township, Bangor county is located in Naqu City, Tibet Autonomous Region, about 16 kilometers away from Qinglong Township, 30 kilometers away from Namuco scenic spot and 82 kilometers away from Bangor county.

Architectural Culture

| Volume and white wall | Prayer flags | Facade | Accounting room |

Climate Analysis

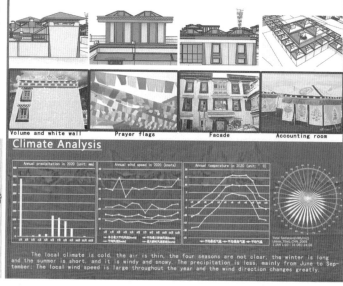

The local climate is cold, the air is thin, the four seasons are not clear, the winter is long and the summer is short, and it is windy and snowy. The precipitation is less, mainly from June to September; The local wind speed is large throughout the year and the wind direction changes greatly.

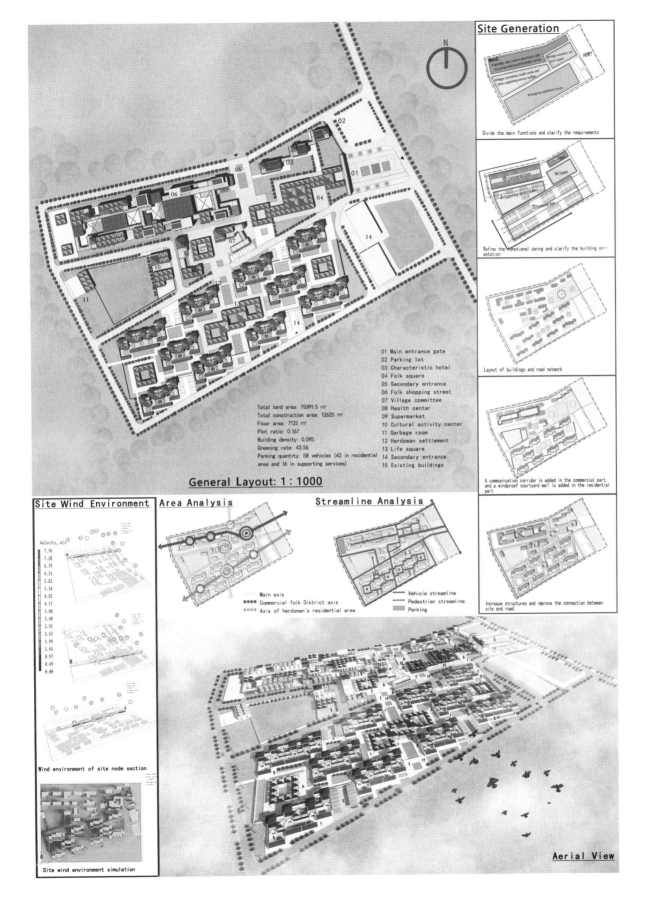

01 Main Entrance Gate

02 Aerial view of the square

02 Square landscape

03 Square landscape

Technical Overview

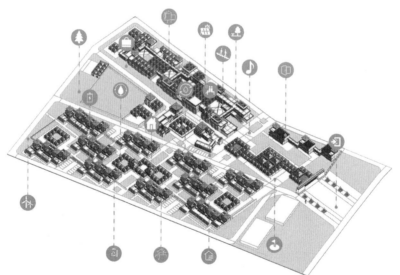

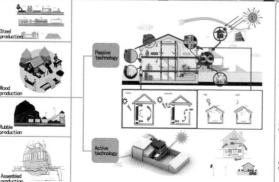

Carbon emission diagram of construction materials in herdsmen's settlements

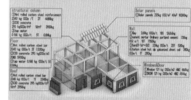

Energy consumption of carbon emission in building construction

Earth moving bulldozer 72H 60kg 11361kg
Excavator 72H 63 kg 11929 kg
Internal pressure of steel wheel 72H 42.95 kg 8133 kg
Road Machines
Foundation works Pile driver 72H
Grouting machine 72H
Frame works Straightentener 72H
Crawler 240H 18.42 kg 10167 kg
crane
Wall works crawler 240H 18.42 kg 10167 kg
Vibrator
Electric welder 10H 154.63kwh 618.52kg
Floor engineering crawler 240H 18.42 kg 10167 kg crane
Vibrator
Electric welder 10H 154.63kwh 618.52kg
total 63,161.04kg

Energy consumption of carbon emissions from building demolition

Crane 120H 18.42kg 5,083.5kg
Floor and Roof Works (Removal) Crawler Type
Crane 120H 18.42kg 5,083.5kg
Other crawler type
Crane 120H 18.42kg 5,083.5kg totol 15,250.5kg

Simulation of Annual Average Energy Consumption in Herdsman Settlements Based on Meteorological Data

Herdsman Settlements

Energy Consumption Items	Unit Energy	Total Energy Consumption	Carbon Emissions
Annual heating energy consumption	45.1kwh/M2	200M2	9020kwh
Annual power supply and energy consumption		1460kwh	1460kwh
Annual domestic water consumption		340T	340T	97.19kg
Annual Solar Panel Power Generation	72kwh/M2	40M2	2880kwh
Annual methane tank heat generation (for heating)	45.1kwh/M2	47%(thermal efficiency)	107.28kwh	45.04kg

Commercial buildings

Energy Consumption Items	Unit Energy	Total Energy Consumption	Carbon Emissions
Annual heating energy consumption		6000M2	
Annual power supply and energy consumption		350,400kwh	350,400kwh
Commercial water use		1533T	1533T	257.7kg
Annual Solar Panel Power Generation	72kwh/M2	5000M2	360000kwh

Estimation of Energy Consumption for Transport of Building Materials

Heavy diesel truck transportation (load 18t)		Transport coefficient	
Types of building materials	Weight of building materials	Transport carbon emissions	Total carbon emissions
C30	2600T	160,600kg	
a steel bar	300T	18,550kg	
doors and windows	350T	19,800kg	277.350kg
Rock wool board	200T	12,000kg	
Solar panels	500T	32,250kg	
other	300T	18,550kg	

Renderings of Herdsmen Settlements

Building Generation **Building Environment Analysis** **Building explosion analysis diagram** Fabricated structure + some local materials

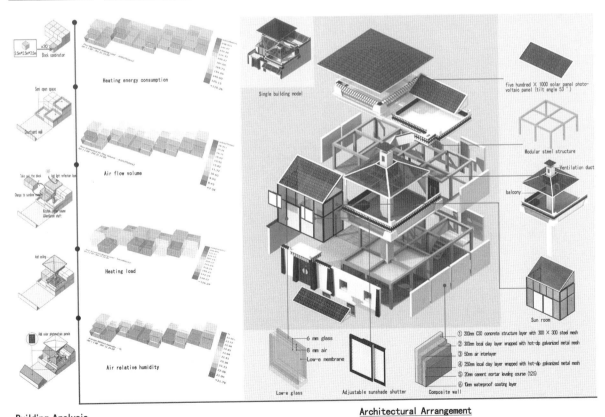

Building Analysis **Architectural Arrangement**

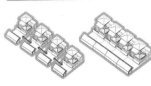

Commercial street renderings

Construction Technology Analysis

Energy utilization strategy
- Active technology: Energy heat pump, Automatic solar folding door, Solar photovoltaic panel, Transparent photovoltaic glass, Trombe wall, Light guide illumination
- Passive technology: Gas tightness, Air flow internal circulation system, Tunnel wind well, Glasshouse, Optimum orientation, Heat storage heavy material

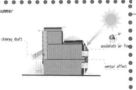

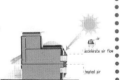
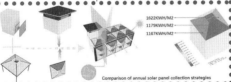

Comparison of annual solar panel collection strategies
- 1622 KWH/M2
- 1179 KWH/M2
- 1167 KWH/M2

Wall structure:
① 200mm C30 concrete structure layer with 300×300 steel mesh
② 300mm local clay layer wrapped with hot-dip galvanized metal mesh
③ 50mm air interlayer
④ 250mm local clay layer wrapped with hot-dip galvanized metal mesh
⑤ 20mm cement mortar leveling course (1:2.5)
⑥ 10mm waterproof coating layer

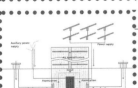
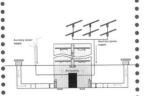
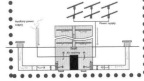

Construction project	Main technologies	Technical advantages
Herdsman settlement	Solar photovoltaic panel power generation technology (combined with shading technology)	Solve the shortage of power supply in Tibet / Help double carbon target / Ladybug calculates the solar altitude angle
	Biogas digester heating system	Solve the problem of power supply for herdsmen in winter / Solve livestock manure
	Carbon saving measures for clay wall (in line with wall thermal)	Respect for regionality / Save building materials and reduce carbon emissions / Increase building air tightness
	Third floor glass window	
Commercial buildings	Solar photovoltaic panel power generation technology (combined with shading technology)	Solve the shortage of power supply in Tibet / Help double carbon target / Ladybug calculates the solar altitude angle
	Venturi effect	Solve indoor ventilation in summer
	Daylight conversion technology	Solve indoor lighting during the day / Reduce carbon emissions
	Transparent photovoltaic glass	Solve the problem of commercial power consumption
apartment	Solar photovoltaic panel power generation technology (combined with shading technology)	Solve the shortage of power supply in Tibet / Help double carbon target / Ladybug calculates the solar altitude angle
	Carbon saving measures for clay wall (in line with wall thermal)	Respect for regionality / Save building materials and reduce carbon emissions
Planning layout	Against the dominant wind direction in winter	Reduce thermal heat dissipation in winter
All carbon simulation technology	According to the national standard GB/T 51366-2019, carbon emissions are calculated from five aspects: construction, operation, materials,	It is concluded that the local clay wall is the best wall material and the carbon emission is the least
Simulation analysis of energy consumption of lady bug and honeybee	According to the designed wall thermal resistance and room layout in combination with EPW	The simulation results are controllable and meet the thermal design standards in cold areas
water	Water filtration device / Rainwater collection device / Reclaimed water treatment	Tibet is short of water, and the per capita water consumption is small. The settlements are equipped with centralized water filtration system to save and reduce emissions
Waste disposal	Waste classification / Biogas digester	Rational utilization of waste, classified treatment, in-situ degradation of construction waste

综合奖 · 优秀奖
General Prize Awarded · Honorable Mention Prize

注 册 号：100276
Registration No：100276

项目名称：阳起风落
Low Carbon Community Which Use Solar and Wind Energy

作　　者：张浩然、刘梓涵
Authors：Zhang Haoran, Liu Zihan

参赛单位：昆明理工大学
Participating Unit：Kunming University of Science and Technology

指导教师：谭良斌
Instructor：Tan Liangbin

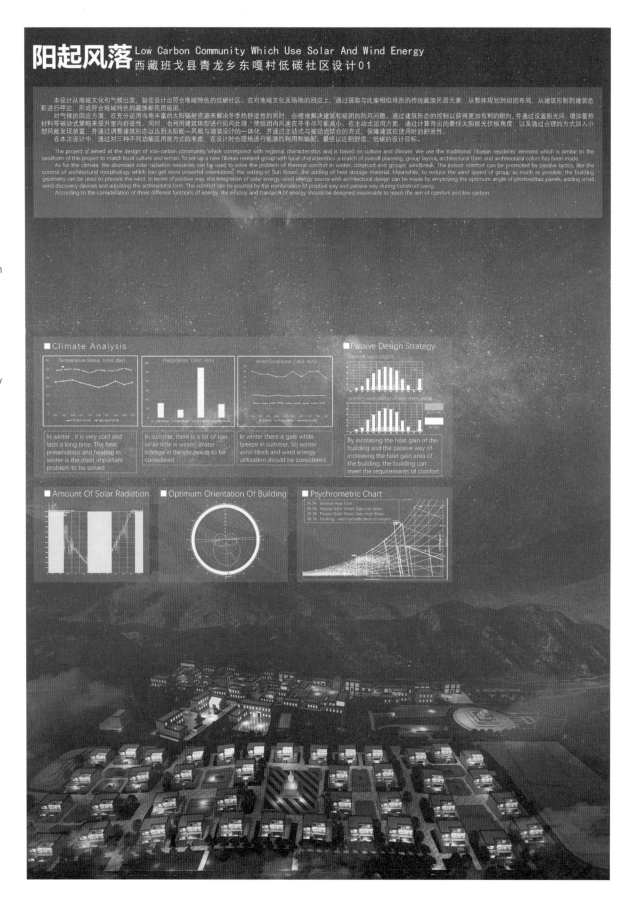

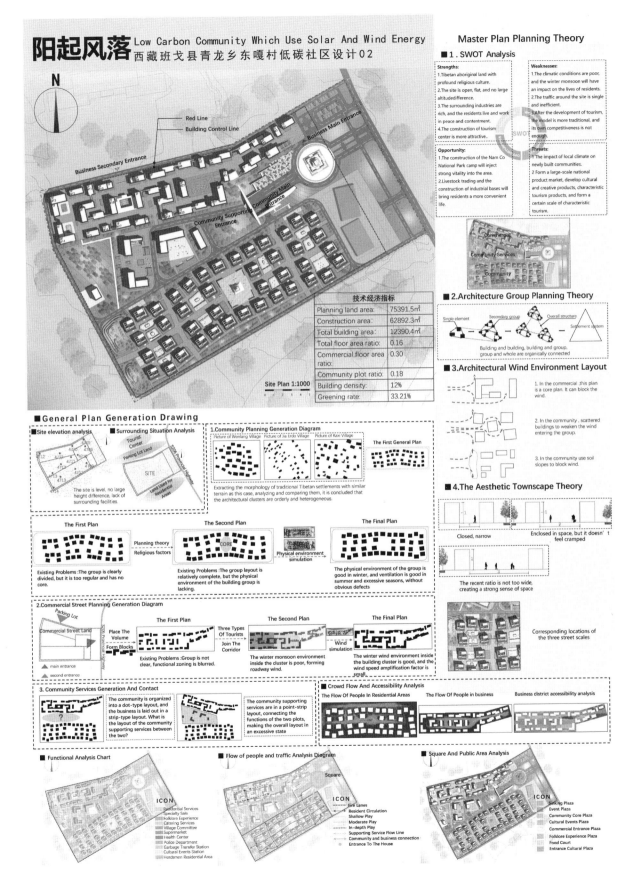

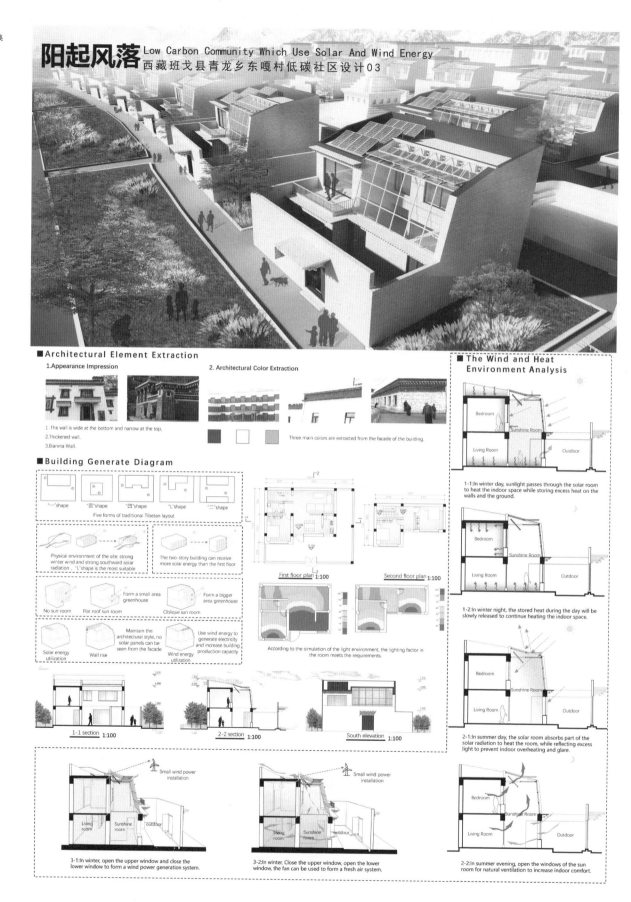

阳起风落 Low Carbon Community Which Use Solar and Wind Energy
西藏班戈县青龙乡东嘎村低碳社区设计 04

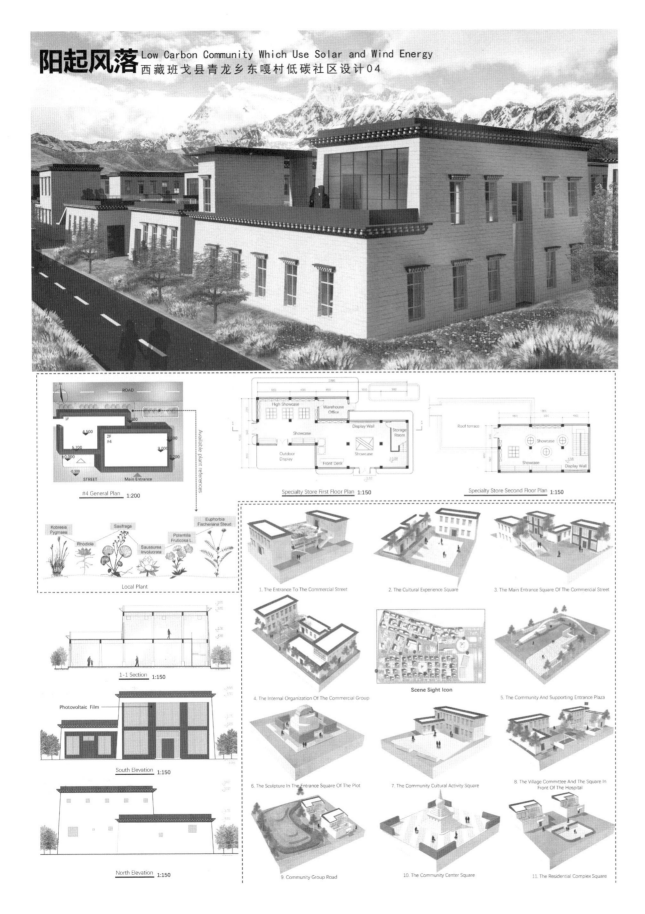

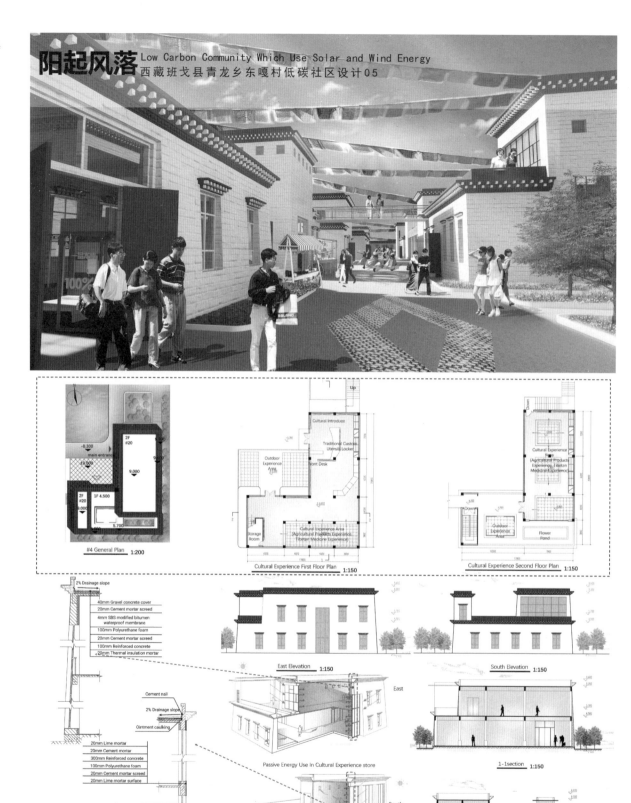

阳起风落 Low Carbon Community Which Use Solar and Wind Energy
西藏班戈县青龙乡东嘎村低碳社区设计 06

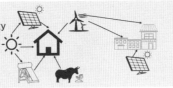

■ Sunshine Shadow Analysis

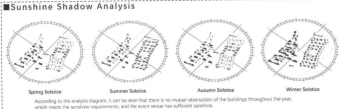

According to the analysis diagram, it can be seen that there is no mutual obstruction of the buildings throughout the year, which meets the sunshine requirements, and the event venue has sufficient sunshine.

■ Wind Environment Analysis

According to the wind environment simulation, there is no large wind field inside the community in winter, which meets the requirements of activities, and there is no strong wind inside the commercial block; in summer and excessive seasons, the internal wind environment of the building cluster is better, which can meet the ventilation requirements.

■ Renewable Energy Utilization

Use solar power, wind power and biogas power generation to meet the daily use of residential houses.

The use of renewable energy

There is a height difference in the square, and at the same time, the surroundings are high, which is used to block the wind. The square on the first floor is used for daily activities of residents, and the second floor is used for stocking livestock.

Square Section · The Square Of The Residential Group

■ Residential Wall Node Diagram

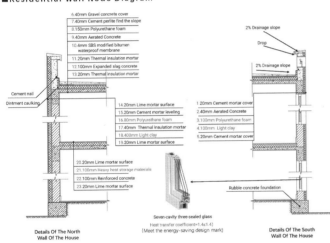

■ Building Energy Efficiency Design

Position	Construction		Thermal Performance	Satisfied
Roof		6.40mm Gravel concrete cover 7.40mm Cement perlite find the slope 8.150mm Polyurethane foam 9.40mm Aerated Concrete 10.4mm SBS modified bitumen waterproof membrane 11.20mm Thermal insulation mortar 12.100mm Expanded slag concrete 13.20mm Thermal insulation mortar	K=0.15≤0.15 (JGJ26-2018 requirements regarding the roof)	Yes
South Wall		1.20mm Cement mortar cover 2.40mm Aerated Concrete 3.100mm Polyurethane foam 4.100mm Light clay 5.20mm Cement mortar cover	K=0.21≤0.25 (JGJ26-2018 on the wall in the requirements)	Yes
North Wall		14.20mm Lime mortar surface 15.20mm Cement mortar leveling 16.80mm Polyurethane foam 17.40mm Thermal insulation mortar 18.400mm Light clay 19.20mm Lime mortar surface	K=0.23≤0.25 (JGJ26-2018 requirements regarding the wall)	Yes

■ Sun And Wind House Structure Node

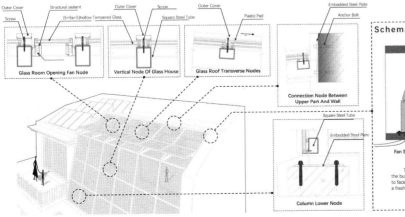

Schematic Diagram Of Wind Power Plant

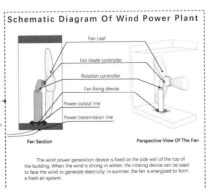

The wind power generation device is fixed on the side wall of the top of the building. When the wind is strong in winter, the rotating device can be used to face the wind to generate electricity; in summer, the fan is energized to form a fresh air system.

综合奖·优秀奖
General Prize Awarded · Honorable Mention Prize

注 册 号：100288
Registration No：100288

项目名称："沐"色·"阳"幡
Bathe in Light Born to the Sun

作　者：李秋儒、张岩、陈文霜、杨晋
Authors：Li Qiuru, Zhang Yan, Chen Wenshuang, Yang Jin

参赛单位：西安建筑科技大学
Participating Unit：Xi'an University of Architecture and Technology

指导教师：何泉
Instructor：He Quan

"沐"色·"阳"幡 1
Bathe in Light Born to the Sun

设计说明

本设计主题为"沐色·阳幡"。"沐"指沐浴阳光，"色"指色彩的光，沐色指在方案建筑中不仅利用了太阳能技术，透过经幡色彩的玻璃，人们也可以感受到彩色的阳光；"阳"指扬帆，有积极向上的生活态度，同时指出阳光的作用及经幡理念的植入。

方案结合主被动太阳能技术，针对基地气候、风向及水源条件，规划建筑群体布局建筑朝向偏东15°方向布局以获得最佳朝向。商业街区部分设置内街，在一个较封闭的环境里给游客提供舒适的环境，故少天气干扰。同时内街顶棚采用彩色光伏玻璃板，不仅发挥了阳光用为建筑集热的作用，而且光伏发电生的同时也可供街区使用。住宅部分分为两个组团，每个组团间设置一个能量生态大棚，以利用及吸收更多的太阳能。作为一个能源互联的核心站，将自身吸收的太阳能转为电能，为住宅用电进行补充；同时，充当了温室大棚的作用，加上循环利用的生物能，村民可在此种植蔬菜。交通部分，规划时根据藏族人民的生活习惯，在场地中设置旁马场、节庆广场，满足人们相聚载舞的需求。

在被动式设计上，建筑物的南向均设置大开窗和阳光廊道或集热墙，冬季可获得太阳热能；在主动式设计上，采用屋面太阳能集热系统、风能和地能能作为辅助热源，进行地板采暖、加热新风、制热水等；在产业化上采用藏族传统建筑材料建造，用土坯砖作为构件主材料，其中民居部分的墙体中间夹杂传统的边玛草，达到建材循环的目的。

The theme of this design is "bathing color · Yang flag". "Bathing" refers to bathing in sunshine. "color" refers to colored light. Bathing color refers to that in the scheme building, people can not only use solar energy technology, but also feel colored sunshine through the glass with the color of prayer flags; "Yang flag" also refers to sailing with a positive attitude towards life, and points out the role of sunshine and the implantation of the concept of prayer flags.
The scheme combines active and passive solar energy technology, and plans the layout of buildings according to the climate, wind direction and water source conditions of the base. Most buildings are arranged 15° south by east to obtain the best orientation. The inner street is set in the commercial block to provide a comfortable environment for tourists in a relatively closed environment with less weather interference. At the same time, the inner street ceiling adopts color photovoltaic glass panels, which not only plays the role of collecting heat for buildings in the sunshine room, but also the electric energy generated by photovoltaic panels can be used for blocks. The residential part is divided into two groups. An energy ecological greenhouse is set between each group to absorb more solar energy. As a core station of energy interconnection, it converts the solar energy absorbed by itself into electric energy to supplement the residential electricity. At the same time, it also acts as a greenhouse. With the recycled bioenergy, villagers can grow vegetables here. For the transportation part, according to the living habits of the Tibetan people, horse farms and festival squares are set in the site to meet the needs of people to gather and dance.
In the passive design, large windows and sunshine corridors or heat collecting walls are set in the south of the building, so that solar heat energy can be obtained in winter; In the active design, the roof solar heat collection system, wind energy and geothermal energy are used as auxiliary heat sources for floor heating, fresh air heating, hot water making, etc. In terms of industrialization, it is built with traditional Tibetan building materials, and adobe bricks are used as the main materials of components, in which the walls of folk houses are mixed with traditional edge grass, so as to achieve the purpose of building material recycling.

● Site Location Analysis
场地分析

Site plan

The project site is located in Dongga village, Qinglong Township, bange County, 90.801°N and 31.091°E. Provincial Highway S206 banluo line passes through the east of the base in a north-south direction.

● Tibetan Cultural Tradition
藏族文化传统

People here have so many recreational activities, they need many squares for singing and dancing. They have strong religious belief and unique characteristics of the architecture.

这里的人们有很多宗教活动，他们需要很多广场来唱歌跳舞。他们有强烈的宗教信仰和独特的建筑特色。

● Year-Round Weather Condition
气候分析

     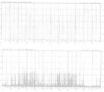

Dongga village, Qinglong Township, bange county is a semi-arid monsoon climate area in the plateau sub cold zone. The climate is cold, the air is thin, the four seasons are not clear, the winter is long and the summer is short, there are many windy and snowy weather. The temperature difference between day and night is large, the annual average temperature is about zero, the annual sunshine time is more than 2850 hours, the annual maximum temperature is 21.9℃, the minimum temperature is -28.6℃, and the precipitation is small.
The annual solar radiation heat gain per square meter is 401.91 kwh/, 58.56 kwh/in winter and 144.59 kwh/in summer.

班戈县青龙乡东嘎村属亚寒带半干旱季风气候区，气候寒冷，空气稀薄，四季不分明，冬长夏短，多风雪天气，昼夜温差大，年平均气温零度左右，年日照时间为2850多小时，年最高气温21.9℃，最低气温-28.6℃，降水量少。

Design Strategy

1 Windproof and Heat Preservation in Heating Period, Insulation and Natural Ventilation in Cooling Period.
2 Passive Energy-saving Technology Mainly, Considering Positive Technology.
3 The Use of Solar Energy, Biomass and Other Renewable Energy Sources.
4. Local Material and Conventional Construction Method.

1.采暖期防风保温，降温期保温自然通风。
2.以被动节能技术为主，考虑积极节能技术。
3.太阳能、生物质能和其他可再生能源的使用。
4.当地材料和常规施工方法。

"沐"色·"阳"幡 2
Bathe in Light Born to the Sun

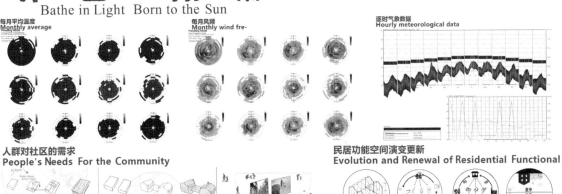

每月平均温度 Monthly average
每月风频 Monthly wind frequency
逐时气象数据 Hourly meteorological data

人群对社区的需求 People's Needs For the Community

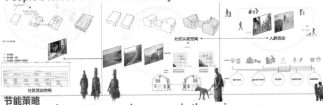

民居功能空间演变更新 Evolution and Renewal of Residential Functional

节能策略 Energy-saving strategy: solar energy is the main energy, geothermal, wind and other multi-complementary new energy system

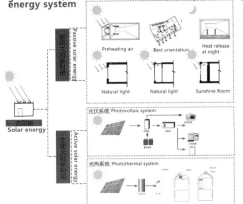

Folk houses have space-time continuity, along with the change of new production mode, the space mode of traditional houses contradicts the modern life style, and with the implementation of the policy of returning farmland to forests, the aquaculture industry launches economic operations such as mushrooms and fungus from high mountain and gorge areas. Farmers need household industry and have new functional requirements. The space function of traditional residential houses can not satisfy the reality of modern mountainous agriculture.

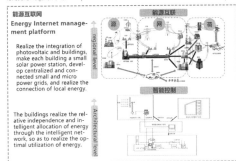

Energy Internet management platform

Realize the integration of photovoltaic and buildings, make each building a small solar power station, develop centralized and connected small and micro power grids, and realize the connection of local energy.

The buildings realize the relative independence and intelligent allocation of energy through the intelligent network, so as to realize the optimal utilization of energy.

消费人群特征 Consumer characteristics

鸟瞰图 Aerial View

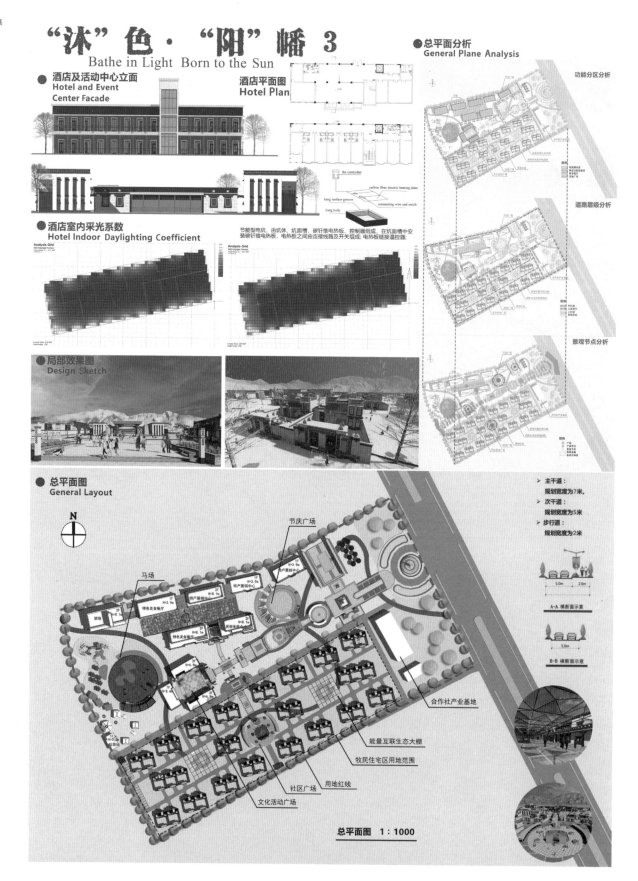

"沐"色·"阳"幡 4
Bathe in Light Born to the Sun

1栋特卖平面 1:350
1# Floor plan of sale 1:350

3栋特卖一层平面 1:350
3# First floor plan of sale 1:350

民俗一层平面 1:350
First floor plan of folklore show 1:350

2栋特卖平面 1:350
2# Floor plan of sale 1:350

3栋特卖二层平面 1:350
3# Second floor plan of sale 1:350

民俗二层平面 1:350
Second floor plan of folklore show 1:350

商业街立面图
Commercial Street Elevation

● 商业街建筑采用地源热泵 Ground Source Heat Pump

冬季供热时，大地作为热泵机组的低温热源，通过垂直系统获取土壤热量为室内提供热量和供应热水。
When heating in winter, the earth is used as the low temperature heat source of the heat pump unit, and the soil heat is obtained indoors through the vertical system. Provide heat and supply hot water.

夏季制冷时，大地为散热场所，室内热量通过垂直系统传入大地，余热被系统回收用来提供热水。
When cooling in summer, the earth is a place for heat dissipation, indoor heat is transferred to the earth through the vertical system, and the waste heat is recovered by the system to provide hot water.

● 酒店采用空气集热器加蓄热地板
Air Collector and Heat Storage Floor

● 商业街的集热呼吸墙
Heat Collecting Breathing Wall of Commercial Street

When there is sunshine in the daytime in winter, most of the solar radiation is absorbed by the surface of the metal collecting plate, and the surface temperature rises. On the one hand, heat is released to the interlayer air, and on the other hand, it is conducted to the room through the wall, that is, radiant heat is supplied. After the air in the interlayer is heated, the temperature rises and forms a natural circulation with the indoor air through the upper and lower air outlets. The hot air continuously enters the room from the upper air outlet and transfers heat to the room. This part of the heat is convective heating.

● 伞状雨水管分析
Analysis of Umbrella Rainwater Pipe

Photovoltaic panels

Venetian blinds reduce sunlight, create shadows

Vegetation layer

Matrix layer

Drainage aquifer

● 光伏采暖大棚
Photovoltaic Heating Greenhouse

● 雨水收集系统
Rainwater Collection System

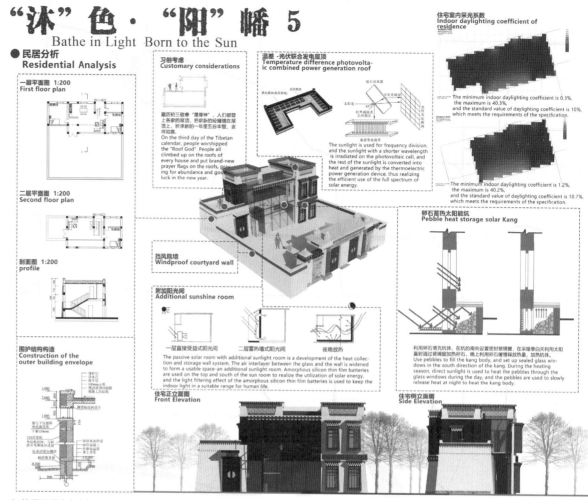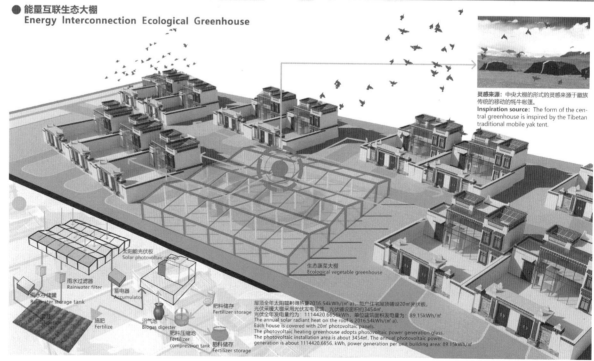

"沐"色·"阳"幡 6
Bathe in Light Born to the Sun

Model Perspective Modeling

Cloud Map of Plane Wind Velocity Distribution at Pedestrian Height in Summer

Under the condition of typical wind speed and direction in summer, there is no eddy or no wind area in the human activity area

Cloud Map of Plane Pressure Distribution at Pedestrian Height in Summer

Under the conditions of typical wind speed and direction in summer, the wind pressure difference between indoor and outdoor surfaces of more than 50% openable external windows is greater than 0.5pa

Cloud Map of Plane Wind Velocity Distribution at Pedestrian Height in Winter

Under the typical wind speed and wind direction in winter, the wind speed at the pedestrian area around the building 1.5m above the ground is less than 5m/s, the wind speed in outdoor rest area and children's entertainment area is less than 2m/s, and the outdoor wind speed amplification factor is small.

Cloud Map of Plane Pressure Distribution at Pedestrian Height in Winter

Under the typical wind speed and direction in winter, except for the first row of buildings facing the wind, the wind pressure difference between the windward surface and the leeward surface of the building shall not be greater than 5pa

Summer Building South Elevation Surface Pressure Cloud

Winter Building South Elevation Surface Pressure Cloud

Site sunshine analysis

- Spring Equinox Day Top View
- Summer Solar Top View
- Autumn Equinox Top View
- Winter Solstice Top View

Annual heat load

According to the calculation results, the annual heat load is 560.5kw

Annual cooling load

According to the calculation results, taking the average value of total cooling value, the annual cooling load is about 98.62kw

The south facade of some houses and hotels on the winter solstice has sufficient sunshine time.

Summary of Full-Year Energy Consumption for Commercial Buildings

Calculated annual solar radiation load

According to the calculation, the estimated annual solar radiation amount on the level surface can reach 2016.54kWh/(a).

Annual Solar Trajectory Map

According to DB simulation, the annual total energy consumption of commercial street is about 693378kwh. Photovoltaic power generation can meet 73% of the total energy consumption requirements of the community throughout the year, and meet the remaining 27% of the energy demand with geo heat pump, photothermal and wind energy systems. The community can achieve the goal of zero carbon emission in the operation process. At the same time, there is a lot of space for photovoltaic panels on the roof of commercial blocks, and the whole community has the potential to realize negative carbon emission.

中水系统 Reclaimed Water System

建筑污水 Construction sewage → 格栅 a grille → 曝气沉沙池 Aeration Sedimentation → 初次沉淀 Primary precipitation → 生物处理 biological treatment → 二次沉淀 Secondary precipitation → 消毒 disinfect → 回用水 Reuse water

填埋 Landfill / 农用 Agriculture / 焚烧 Burn / 沤肥 Waterlogged compost

初始污泥 Initial sludge / 回流污泥 Return sludge / 剩余污泥 excess sludge / 污泥 Initial sludge

洗沙间 Sand washing room → 混合污泥 Mixed sludge → 泥浆浓缩 Mud concentration → 污泥消化 Sludge digestion → 脱水机房 Dehydration machine room

综合奖·优秀奖
General Prize Awarded · Honorable Mention Prize

注 册 号：100352
Registration No：100352

项目名称：行寻圣湖，游居光筑
Green and Blue—A Design of Dongga Low-Carbon Community as the Gate of Holy Lake Namtso

作　　者：张子裕、许碧康、宁娇荣、蔡雨辰
Authors：Zhang Ziyu, Xu Bikang, Ning Jiaorong, Cai Yuchen

参赛单位：南京工业大学
Participating Unit：Nanjing University of Technology

指导教师：胡振宇、王一丁、薛春霖
Instructors：Hu Zhenyu, Wang Yiding, Xue Chunlin

行寻圣湖，游居光筑　Green and Blue
A Design of Dongga Low-carbon Community as the Gate of Holy Lake Namtso

设计说明

"行寻圣湖，游居光筑"是贯穿本次设计的概念。基地被视为游客前往圣湖——纳木措沿途中的"中转站"，为游客提供餐饮、购物、休息、休闲、聚会和祭祀等服务，同时为当地牧民提供完善的住宅及配套基础设施。

总体规划以"两轴六心四片"的结构进行，其中，商业片区以太阳能廊道为动线，串联组织起不同的功能空间。在单体建筑设计方面，建筑材料、造型、细部等借鉴了传统藏式建筑的做法，而建筑构造、结构、设备则融合了现代太阳能技术，由此探索了新藏式太阳能建筑一体化设计的路径。

Green and Blue: An Introduction to the Design

"Green and Blue" is the concept throughout the design which Green refers to the green building and technologies and Blue symbolizes the photovoltaic panel and Lake Namtso.

Located in Dongga, Bange County, the low-carbon community is designed as the "transfer station" on the way to the Holy Lake or Lake Namtso. It provides restaurants, shops, folk products mall, hotel, Tarchin plaza, parking lot and other facilities for the tourists, and also outlines a residential quarter and service buildings for the local herdsmen.

The master plan defines its form by the structure of "two axes—six centers—four areas". In the business area, a long solar corridor is arranged to link the different functional spaces. On the whole, the new style of modern solar building integrated with traditional Tibetan architecture features is explored through the tectonics of materials, structure, shapes, details and techniques.

Location Analysis

Site

Climatic Analysis

Humanistic Analysis

Cultural Custom

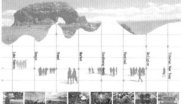

SWOT Analysis

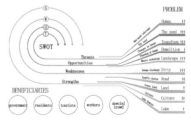

行寻圣湖，游居光筑
Green and Blue
A Design of Dongga Low-carbon Community as the Gate of Holy Lake Namtso

Legend:
1. Community main entrance
2. Tianmen square
3. Parking
4. Memory to Tarchin
5. Souvenir shop
6. Bus stop
7. Impression of Lake Namtso
8. Traces of Mountains
9. Folk experience pavilion
10. Supermarket
11. Cultural activity station
12. Guard office
13. Catering store
14. Villagers' committee
15. Culture forum
16. Commercial Plaza
17. Restaurant
18. Hotel
19. Garbage station
20. Hospital
21. Residential quarter
22. Resident service center
23. Cooperative industrial base
24. Tibetan style garden

Overall Analysis
- Functional Division
- Space Structure
- Traffic Analysis
- Landscape

Traditional Architectural Style
- yellow — unused
- red — use
- white — use
- window — Trapezoidal window
- wall — Contracture wall
- eave wall — Bianma eaves wall

Software Simulation
Wet and dry bulb temperature
Dry Bulb Temperature (C) - Hourly
1 JAN 1:00 - 31 DEC 24:00

Dew Point Temperature (C) - Hourly
1 JAN 1:00 - 31 DEC 24:00

Sunlight Hours Simulation
According to the results, the sunlight hours is more than 2 hours in the winter solstice.

Winter solstice
According to the results, the sunlight hours is more than 2 hours in the major cold.

Major Cold

Conclusion
Dry and wet bulb temperature simulation is used to estimate energy consumption of HVAC buildings. Degree-days and temperature frequency are calculated. After simulation, it is found that the temperature difference between dry and wet spheres is large, so it can be known that the air humidity is small and the water evaporation is large.

Psychrometric chart combined with passive strategy

Sight Analysis

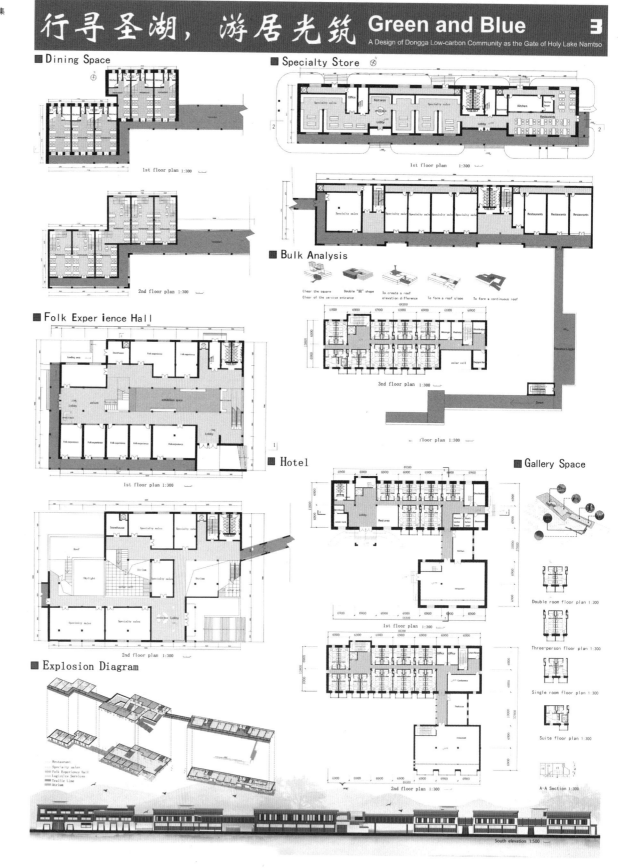

行寻圣湖，游居光筑 Green and Blue

A Design of Dongga Low-carbon Community as the Gate of Holy Lake Namtso

■ Commercial Section

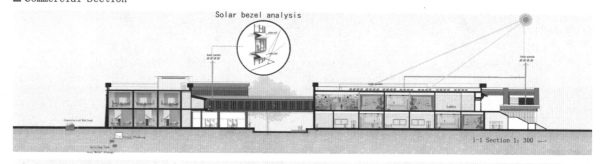

■ Photovoltaic System

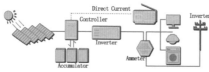

■ Solar Water Heater

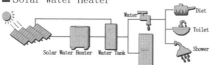

■ Solar Parking

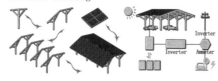

■ Solar Oxygen System

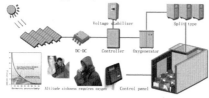

■ Combined Solar and Biogas System

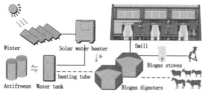

■ Building Construction & Details

Cornice

The cornice is symbolized by white round or square rafters, which are repeated horizontally to form a horizontal composition on the facade.

Eight Auspicious Patterns

Goldfish Vase Conch Wheel
Knot Pennant Umbrella Lotus

Tibetan traditional symbols will be painted on the drawings, which can both block solar radiation and reduce the impact of harsh natural weather on the structure

Door and Window

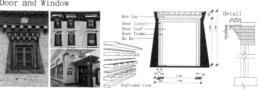

■ Hotel Section

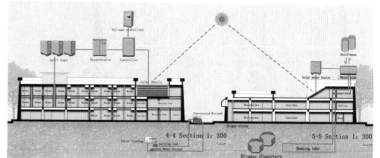

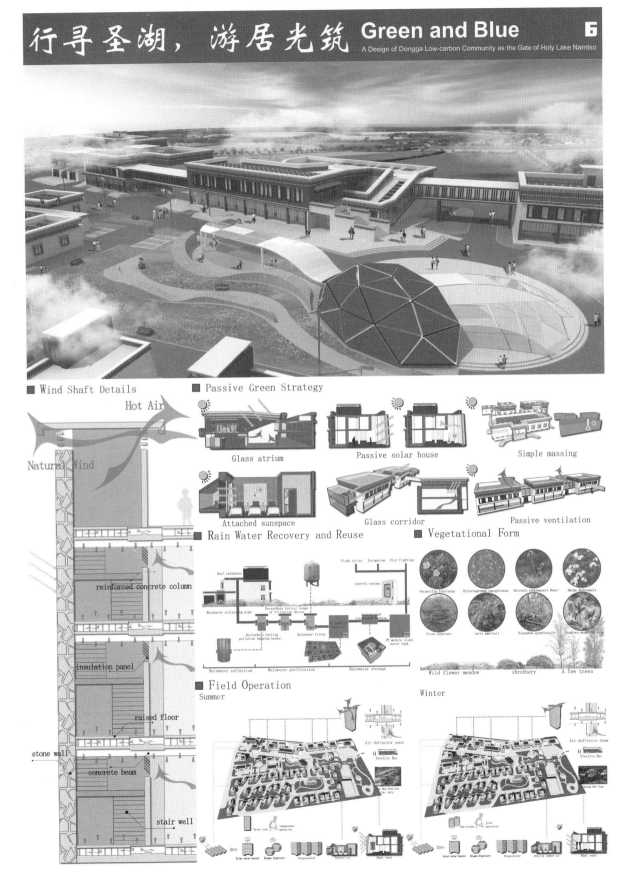

综合奖·优秀奖
General Prize Awarded · Honorable Mention Prize

注 册 号：100358
Registration No：100358

项目名称：坛城·融光
　　　　　Glory City

作　　者：杜　舰、姜　雨、邓立瑞、
　　　　　赵雨濛、林子琪、葛祎明

Authors：Du Jian, Jiang Yu, Deng Lirui,
　　　　　Zhao Yumeng, Lin Ziqi,
　　　　　Ge Yiming

参赛单位：东南大学
Participating Unit：Southeast University

指导教师：沈宇驰、王　伟
Instructors：Shen Yuchi, Wang Wei

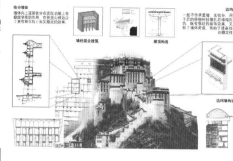
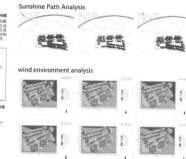

项目地位于班戈县青龙乡东嘎村，北纬90.801°，东经31.091°距纳木措景区约30公里，距青龙乡16公里，距班戈县82公里。省道S206班洛线从基地东侧南北向穿过。

该项目将计划建设为集居住、旅游接待为一体化的低碳社区，目前已建成纳木错国家公园露营区游客中心、畜牧品交易中心、合作社产业基地等建筑。

The project site is located in Dongga village, Qinglong Township, bange County, 90.801° n, 31.091° e, about 30km away from Namucuo scenic spot, 16km away from Qinglong Township and 82km away from bange county. Provincial Highway S206 banluo line passes through the east of the base in a north-south direction.

Sunshine Path Analysis
wind environment analysis
humidity analysis

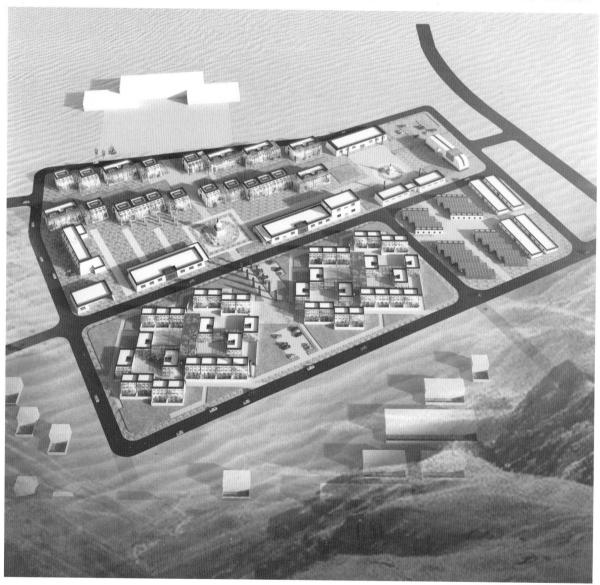

设计说明：

本设计以"坛城 融光"为主题，力图创造传统藏地人文精神与现代绿色生态技术的有机耦合，"坛城"既象征着藏传佛教宇宙世界观的本源，也代表着青藏地区最具民族特色的城镇空间结构。我们以坛城作为基本空间原型，以适应藏区生产条件与运输条件的、低技的被动式绿色生态方法为出发点，利用"阳光房体系"，克服高原严峻的自然环境，将充沛的阳光资源汇集起来，营造"高原绿洲"，创建"阳光坛城"。

居住、商业板块的阳光房系统，结合基于藏区传统玛尼墙创新改造而成的蓄热墙体、基于传统藏式窗户转译而成的保温窗套以及现代化的可调节遮阳和采光天窗等技术，极好适应严峻的高原气候以及完成建筑的在地性表达。此外，通过汽车营地的光伏发电系统，为整个场地提供绿色能源，完成太阳能热水系统、污水无害化处理系统与垃圾处理系统的构建，为当地牧民生活以及纳木措北岸游客旅游营造绿色生态的宜居环境。

The design is based on the theme of "Glory City", trying to create an organic coupling between traditional Tibetan humanistic spirit and modern green ecological technology. "Mandala" not only symbolizes the origin of Tibetan Buddhism cosmic world view, but also represents the most ethnically characteristic urban spatial structure in Qinghai-Tibet area. We take "Mandala" as the basic space prototype, and take the low-tech passive green ecological method that adapts to the production and transportation conditions in Tibetan areas as the starting point. We use the "Conservatory System" to overcome the harsh natural environment of the plateau and bring plenty of sunshine. Collect resources to create a "plain oasis" and create a "sunshine mandala".

Conservatory system of the residential and commercial sectors, combined with the thermal storage wall based on the innovative transformation of the traditional Tibetan Mani wall, the thermal insulation window cover based on the traditional Tibetan window translation, and modern adjustable sunshade and daylight skylights and other technologies. It adapts well to the severe plateau climate and completes the local expression of the building. In addition, the photovoltaic power generation system of the car camp provides green energy for the entire site, completes the construction of a solar hot water system, a harmless sewage treatment system and a garbage disposal system, and creates a green ecology for the life of local residents and tourists on the north bank of Nam Co. Livable environment.

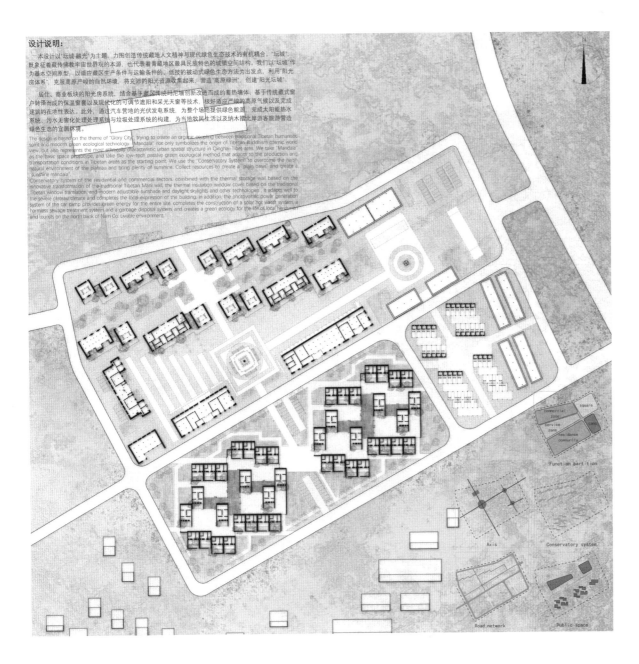

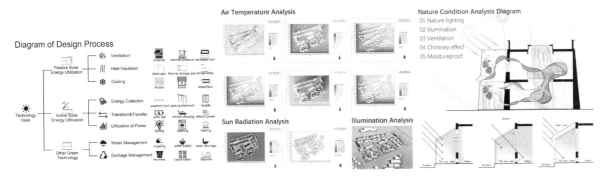

2021 台达杯国际太阳能建筑设计竞赛获奖作品集

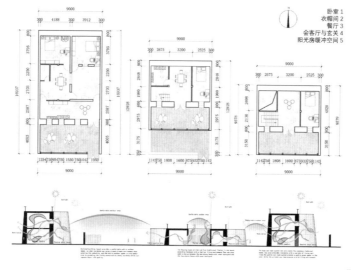

air temperature analysis

居住区的气温随太阳升起逐渐升高，且明显可看出组团内部的温度较外部要高。

The temperature in the residential area gradually rises as the sun rises, and it can be clearly seen that the temperature inside the cluster is higher than the outside.

wind environment analysis

居住区内部的风环境因组团的布局形式有明显改善，风速可较外部降低一倍。

The wind environment inside the residential area has been significantly improved due to the group layout, and the wind speed can be doubled compared to the outside.

sun radiation analysis / relative humidity analysis

居住区的太阳辐射整体较强，阳光房利用很合理，内部的相对湿度也较为稳定。

The overall solar radiation in the residential area is relatively strong, the use of the sun room is reasonable, and the relative humidity inside is relatively stable.

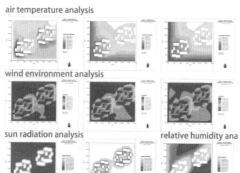

沿南北向的剖轴测图可清晰地看出，两种形式的住宅都充分发挥了阳光房、玛尼墙、庭院蓄热、气体交换的功能，通过被动式的技术获得了适宜的居住条件。

The cross-sectional isometric drawing along the north-south direction clearly shows that the two types of houses have fully utilized the functions of Conservatory, the mani wall, the courtyard heat storage, and the gas exchange, and have obtained suitable living conditions through passive means.

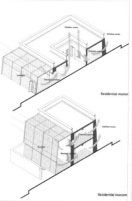

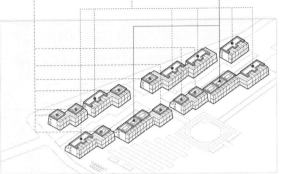

场地划分	体量敲定	"锯齿状"相错	阳光房走廊-绿化
Two axes intersect, dividing the site into four zones.	Settled into a long volume	Zigzag form	Sun room corridor - Green

1. 自然通风 natural ventilation
2. 热压通风 thermal pressure ventilation
3. 绿化 greening
4. 室内商业街 interior pedestrian street
5. 日间 daytime
6. 晚间 night
7. 光照 sunlight

村委会 2F：
建筑坐落于广场尽头面向广场，具有村委会象征的轴对称的形体。

卫生院 2F：
与商业建筑一起，面对场地入口形成对外完整界面。

文化活动站 1F：
沟通南面居民区和居民活动广场。

警务处 1F：
与商业、广场、居民区均联系便捷，便于治安管理。

Village Committee 2F:
The building is located at the end of the square facing the square and has an axisymmetric shape symbolized by the village committee.

Health Center 2F:
Together with the commercial building, it forms a complete external interface entrance of the site.

Cultural Activity Station 1F:
Communicate with the southern residential area and residential activity square.

Police Force 1F:
It has convenient connections with business, squares, and residential areas, and is convenient for public security management.

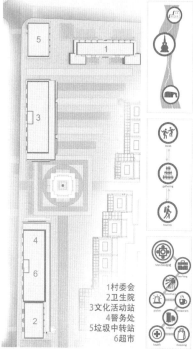

1 村委会
2 卫生院
3 文化活动站
4 警务处
5 垃圾中转站
6 超市

交通文化广场可分别通往其他三个功能地块。一座白塔位于文化广场中央，有同时引向商业、坛城、汽车营地三条轴线上之妙，使西藏特有的神圣氛围在场地入口得以充分烘托。

The traffic culture square can lead to the other three functional plots. A white pagoda is located in the center of the culture square, and it is located on three axes at the same time, so that the unique sacred atmosphere of Tibet can be fully emphasized at the entrance of the site.

贴墙面布置一排转经轮，极强的透视感将视线引向坛城广场，亦朝向虔诚的教徒们心中的纳木措。

A row of prayer wheels are arranged on the wall, and the strong sense of perspective leads the sight to the Mandala Square, and also to the Namco in the hearts of the devout believers.

 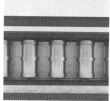 

坛城广场是整个场地的中心，是"坛城"中最高地位的象征。洁白佛塔是信仰的终极表达与完美寄托。人群流线于此汇集，可供举行盛大的节日庆典。

The Mandala Square is the center of the entire site and the symbol of the highest status in the "Mandala." The white stupa is the ultimate expression and perfect sustenance of faith. This square is a place where crowds gather and can be used for grand festivals.

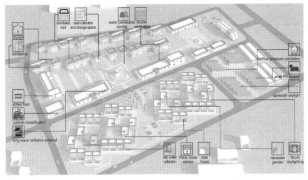
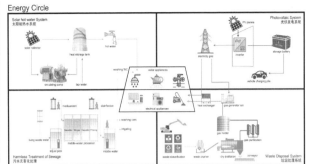

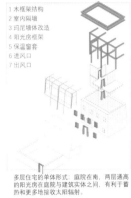
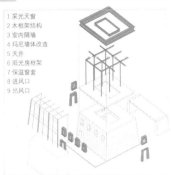

单层住宅的单体形式:阳光房在南,建筑中间置入庭院,庭院方向面对组团内部,有利于蓄热和气体交换。

The single form of single-storey house: Conservatory is in the south, the courtyard is placed in the middle of the building, and the courtyard faces the inside of the cluster.

多层住宅的单体形式:庭院在南,两层通高的阳光房在庭院与建筑实体之间,有利于蓄热和更多地接收太阳辐射。

The single form of multi-storey houses: the courtyard is in the south, and the two-storey Conservatory is between the courtyard and the building entity.

商业建筑的单体形式:阳光房在南,建筑采用木结构框架和玛尼墙,北边为灰空间的形式,顶部设计有天窗。

The single form of the commercial building: Conservatory is in the south, the building uses a wooden frame and mani walls, and the north is in the form of a gray space, with a skylight on the top.

营地的单体形式:朝南的停车位为钢框架结构,上面铺设有太阳能光伏板,北部为可休息的小建筑,旁边有充电桩。

The single form of the camp: the south-facing parking space is a steel frame structure with solar photovoltaic panels laid on it, and the north is a small building for rest, with charging piles next to it.

夏季通风:烟囱效应
ventilate in summer: stack effect

冬季蓄热:暖房
thermal storage in winter: conservatory

居住用电 汽车充电桩
for life   Charging Station

可供设备充电
for device

综合奖·优秀奖
General Prize Awarded·
Honorable Mention Prize

注 册 号：100371
Registration No：100371

项目名称：光环·藏腔
Light Ring· Hot Cavity of Tibetan

作 者：王妍淇、张 雯、凌嘉敏
Authors：Wang Yanqi, Zhang Wen, Ling Jiamin

参赛单位：重庆大学
Participating Unit：Chongqing University

指导教师：周铁军、张海滨
Instructors：Zhou Tiejun, Zhang Haibin

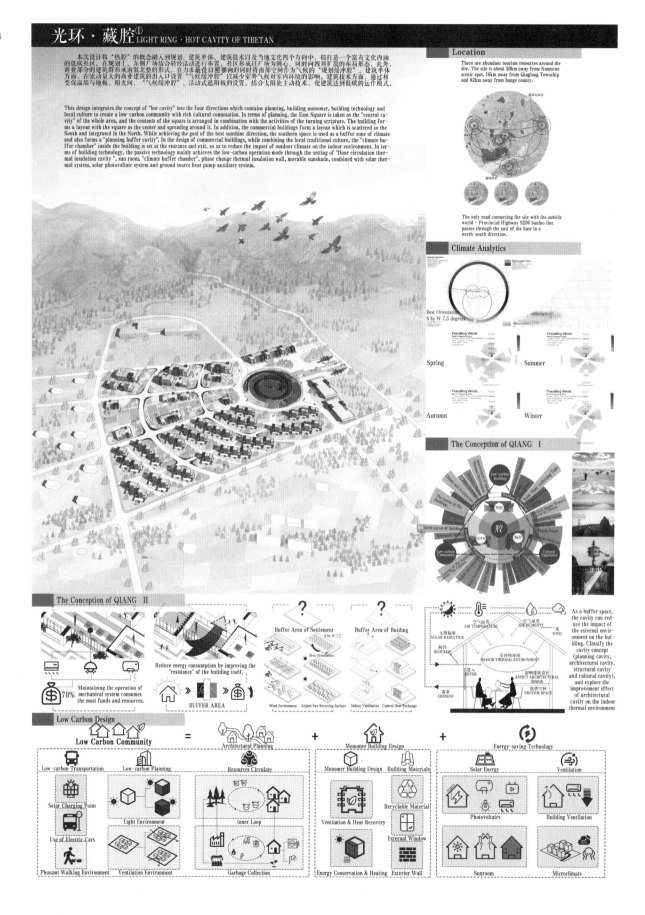

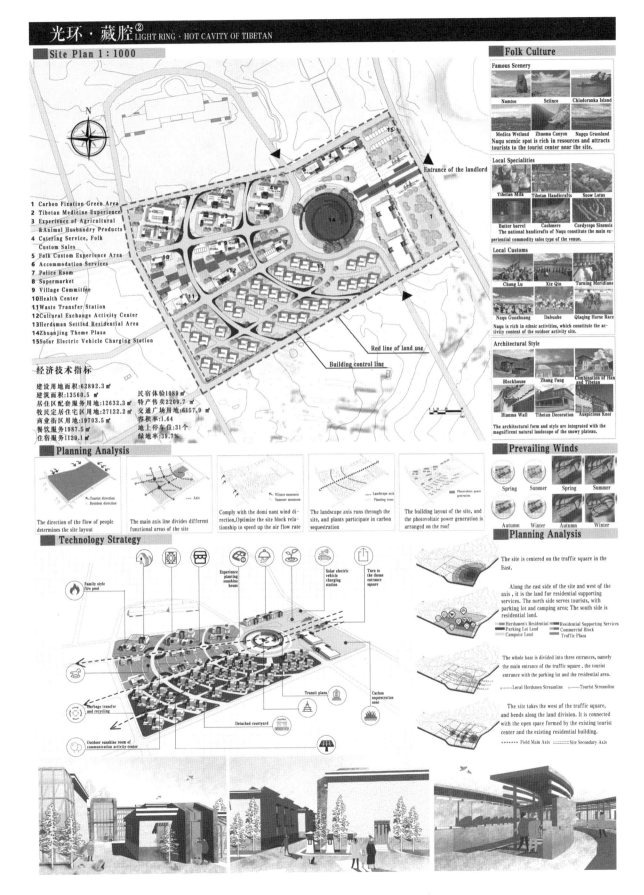

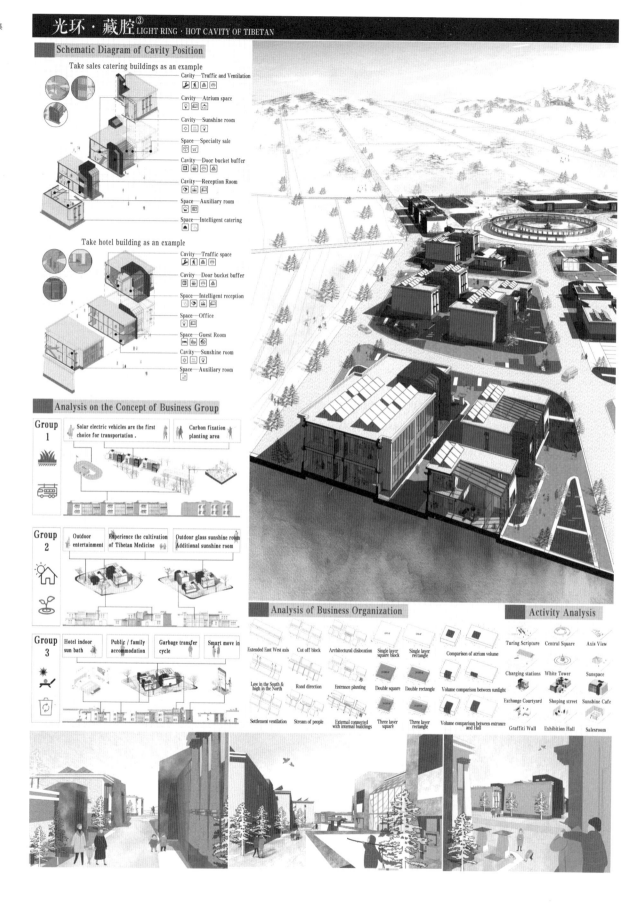

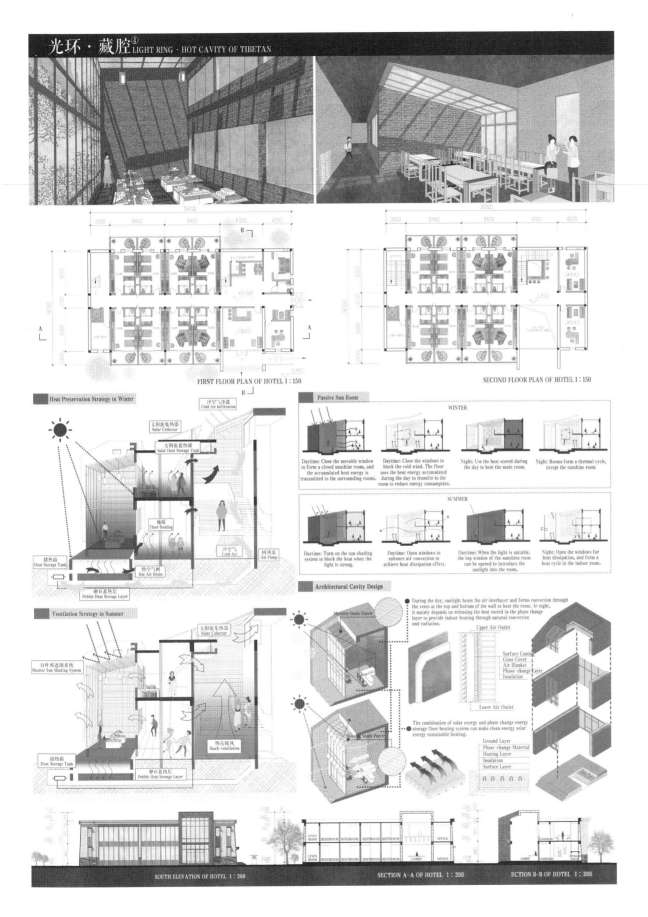

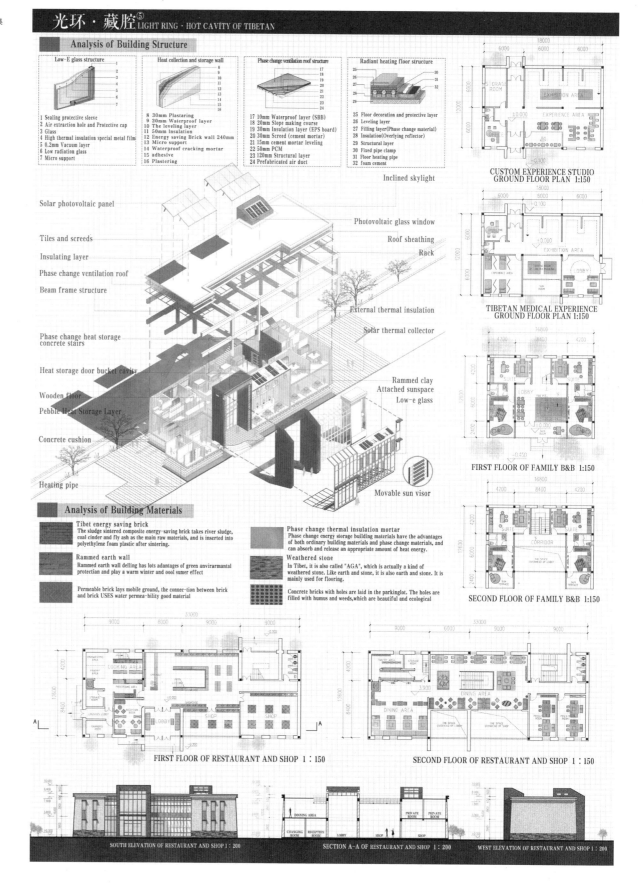

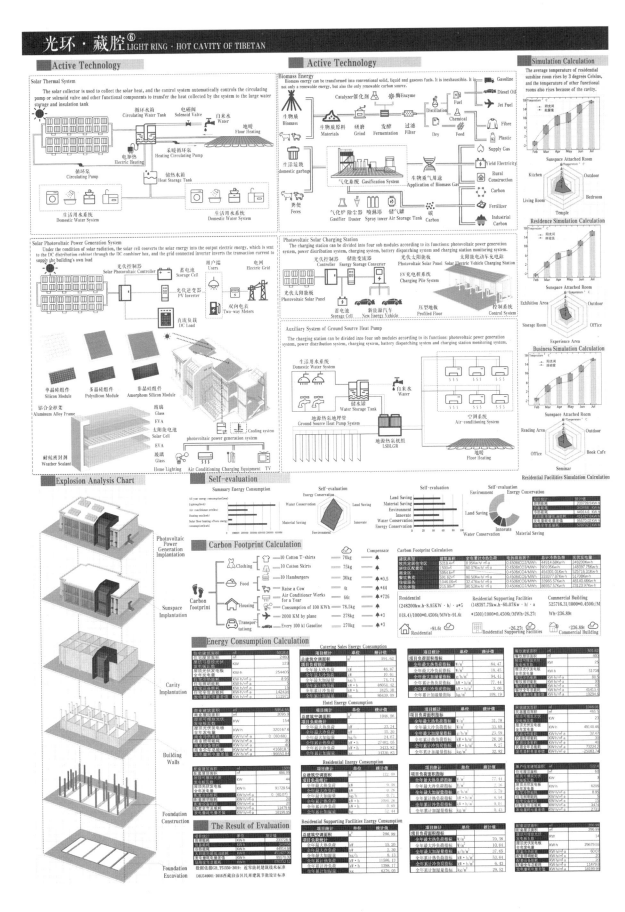

综合奖·优秀奖
General Prize Awarded·Honorable Mention Prize

注 册 号：100382
Registration No：100382
项目名称：回聚
　　　　　Cycling and Converging
作　　者：陈彦儒、陈潇语、李匀思
Authors：Chen Yanru, Chen Xiaoyu, Li Yunsi
参赛单位：福州大学
Participating Unit：Fuzhou University
指导教师：邱文明
Instructor：Qiu Wenming

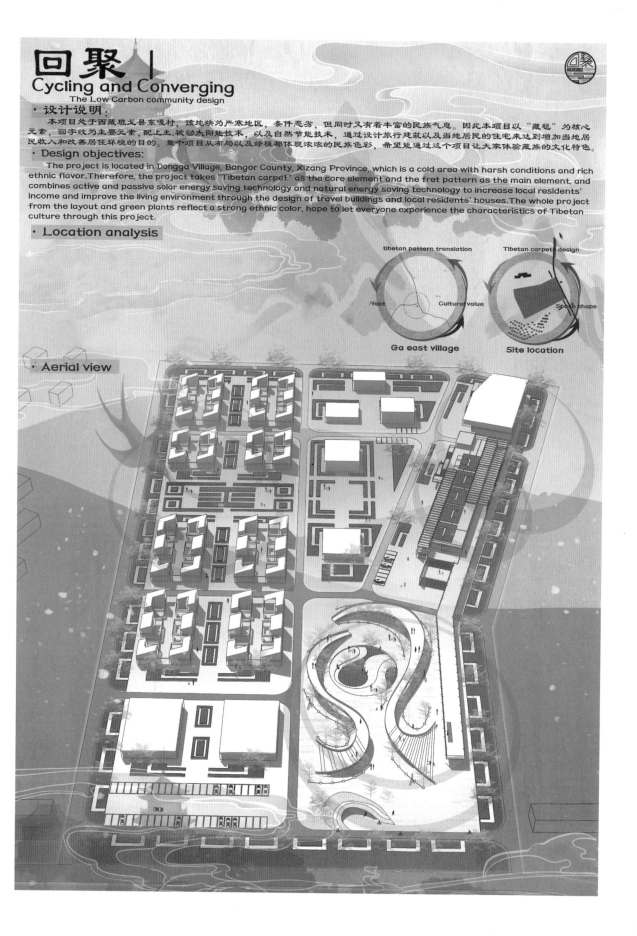

回聚 Cycling and Converging
The Low Carbon Community Design

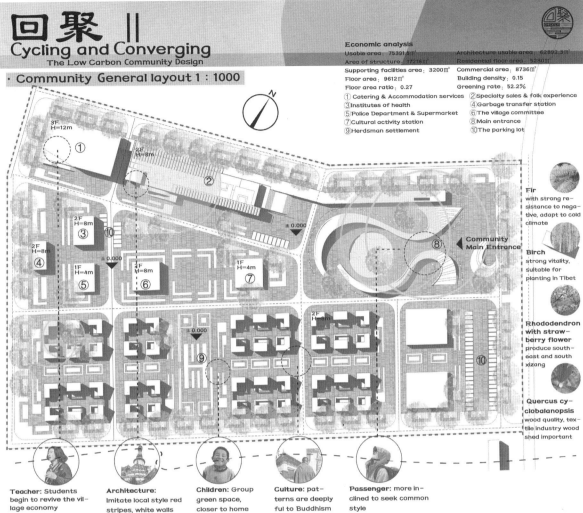

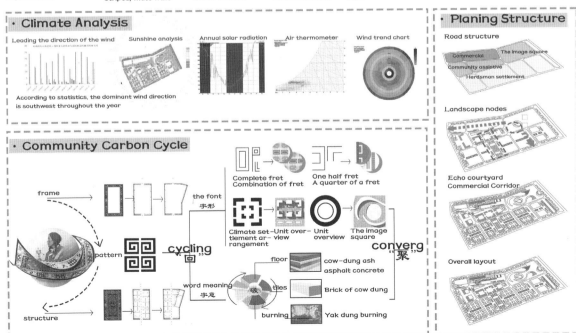

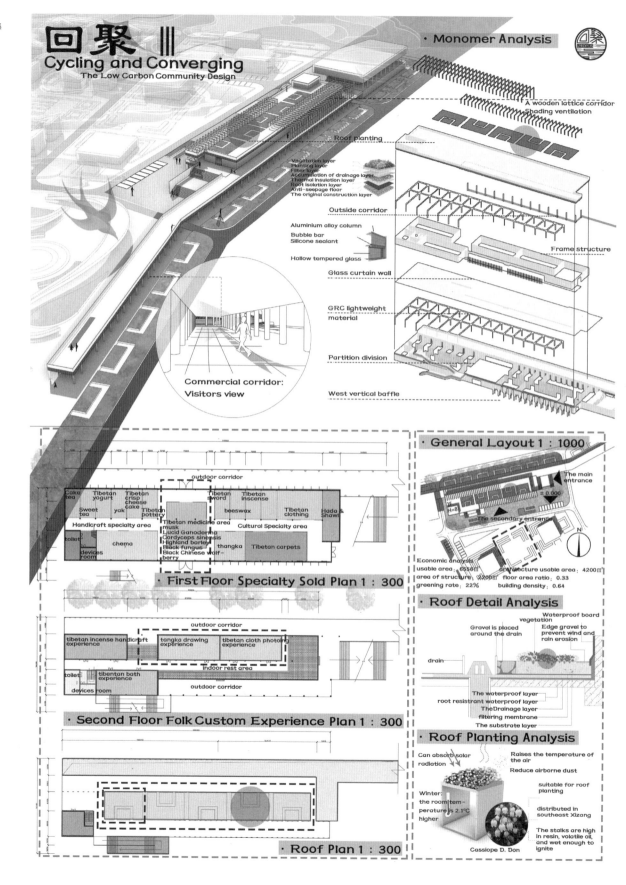

回聚 IV
Cycling and Converging
The Low Carbon community design

- South elevation 1:300

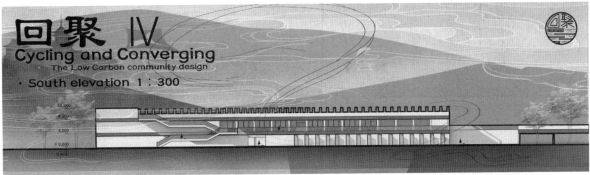

- **Herdsmen Community Residential Area Single Design Part**

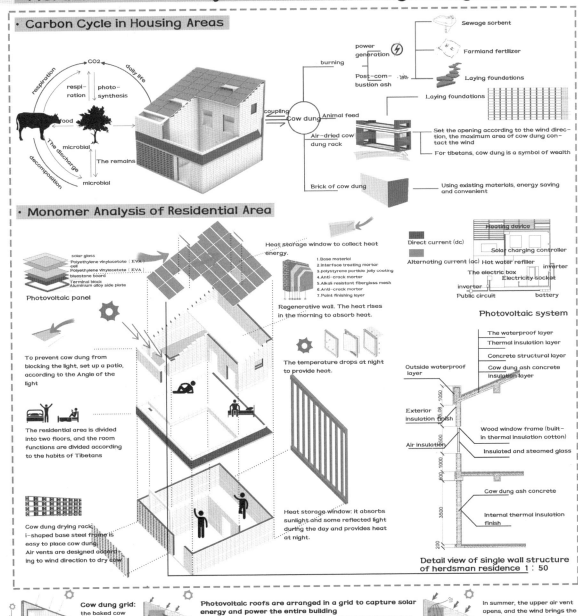

综合奖·优秀奖
General Prize Awarded·
Honorable Mention Prize

注 册 号：100388
Registration No：100388
项目名称：匡郭／织轨
Outer-Internal Weaving
作　　者：刘锐捷、郑仲意、杨雪娴
Authors：Liu Ruijie，Zheng Zhongyi，
　　　　　Yang Xuexian
参赛单位：山东建筑大学
Participating Unit：Shandong Jianzhu
　　　　　University
指导教师：房　涛
Instructor：Fang Tao

Design Description

本设计为西藏地区集居住、旅游接待一体化的低碳社区设计。方案从地域气候及文化出发，采用"外围内织"的设计理念，围合的建筑体量优化了场地热工环境和风环境，并通过羊毛算法程序合理组织内部道路，达成最短路径以达成降碳目标。此外还应用了阳光间、太阳能墙等被动式策略来提升室内舒适性，并通过计算得出最佳太阳能光伏板角度并通过调整建筑形态达到太阳能建筑设计一体化，旨在设计出具有西藏地域性特色的绿色低碳社区。

This design is a low-carbon community design integrating residence, tourism and reception in Tibet. Starting from the regional climate and culture, the plan adopts the design concept of "outer-internal weaving". The enclosed building volume optimizes the thermal environment and wind environment of the site, and rationally organizes the internal roads through the wool algorithm program to achieve the shortest path to achieve carbon reduction Target. In addition, passive strategies such as sunshine room and solar wall are applied to improve indoor comfort; and the best solar photovoltaic panel angle is calculated through calculation and the building form is adjusted to achieve the integration of solar building design, aiming to design a design with regional characteristics of Tibet Green low-carbon community.

Wind Environment Analysis

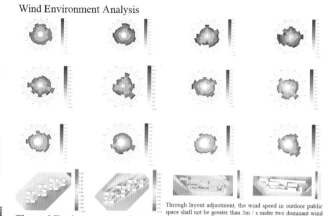

Through layout adjustment, the wind speed in outdoor public space shall not be greater than 5m / s under two dominant wind directions

Thermal Environment Analysis

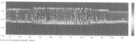

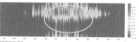

Daytime radiation up to 700wh / ㎡. The annual sunshine is stable and does not change with seasons, It has a good foundation for solar energy utilization.

Most of the year Below 13 ℃, The maximum temperature is 29 degrees, The thermal insulation of site and building is Top priority.

The temperature difference between day and night is more than 13 degrees.

Sunshine Analysis

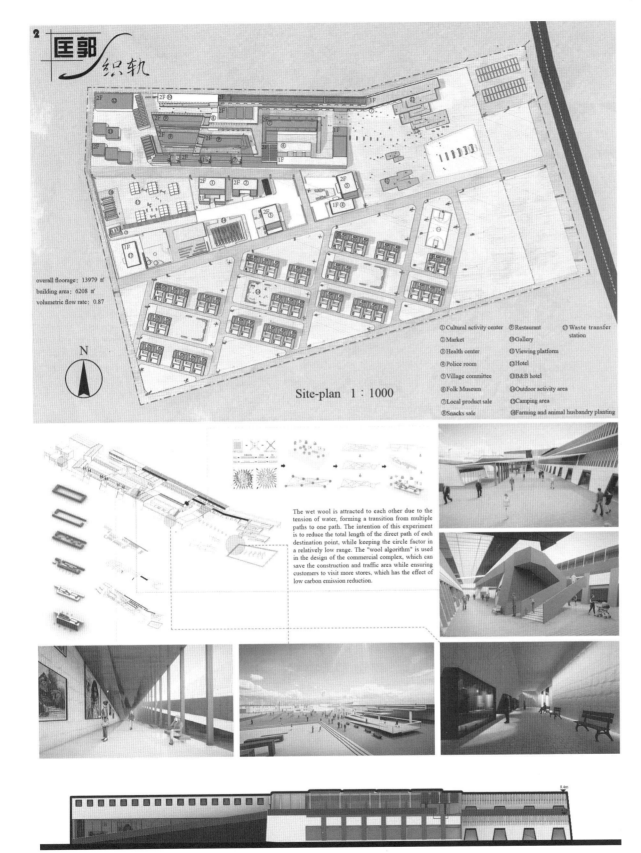

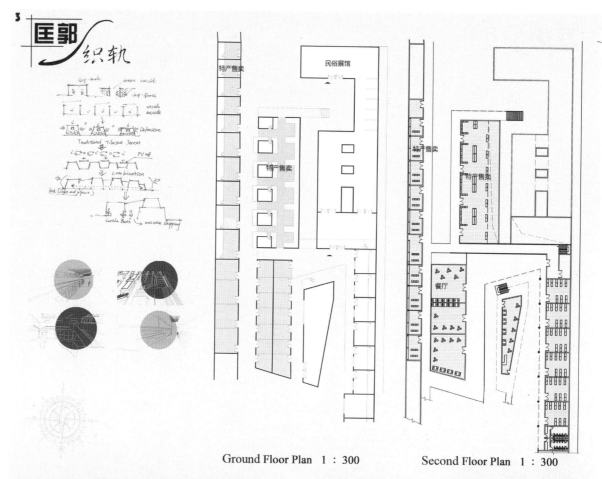

Ground Floor Plan 1:300   Second Floor Plan 1:300

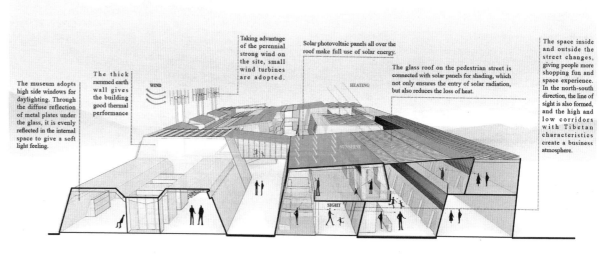

The museum adopts high side windows for daylighting. Through the diffuse reflection of metal plates under the glass, it is evenly reflected in the internal space to give a soft light feeling.

The thick rammed earth wall gives the building good thermal performance.

Taking advantage of the perennial strong wind on the site, small wind turbines are adopted.

Solar photovoltaic panels all over the roof make full use of solar energy.

The glass roof on the pedestrian street is connected with solar panels for shading, which not only ensures the entry of solar radiation, but also reduces the loss of heat.

The space inside and outside the street changes, giving people more shopping fun and space experience. In the north-south direction, the line of sight is also formed, and the high and low corridors with Tibetan characteristics create a business atmosphere.

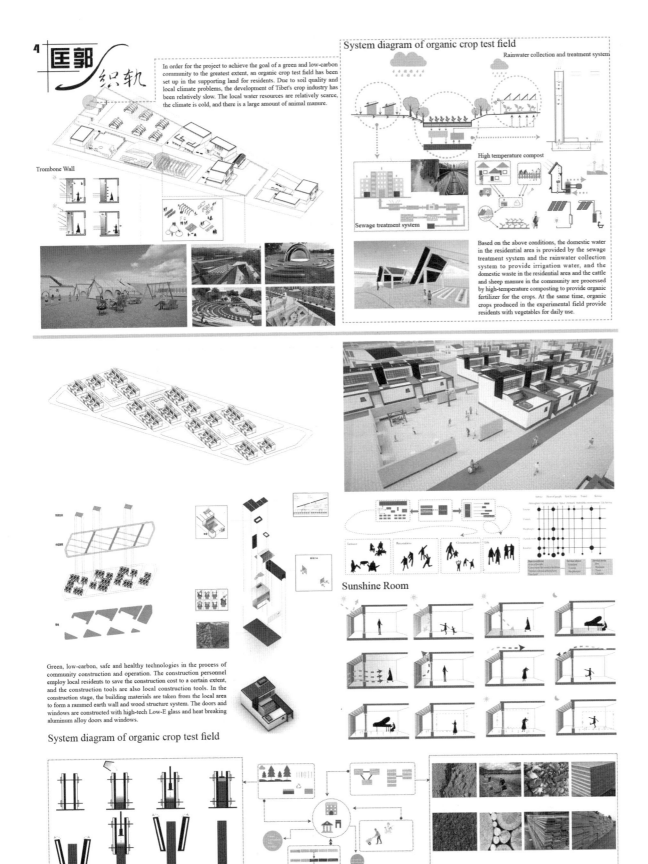

综合奖·优秀奖
General Prize Awarded · Honorable Mention Prize

注 册 号：100389
Registration No：100389

项目名称：聚·星空之下
Tribe Under the Stars-Low Carbon Community Design in Tibet

作　　者：孙之桐、肖　瑶、沈冰珏、励姿玮
Authors：Sun Zhitong, Xiao Yao, Shen Bingjue, Li Ziwei

参赛单位：南京工业大学
Participating Unit：Nanjing University of Technology

指导教师：薛　洁、刘　强
Instructors：Xue Jie, Liu Qiang

Tribe Under the Stars II

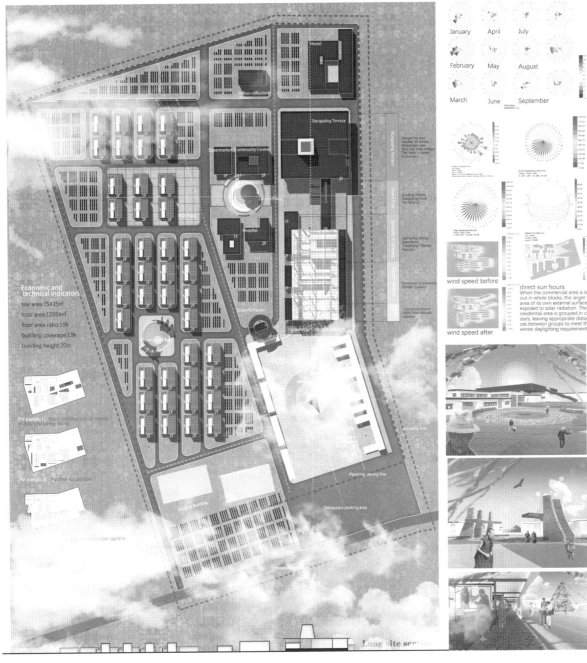

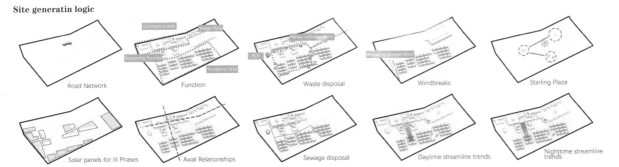

Tribe Under the Stars III

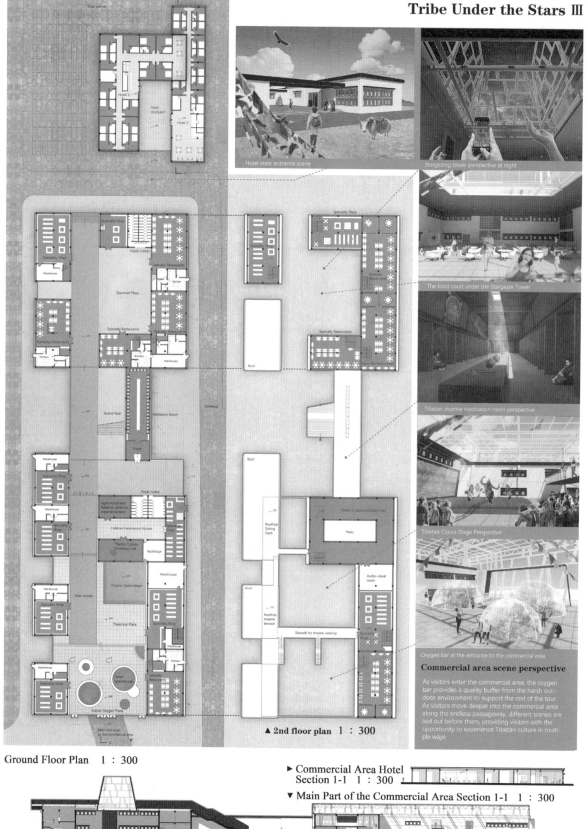

Ground Floor Plan 1:300

▲ 2nd floor plan 1:300

Commercial area scene perspective

As visitors enter the commercial area, the oxygen bar provides a quality buffer from the harsh outdoor environment to support the rest of the tour. As visitors move deeper into the commercial area along the endless passageway, different scenes are laid out before them, providing visitors with the opportunity to experience Tibetan culture in multiple ways.

► Commercial Area Hotel Section 1-1 1:300
▼ Main Part of the Commercial Area Section 1-1 1:300

Tribe Under the Stars IV

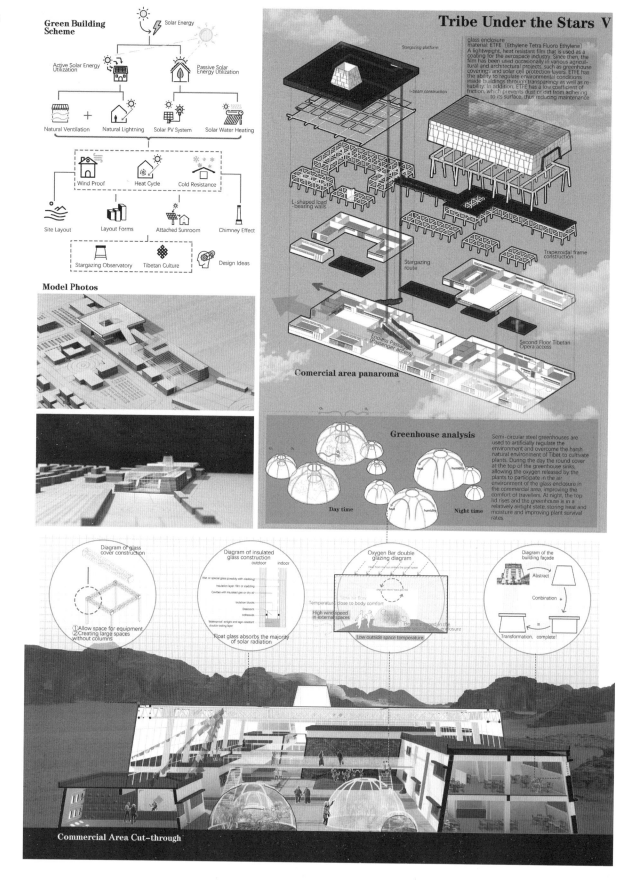

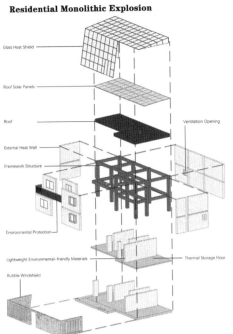
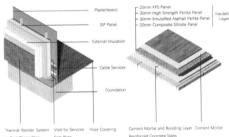
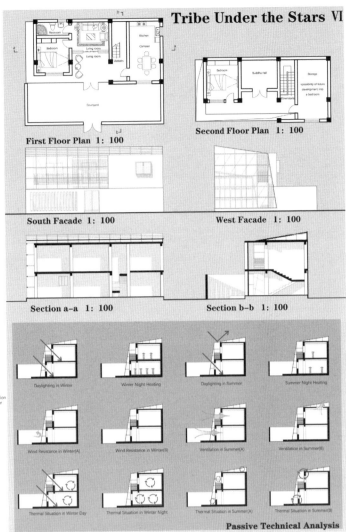

综合奖·优秀奖
General Prize Awarded · Honorable Mention Prize

注 册 号：100400
Registration No：100400
项目名称：廊桥日暖渡游人
　　　　　Ferrying Corridor
作　　者：杨华婷、吴相礼、顾迦艺、字月婷
Authors：Yang Huating, Wu Xiangli, Gu Jiayi, Zi Yueting
参赛单位：南京工业大学
Participating Unit：Nanjing University of Technology
指导教师：林杰文
Instructor：Lin Jiewen

廊桥日暖渡游人
Ferrying Corridor

Design Description

设计以"廊桥"为主题，结合温暖的阳光，旨在让放牧的人和游玩的人停留，赋予场地自然的物质与情感属性。

因此，本方案以被动式技术为设计的主要出发点，结合人群活动行为、西藏传统建筑形式、游牧文化，重点解决建筑的冬季保温采暖问题。通过场地布局、建筑形态的设计，削弱恶劣气候因素带来的影响。对不同功能区域采用不同形式的阳光间，运用特朗勃墙、相变材料和蓄热材质来提升室内舒适性。计算最佳的太阳能光伏板角度，并结合建筑形成太阳能建筑设计一体化。

The theme of the project is "corridor",combined with warm sunshine to stay both nomatic peole and tourists, which endows the site with natural material and emotional attributes.

Therefore, this skeme takes passive technology as main starting point,combined crowd activity behabior,traditional architectural form in Tibet, nomadic culture, focusing on solving the problem of the winter heat preservation. Through the layout of the site and architectural form design, the negative impact of severe climate factors can be weakened. Different forms of sunlight rooms used in different functional areas and passive technologies such as Trump Wall, phase change materials and heat storarage body are applied to improve indoor comfort. The integration of solar energy and architectural design are achieved by calculating the best angle of solar photovoltaic panels and adjusting the architectural form.

Economic and Technical Index

Site area	75391.5 ㎡	Plot Ratio	1.163
Gross oor Area	12659 ㎡	Building density	0.144
Land occupation for building	10884 ㎡	Highest height of building	11.5m

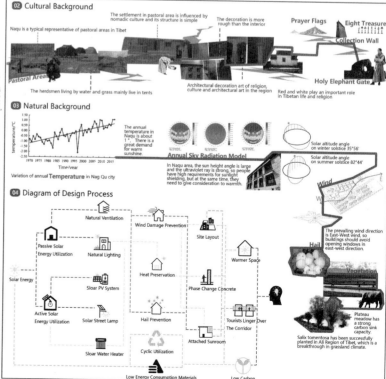

廊桥日暖渡游人

Ferrying Corridor

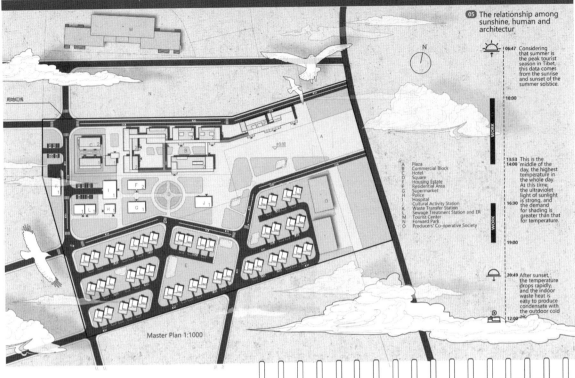

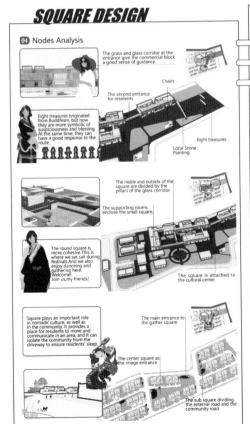

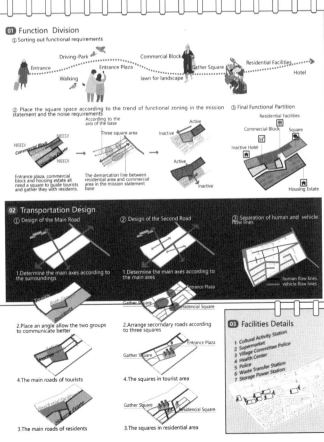

廊桥日暖渡游人　　Ferrying Corridor

03 What's in the houselets?

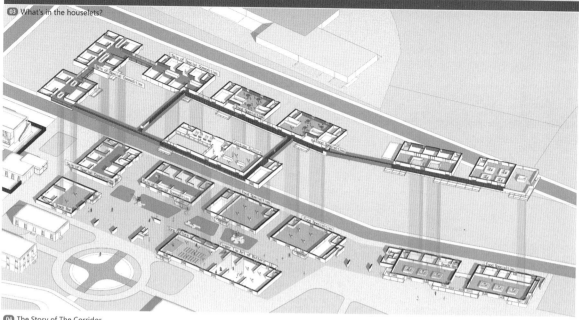

04 The Story of The Corridor

COMMERCIAL BLOCK

01 Block Generation

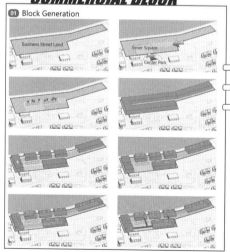

02 Modularization

① Shape generation
② Fabricated frame system
③ Modularization

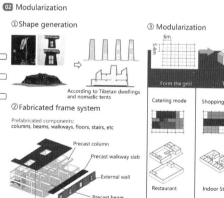
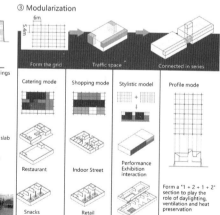

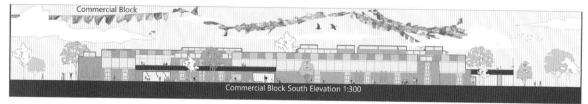

Commercial Block South Elevation 1:300

廊桥日暖渡游人　　　　　Ferrying Corridor

03 Art of Light

HOTEL DESIGN

01 Function and Streamline Analysis

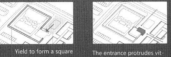

Yield to form a square

A fence to keep out wild animals

The entrance protrudes vitreous blocks to cater to the flow of people.

The south side of the guest room is staggered with sun room to improve the thermal insulation effect.

The north and south facades of the two buildings echo each other with entrances and sun rooms.

MUSEUM of CULTURE

01 Block Generation

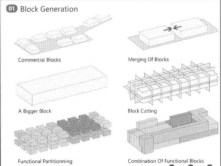

Commercial Blocks　　Merging Of Blocks

A Bigger Block　　Block Cutting

Functional Partitionning　　Combination Of Functional Blocks

HOUSING ESTATE DESIGN

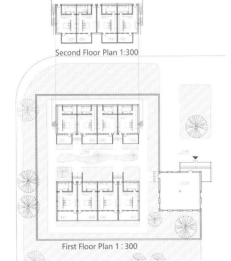

Second Floor Plan 1:300

First Floor Plan 1:300

01 Function and Streamline Analysis
① Relationship with Old Housing Estate

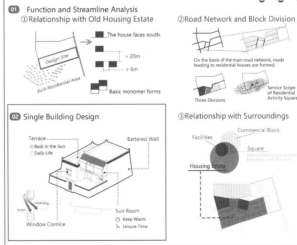

The house faces south.
> 20m
> 6m
Basic monomer forms

② Road Network and Block Division

On the basis of the main road network, roads leading to residential houses are formed.

Three Divisions　Service Scope of Residential Activity Square

02 Single Building Design

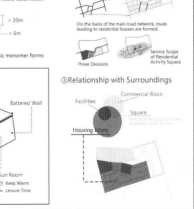

Terrace — Bask in the Sun / Daily Life
Battered Wall
morning / noon
Window Cornice
Sun Room — Keep Warm / Leisure Time

③ Relationship with Surroundings

Commercial Block
Facilities
Square
Housing Estate

Hotel　　　　Dwelling House

South Elevation 1:300　　South Elevation 1:150　　Section 1:150

廊桥日暖渡游人 Ferrying Corridor

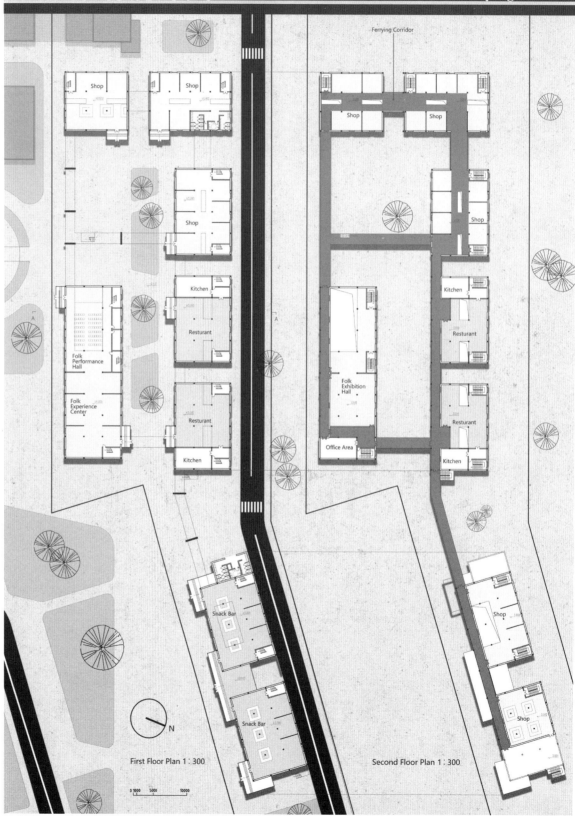

First Floor Plan 1:300

Second Floor Plan 1:300

廊桥日暖渡游人　　Ferrying Corridor

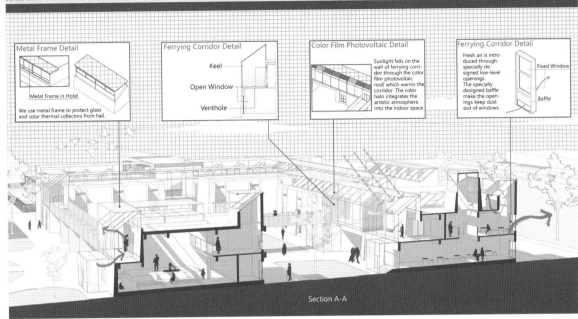

Section A-A

TECHNOLOGY STRATEGY

01 Application of Technology

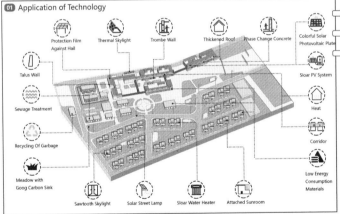

02 Commercial Unit

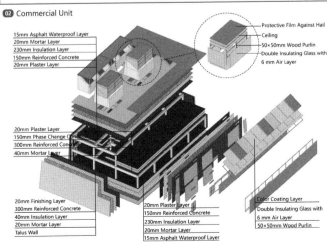

03 Sunroom in the Second Floor

04 Trump Wall in the First Floor

05 Sunroom in the First Floor

06 Different Size of the Sunroom

07 Shadow Analysis

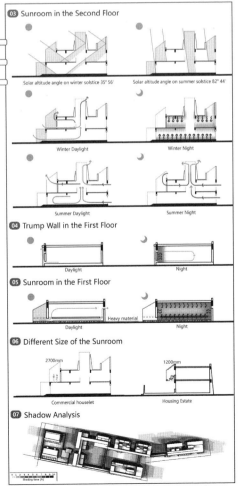

综合奖·优秀奖
General Prize Awarded·
Honorable Mention Prize

注 册 号：100407
Registration No：100407
项目名称：风拂经轮
The Wind Blows the Wheel
作　　者：冯慧玉、宋俊慷、李砖砖、刘西南
Authors：Feng Huiyu, Song Junkang, Li Zhuanzhuan, Liu Xinan
参赛单位：新疆大学
Participating Unit：Xinjiang University
指导教师：滕树勤、王万江、潘永刚
Instructors：Teng Shuqin, Wang Wanjiang, Pan Yonggang

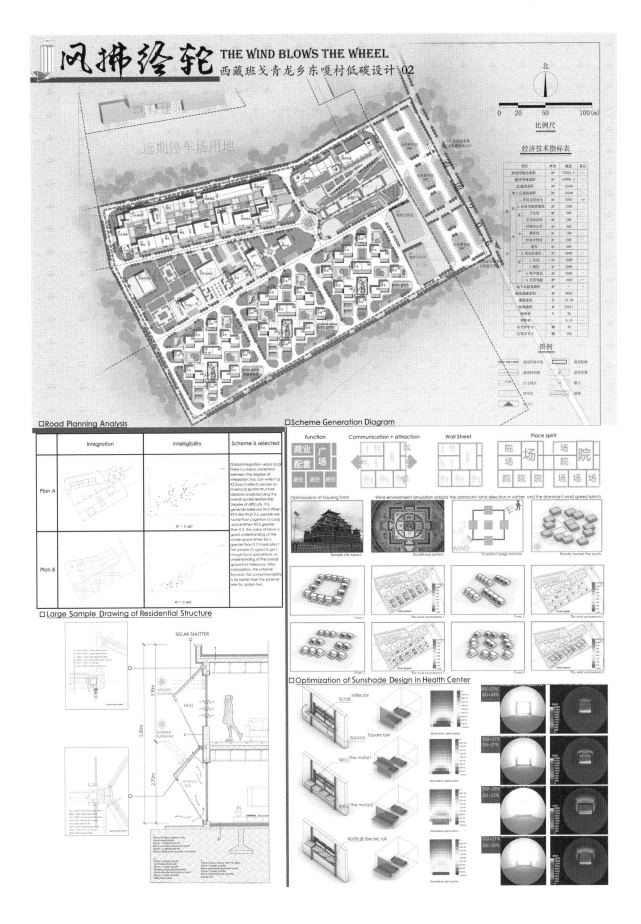

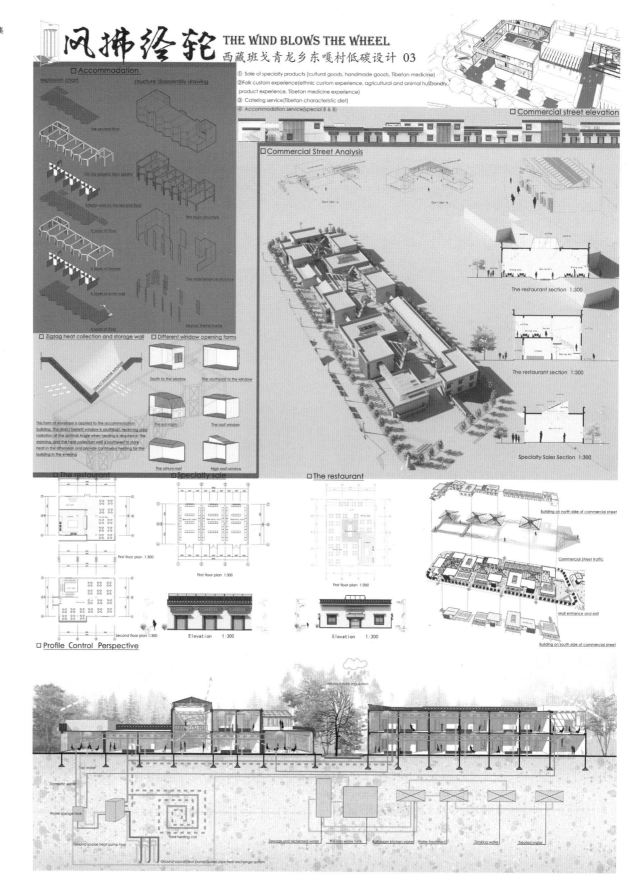

风拂经轮 THE WIND BLOWS THE WHEEL
西藏班戈青龙乡东嘎村低碳设计 04

□ Waste Disposal System Flow Design

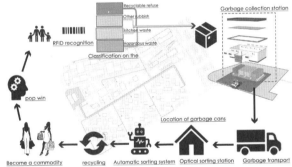

The community garbage collection center will collect the sorted garbage from each garbage point to the garbage station in the form of pipeline

□ Traffic Plaza Design

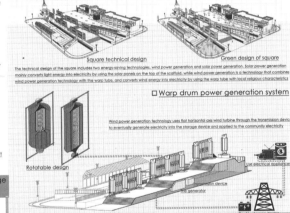

Square technical design / Green design of square

The technical design of the square includes two energy-saving technologies, wind power generation and solar power generation. Solar power generation mainly converts light energy into electricity by using the solar panels on the top of the scaffold, while wind power generation is a technology that combines wind power generation technology with the warp tube, and converts wind energy into electricity by using the warp tube with local religious characteristics

□ Warp drum power generation system

Wind power generation technology uses flat horizontal axis wind turbine through the transmission device to eventually generate electricity into the storage device and applied to the community electricity

□ Solar power generation technology

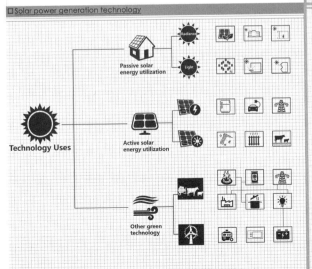

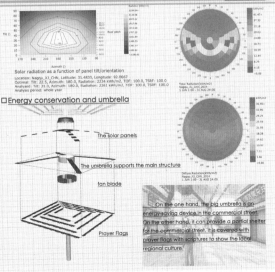

Solar radiation as a function of panel tilt/orientation
Location: Naqqu_XJ_CHN, Latitude: 31.4833, Longitude: 92.0667
Optimal: Tilt: 22.5, Azimuth: 180.0, Radiation: 2234 kWh/m2, TOF: 100.0, TSRF: 100.0
Analysed: Tilt: 31.0, Azimuth: 180.0, Radiation: 2261 kWh/m2, TOF: 100.0, TSRF: 100.0
Analysis period: whole year

□ Energy conservation and umbrella

On the one hand, the big umbrella is an energy-saving device in the commercial street. On the other hand, it can provide a partial shelter for the commercial street. It is covered with prayer flags with scriptures to show the local regional culture.

综合奖·优秀奖
General Prize Awarded · Honorable Mention Prize

注 册 号：100412
Registration No：100412
项目名称：风回路转
　　　　　Wind Back
作　　者：李若天、李金洋、金在纯、
　　　　　邹庆泽
Authors：Li Ruotian, Li Jinyang,
　　　　　Kim Jae-Chun, Zou Qingze
参赛单位：苏州科技大学
Participating Unit：Suzhou University of
　　　　　　　　　Science and Technology
指导教师：刘长春、陈守恭、高　姗
Instructors：Liu Changchun,
　　　　　　Chen Shougong, Gao Shan

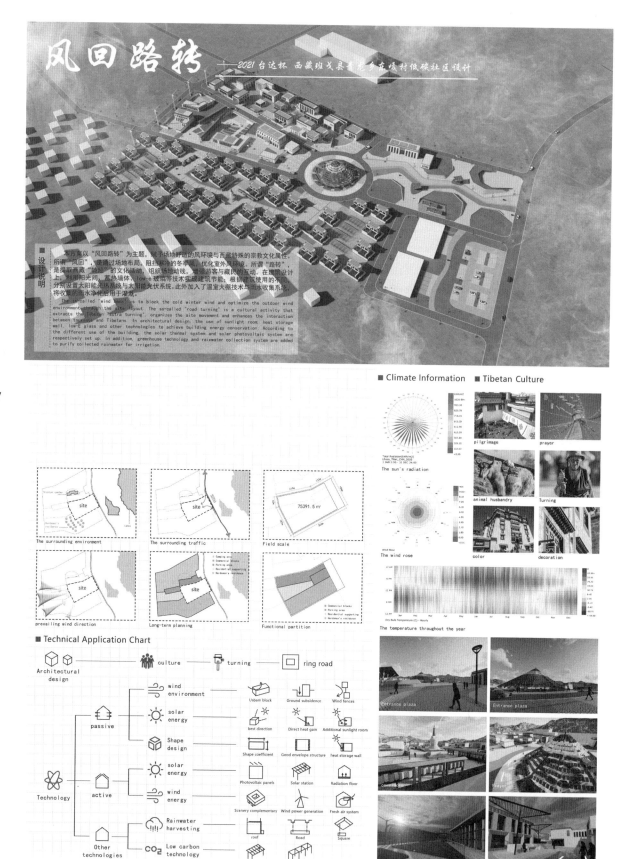

■ Environment Simulation of Wind and Sun

■ Crowd Activity and Sloar Energy

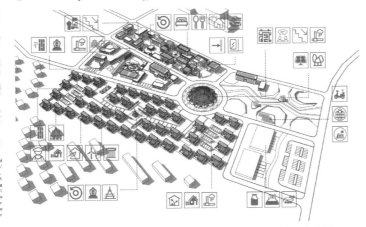

■ Design Analysis

The site area is 75391.5 square meters. Construction land area of 62892.3 square meters

Five functional areas: herdsmen's residential area, supporting residential facilities, commercial blocks, traffic square, and livestock products trading area

Form a preliminary motor vehicle traffic system

Tibetan culture type: scripture turning. Combined with the walking behavior of warp turning, a warp turning loop of three layers is formed

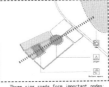

Three ring roads form important nodes: prayer flags and cultural event squares

The boundary of the site uses a continuous volume to block the cold winter wind and make the wind environment inside the site relatively comfortable

The adjacent buildings are connected by the wall and form a wind barrier to meet the functional needs of the building and prevent the winter wind

The site sinks and the surrounding buildings are raised to block the winter winds and create a more comfortable outdoor wind environment

The second meridian road is also the sightseeing route of tourists and the life route of herdsmen

Combining with the climatic characteristics of Bangor County, Xizang Province, and adding technical factors, the architectural deepening design was carried out

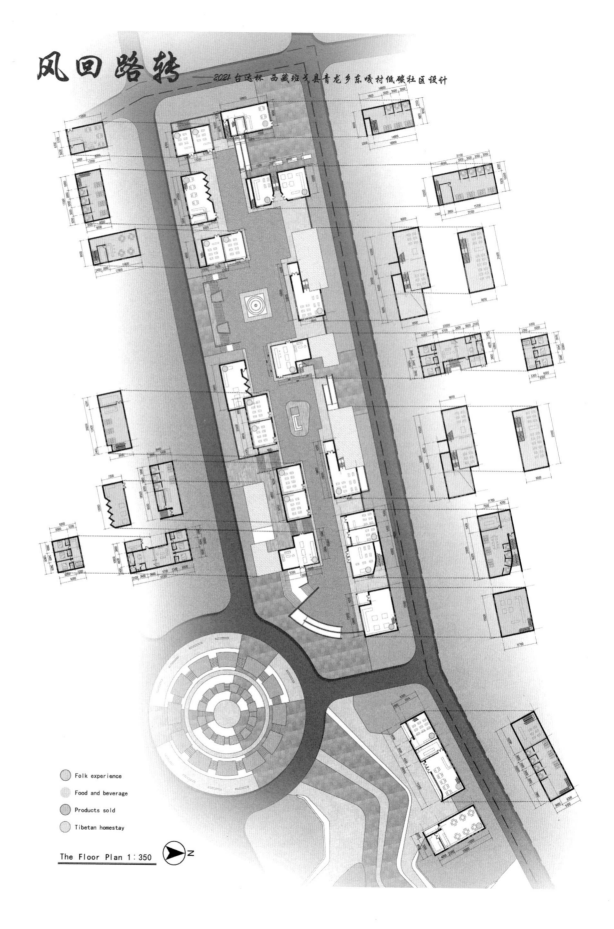

风回路转
2021 台达杯 西藏班戈县青龙乡东喀村低碳社区设计

■ Building Plan

Basement Plan
1. Commercial sales

Ground Floor Plan
1. Reception room
2. Restaurant
3. Kitchen
4. Bathroom
5. Living room
6. Guest Rooms

First Floor Plan
1. Guest Rooms
2. Living space
3. Sunshine Room

■ B&B Single Unit Design

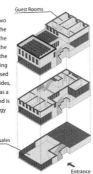

B&B

The location of the B&B is within the commercial area of the site, with two locations, one at the entrance of the visitor centre and the other at the entrance to the commercial street plaza, making it easy for guests to move in. As the commercial street is sunken, the ground floor of the building in the sunken plaza is used as a sales and catering function, while the remaining two floors are used as residential units for the B&B. The living room is used as a sun room to connect the residential units on the north and south sides, and is used as the main B&B communal space. The living room is used as a sun room to connect the living units on the north and south sides, and is used as the main B&B communal space to minimise heating energy consumption.

■ B&B Analysis

Guest Rooms

Commercial sales

Entrance

■ SUNROOM DESIGN DIAGRAM

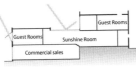
Daylighting in Summer

Daylighting in Winter

Ventilation in Summer

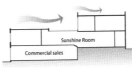
Ventilation in Winter

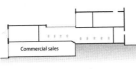
Thermal Situation in Summer

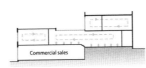
Thermal Situation in Winter

SUMMER
In summer open windows and sun shading devices in the sunroom to reduce the entry of sunlight and open the windows of the sunroom, rooms and ventilated roofs to keep them open to achieve indoor and outdoor air circulation.

Minimise the gathering of heat indoors in summer through natural ventilation and shading devices.

WINTER
Open the sunshade in winter to let the sun light into the room as much as possible. Reduce the opening of windows so that indoor the heat circulating inside does not dissipate. At the same time In the winter, the heat stored in the walls and floors is kept inside during the day. heat stored in the walls and floors heat is released during the night to enhance indoor thermal comfort.

■ Elevation

South Elevation | East Elevation | West Elevation

B&B Energy Saving Technology

Solar PV Panels + Solar Heater

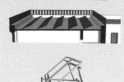

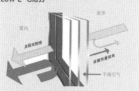

The solar photovoltaic panels and collector panels are placed on the roof by means of brackets, which make it easy to adjust the right angle to make the best use of the solar energy.

Low-E Glass

The windows are made of triple-glazed, two-cavity Low-e glass with a built-in air layer of 12mm thick, which enhances the thermal performance of the windows and doors and greatly improves the heat dissipation of the building.

Thermal insulation and decorative one-piece panels

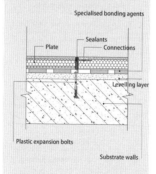

Substrate walls
Specialised bonding agents
Insulation layer

Sealants
Special fasteners
Calcium silicate panels
Faux stone

The insulation is made of 60mm polyurethane and the heat transfer coefficient of the external walls is 0.21, which meets the local energy saving design standards.

B&B Analysis

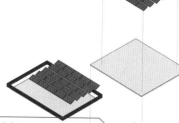

Solar PV Panels + Solar Heater
The solar photovoltaic panels and collector panels are installed at an angle of 22.8°, for more efficient use.

Solar Water Heating Systems
The solar water heating system provides domestic hot water and is also capable of meeting certain heating requirements.

Sunroom
The use of public spaces as sunrooms can reduce the building's heating energy consumption and, in combination with thermal storage walls, can meet the heating requirements to some extent.

Fresh Air Systems
If no fresh air is fed into the occupants may feel suffocated. This is why the use of fresh air systems is so necessary. Efficient The heat recovery efficiency of fresh air systems can reach 90%, which can significantly reduce heating energy consumption.

Low-E Glass
Low-E glass saves energy by reducing the amount of heat lost from the building.

Thermal insulation and decorative one-piece panels

Solar Water Heating Systems

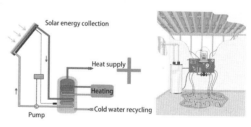

Hot water for domestic use is supplied through solar hot water, while hot water is supplied to the radiant floor for indoor heating. When the sun is not shining enough, hot water can be supplied through auxiliary energy sources.consumption.

Fresh Air Systems

B&B Rendering

风回路转
2021台达杯 西藏班戈县青龙乡东咳村低碳社区设计

■ Other Technologies

Greenhouses
In order to create the concept of low-carbon community, starting from changing the way of life and production of residents, intelligent greenhouse greenhouses will be built in each living group. When the temperature is too low, the greenhouse will be heated by solar water heating system; when the temperature is too high, the skylight on the roof will be opened for ventilation, so as to achieve self-sufficiency of resources as much as possible.

Solar shed
In order to make full use of the site's solar energy resources, solar panels are added to the roof of the parking shed for photovoltaic power generation.

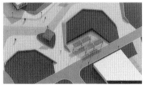

Solar charging station
Two solar charging stations are designed for green travel and carbon emission reduction for residents and tourists in need.

Sinking square
The activity site is settled to effectively improve the wind environment in the activity area. It also helps rainwater harvesting.

■ Retaining structure

Thermal insulation wall and low-E glass enhance the thermal insulation performance of the building itself.

■ Ventilated roof

by controlling the opening and closing of roof skylights, to meet the specific period of natural ventilation.

■ Additional sunlight room
Combined with heat storage wall to reduce building heating energy consumption.

■ Fresh air system
The fresh air system ensures that indoor heat is not lost as much as possible and provides fresh air outdoors.

■ Wind-solar complementary
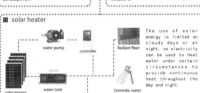
Solar photovoltaic panels are arranged on the roof of each house, and the four corners of the roof are highlighted, which can not only meet the needs of local residents for architectural decoration, but also serve as wind power generation devices. The wind-wind complementary power generation system can provide 7-8 KWH of electricity per household per day

■ solar heater
The use of solar energy is limited on cloudy days or at night, so electricity can be used to heat water under certain circumstances to provide continuous heat throughout the day and night.

■ Biogas utilization

The carbon produced by biogas combustion is far less harmful to the environment than direct emissions.

Daily life of boiling water, cooking and so on | Biogas is used to heat the house | Biogas is burned to heat cold water | Residue recovery and reuse

■ Residential Design

first floor | second floor | south facade | eastern facade

■ Model Photos

   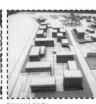

overlooking | Herders settlements | Commercial blocks | The entrance

暖厅·聚能
Warming Hall · Gathering Energy

综合奖·优秀奖
General Prize Awarded · Honorable Mention Prize

注 册 号：100436
Registration No：100436
项 目 名 称：暖厅·聚能
　　　　　　Warming Hall · Gathering Energy
作　　　者：杨瑞航、曾俊鸿、蒋 铁
Authors：Yang Ruihang, Zeng Junhong, Jiang Tie
参 赛 单 位：重庆大学
Participating Unit：Chongqing University
指 导 教 师：张海滨、周铁军
Instructors：Zhang Haibin, Zhou Tiejun

CLIMATE ANALYSIS

Optimal Orientation

Prevailing Trends

REGIONAL ELEMENTS

Bunker

Stone

Sutra Banner

Buddhist Culture

LOCATION

SITE ANALYSIS

Functions are set along the strip

Site is adjacent to major roads

The existing buildings are closely related to the site

Key axes intersect to form points of interest

The main prevailing wind direction is parallel to the distribution of functional areas

The land is flat with almost no height difference

Bangor County belongs to the semi-arid monsoon climate of the highland sub-freezing zone. It's cold and dry, with a long winter of half a year and small annual temperature difference, large temperature difference between day and night. Its annual average temperature is about 0°C, annual maximum temperature of 21.9°C and minimum temperature of -28.6°C. The highest daily temperature is around 6:00 p.m. The annual sunshine time in Bangor County is about 2,850 hours, with very little diffuse solar radiation, and the main radiation is direct solar radiation, which is rich in solar resources. The dominant wind direction is W (westerly)

INTRODUCTION

本方案以"曼荼罗"为主题，赋予场地文化元素和宗教特色。

因此，本方案以被动式技术为设计出发点，结合人群活动行为、西藏传统民居形式，重点加强建筑冬季保温采暖、夏季自然通风处理。利用阳光房、蓄热墙、灰空间、相变蓄热材料等来实现建筑节能。同时，在主动式技术上，根据建筑使用的不同，分别设置太阳能光伏系统和太阳能光热系统。结合建筑设计，集中设计阳光间和传统藏族建筑之间的结合。此外，加入了雨水收集系统，将收集的雨水用于景观环境和冲厕等非生活用水。

The theme of this project is "Mandala", which gives the site cultural elements and religious characteristics.
Therefore, this project takes passive technology as the starting point of design, combines the activity behavior of people and traditional Tibetan dwelling form, and focuses on strengthening the building heat preservation and heating in winter and natural ventilation in summer. Sunroom, heat storage wall, gray space and phase change heat storage materials are used to realize building energy saving. Meanwhile, in the active technology, solar photovoltaic system and solar photothermal system are set up respectively according to the different building use, combined with the architectural design. The combination between the centralized design sunroom and traditional Tibetan architecture. In addition, a rainwater collection system is incorporated to use the collected rainwater for non-domestic water such as landscape environment and toilet flushing.

Design Ideas

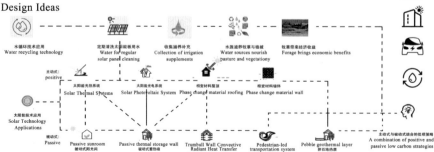

水循环技术应用 Water recycling technology
定期清洗太阳能板用水 Water for regular solar panel cleaning
收集灌溉补充 Collection of irrigation supplements
水源涵养牧草与植被 Water sources nourish pasture and vegetation
牧草带来经济收益 Forage brings economic benefits

太阳能技术应用 Solar Technology Applications

主动式：positive
太阳能光伏系统 Solar Thermal Systems
太阳能光电系统 Solar Photovoltaic System
相变材料屋顶 Phase change material roofing
相变材料墙体 Phase change material wall

被动式 Passive
被动式阳光间 Passive sunroom
被动式蓄热墙 Passive thermal storage wall
特朗勃墙对流辐射传热 Trumbull Wall Convective Radiant Heat Transfer
步行系统主导的交通体系 Pedestrian-led transportation system
卵石地热盘 Pebble geothermal layer

主动式与被动式结合的优碳策略 A combination of positive and passive low carbon strategies

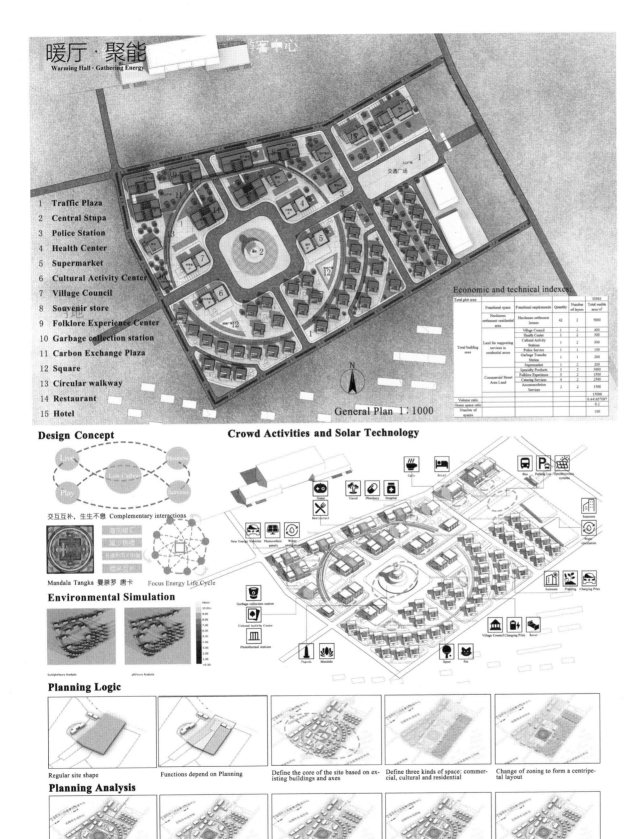

暖厅·聚能
Warming Hall · Gathering Energy

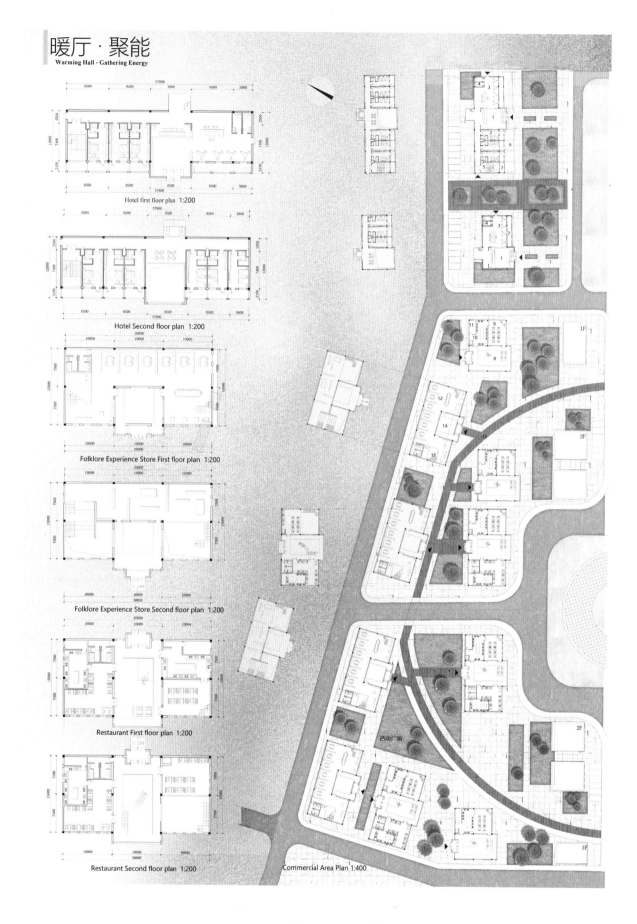

Hotel first floor plan 1:200

Hotel Second floor plan 1:200

Folklore Experience Store First floor plan 1:200

Folklore Experience Store Second floor plan 1:200

Restaurant First floor plan 1:200

Restaurant Second floor plan 1:200

Commercial Area Plan 1:400

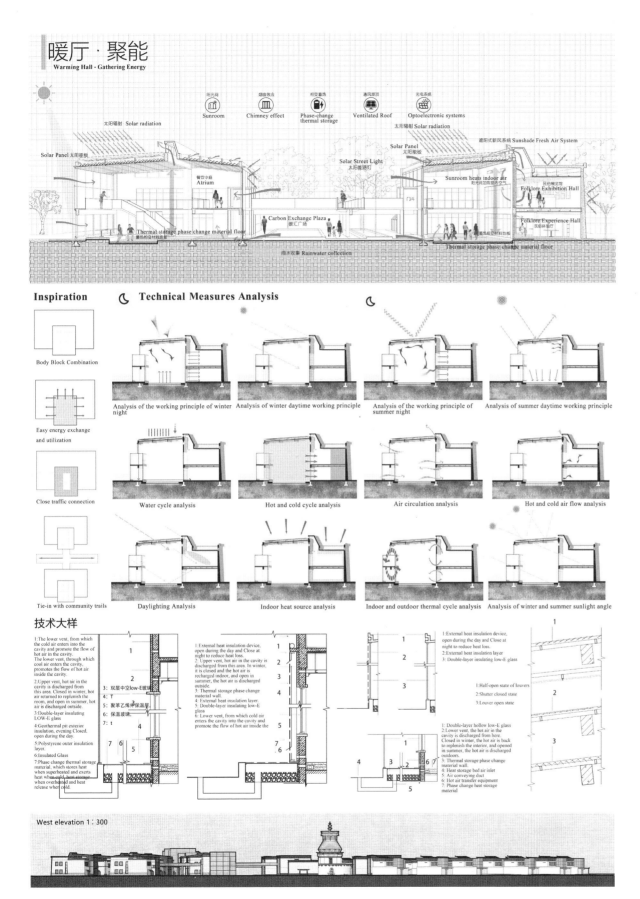

暖厅·聚能
Warming Hall · Gathering Energy

Constructing Process

- Beam and column foundations
- Wall construction
- Windows, doors and atrium elements
- Atrium beams
- solar house
- Sunshade louvers with solar panels

Analysis of Technical Measures

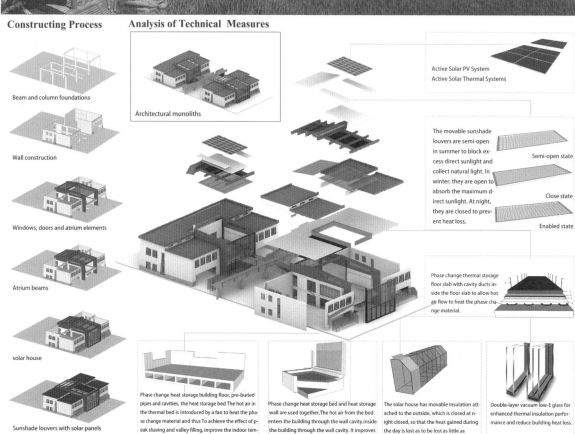

Architectural monoliths

Active Solar PV System
Active Solar Thermal Systems

The movable sunshade louvers are semi-open in summer to block excess direct sunlight and collect natural light. In winter, they are open to absorb the maximum direct sunlight. At night, they are closed to prevent heat loss.

- Semi-open state
- Close state
- Enabled state

Phase change thermal storage floor slab with cavity ducts inside the floor slab to allow hot air flow to heat the phase change material.

Phase change heat storage building floor, pre-buried pipes and cavities, the heat storage bed The hot air in the thermal bed is introduced by a fan to heat the phase change material and thus To achieve the effect of peak shaving and valley filling, improve the indoor temperature variation.

Phase change heat storage bed and heat storage wall are used together,The hot air from the bed enters the building through the wall cavity,inside the building through the wall cavity. It improves the indoor the-rmal environment.

The solar house has movable insulation attached to the outside, which is closed at night closed, so that the heat gained during the day is lost as to be lost as little as possible.

Double-layer vacuum low-E glass for enhanced thermal insulation performance and reduce building heat loss.

South elevation 1:300

暖厅·聚能
Warming Hall · Gathering Energy

Construction Material Analysis

Material type	Material Evaluation	Material type	Material Evaluation
Square steel pipe	Each ton of steel consumes 0.7 tons of coal, the building consumes about 40kg per square meter, 15,000 square meters consumes 600t, and Produce 420t of coal, 113t of CO_2, need to plant 1000 trees	Local wood	Using local wood reduces the amount of energy consumed during the transportation of wood and reduces CO_2 emissions.
Local stone	Stone is used in the building's exterior wainscoting, and local stone is chosen for the interior pebble heat collection layer in the lobby. Reduction of CO_2 emissions during transportation	Porous concrete	Each square meter consumes about $0.35m^3$ of concrete, and one ton of concrete consumes 0.7 degrees of electricity, emitting a total of 8,400 tons of carbon dioxide, which can be neutralized by planting trees and photovoltaic power generation.
Double-layer LOW-E glass	Compared with traditional glass, coated glass has good heat insulation and light transmission, and the use of double-side glass in building facades can further improve the thermal insulation performance.	Polystyrene insulation	Insulation panels prevent heat loss and reduce indoor heating supply, thus reducing electricity consumption and approximately 1,900 tons of CO_2 emissions.
Fly ash bricks	Made from industrial waste and slag produced in the process of coal power generation, each 10,000 bricks consume about 1000 kWh of electricity energy.	Phase change heat storage material	Phase change materials can absorb and release large amounts of heat, and phase change gypsum board can reduce energy consumption by approximately 22.5% per room compared to ordinary gypsum board.

Energy Consumption Calculation Results

Energy consumption table for restaurant buildings

Energy consumption table for hotel buildings

Cultural specialties building energy consumption table

Note: Without considering the effectiveness of passive technical measures premise, the difference between annual power generation and power consumption of commercial buildings is -41,746.0 KW/h, while the residential solar power generation electricity is about 243,682.6KW/h per year, and the annual residential The surplus power can be added to the commercial building part to fill the gap. can be added to the commercial part of the building to fill the After filling the energy consumption gap in the commercial part, it can also save 48252.8KW/h of electricity per year. The goal of zero energy consumption for community operation can be achieved. The goal of zero energy consumption is achieved.

- 49% Materials and construction energy efficiency as a percentage
- 36% Proactive energy efficiency as a percentage
- 15% Passive energy efficiency as a percentage

Total community energy consumption calculation table

换碳分析

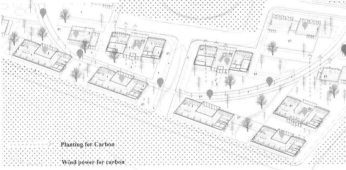

Planting for Carbon
Wind power for carbon

 Bioenergy-for-carbon
 Wind-solar for carbon
 Geothermal Energy for Carbon

Site 1-1 profile

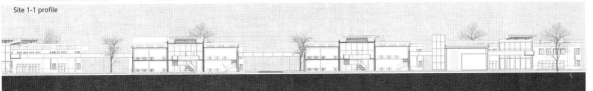

综合奖·优秀奖
General Prize Awarded · Honorable Mention Prize

注 册 号：100476
Registration No：100476

项目名称："盒""光"同尘——西藏班戈县青龙乡东嘎村低碳社区设计
Sunshine, Box, Single Building, Group, Community

作 者：李世萍、李舒祺、卓金明、许浩川、周超超、苏湘茗
Authors：Li Shiping, Li Shuqi, Zhuo Jinming, Xu Haochuan, Zhou Chaochao, Su xiangming

参赛单位：西安建筑科技大学
Participant Unit：Xi'an University of Architecture and Technology

指导教师：李 涛、张 斌、朱新荣
Instructors：Li Tao, Zhang Bin, Zhu Xinrong

"盒" "光" 同尘

篇二：冰开得暖 "光"

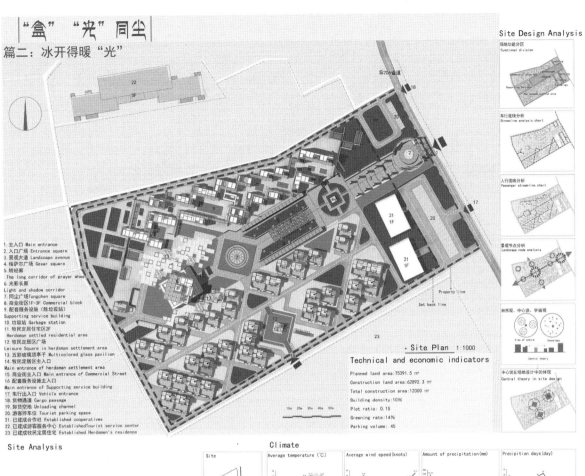

1. 主入口 Main entrance
2. 入口广场 Entrance square
3. 景观大道 Landscape avenue
4. 格萨尔广场 Gesar square
5. 转经廊 The long corridor of prayer wheel
6. 光影长廊 Light and shadow corridor
7. 同尘广场 Tongchen square
8. 商业街区1F-3F Commercial block
9. 配套服务设施（除垃圾站）Supporting service building
10. 垃圾站 Garbage station
11. 牧民定居住宅2F Herdsman settled residential area
12. 牧民定居区广场 Leisure Square in herdsman settlement area
13. 五彩玻璃顶亭子 Multicolored glass pavilion
14. 牧民定居区主入口 Main entrance of herdsman settlement area
15. 商业街主入口 Main entrance of Commercial Street
16. 配套服务设施主入口 Main entrance of Supporting service building
17. 车行出入口 Vehicle entrance
18. 货物通道 Cargo passage
19. 卸货空地 Unloading channel
20. 游客停车位 Tourist parking space
21. 已建成合作社 Established cooperatives
22. 已建成游客服务中心 Established Tourist service center
23. 已建成牧民定居住宅 Established Herdsmen's residence

Site Design Analysis

- 场地功能分区 Functional division
- 车行流线分析 Streamline analysis chart
- 人行流线分析 Passenger streamline chart
- 景观节点分析 Landscape node analysis
- 自然观、中心说、宇宙观 View of nature / Cosmology / Central theory
- 中心说在场地设计中的体现 Central theory in site design

Site Plan 1:1000

Technical and economic indicators
- Planned land area: 75391.5 ㎡
- Construction land area: 62892.3 ㎡
- Total construction area: 12000 ㎡
- Building density: 10%
- Plot ratio: 0.15
- Greening rate: 14%
- Parking volume: 45

Site Analysis

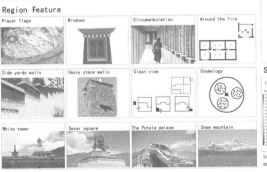

Urban texture of Bangor County:
(1) the texture of the building is extracted from the nearest Bangor county to obtain the best orientation of the base.
(2) According to the traditional wind protection measures, an enclosure is set on the west side of the site to reduce the wind speed of the site.

Region Feature

Climate

Average temperature (℃) / Average wind speed (knots) / Amount of precipitation (mm) / Precipitation days (day)

Wind chart within 1500 days — Dominant wind direction: westerly
Radition 3D chart
Psychrometric chart

Conclusion and Strategy

1. The annual average temperature is low, so thermal insulation is very important;
2. The annual average wind speed is high, so the site wind protection is very important;
3. The rainfall is small, so rainwater recovery can be considered;
4. There are few precipitation days in the whole year, and rainwater recycling measures are taken;
5. The prevailing wind direction throughout the year is westerly
6. The solar radiation is strong, and the time proportion suitable for direct benefit solar heating is large;
7. Only 5.0% of the time in the whole year belongs to the comfortable period, and 57.0% of the time needs to adopt active heating methods such as stove, air conditioner or boiler heating. According to the length of effective time, the applicable passive building strategies are passive solar direct heat gain + low heat storage (25.9%), internal heat gain (24.5%), sun shading (1.5%), simple humidification (1.2%), direct evaporative cooling (0.6%), and wind protection of outdoor space (0.1%).

Shadow Analysis

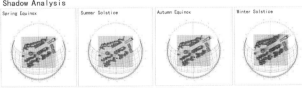

Spring Equinox / Summer Solstice / Autumn Equinox / Winter Solstice

Site Wind Simulation

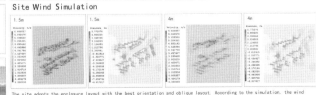

1.5m / 1.5m / 4m / 4m

The site adopts the enclosure layout with the best orientation and oblique layout. According to the simulation, the wind speed and pressure of the outdoor activity site planned are small.

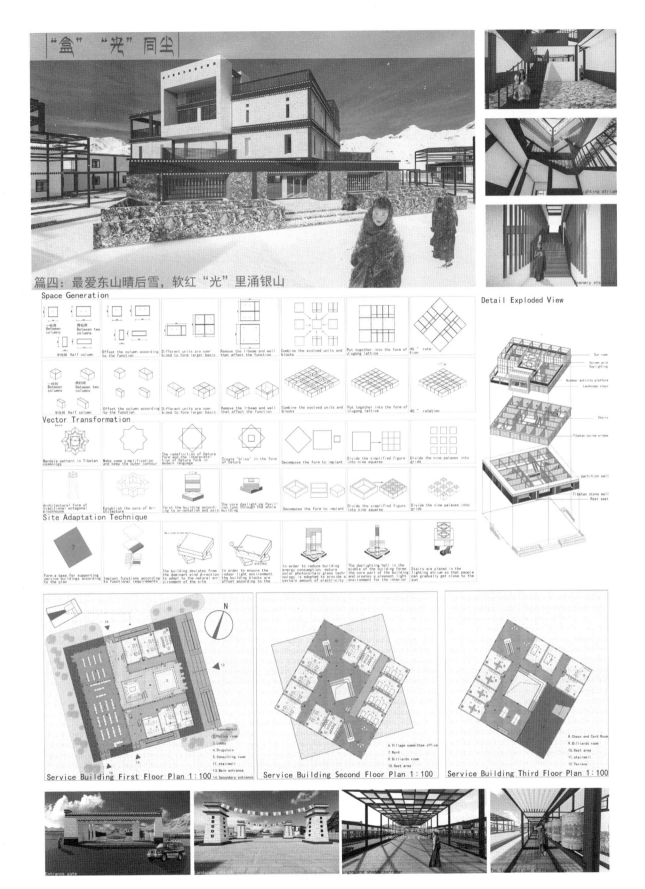

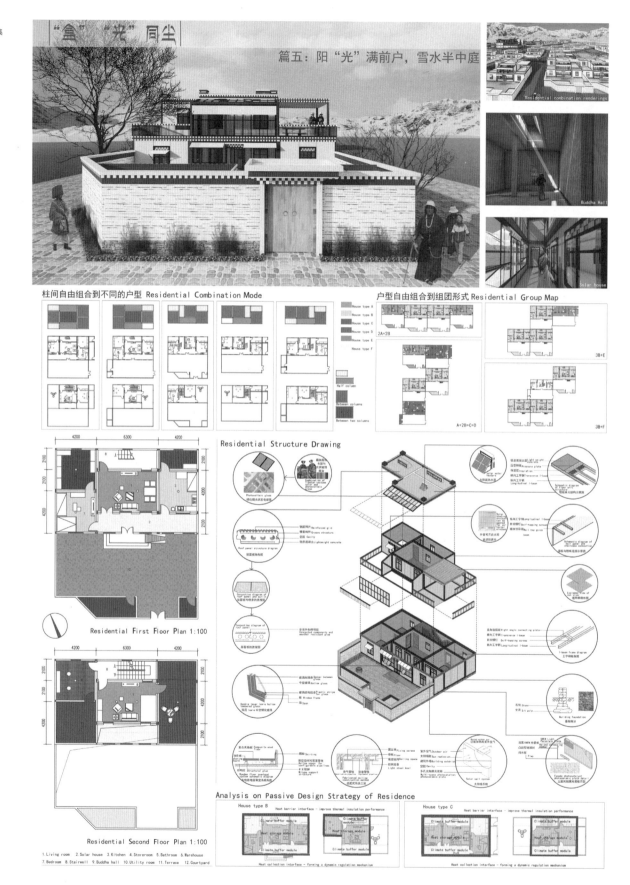

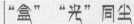

"盒" "光" 同尘

篇六：斜"光"到晓穿朱户

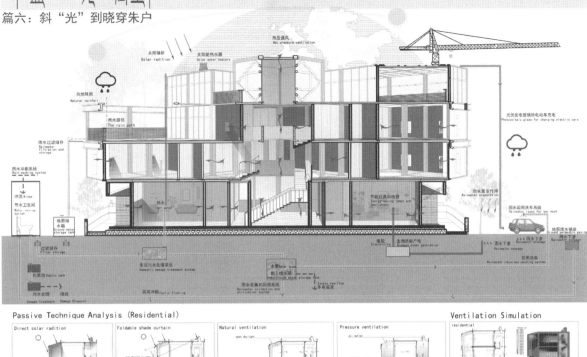

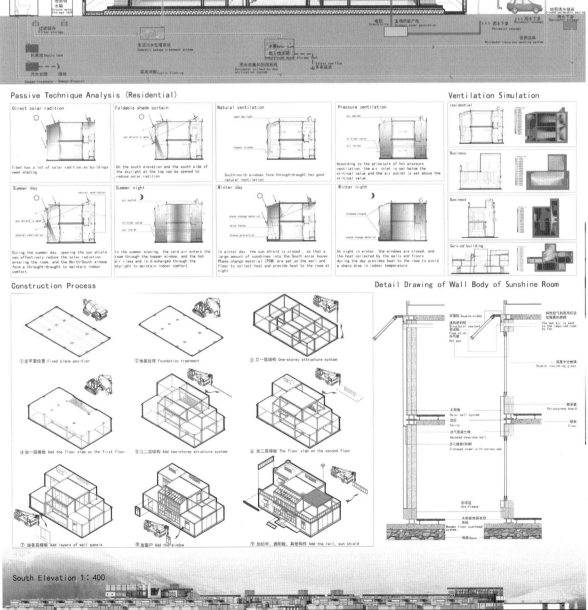

South Elevation 1:400

综合奖·优秀奖
General Prize Awarded·
Honorable Mention Prize

注 册 号：100498
Registration No.：100498
项目名称：风涌墙阻
A Wall to Keep Out the Cold Wind
作　　者：孙思雨、邓凯元、刘乃歌、
李祎平、张淑婧
Authors：Sun Siyu, Deng Kaiyuan,
Liu Naige, Li Yiping,
Zhang Shujing
参赛单位：吉林建筑大学
Participating Unit：Jilin University of Architecture
指导教师：赫双龄、周春艳
Instructors：He Shuangling, Zhou Chunyan

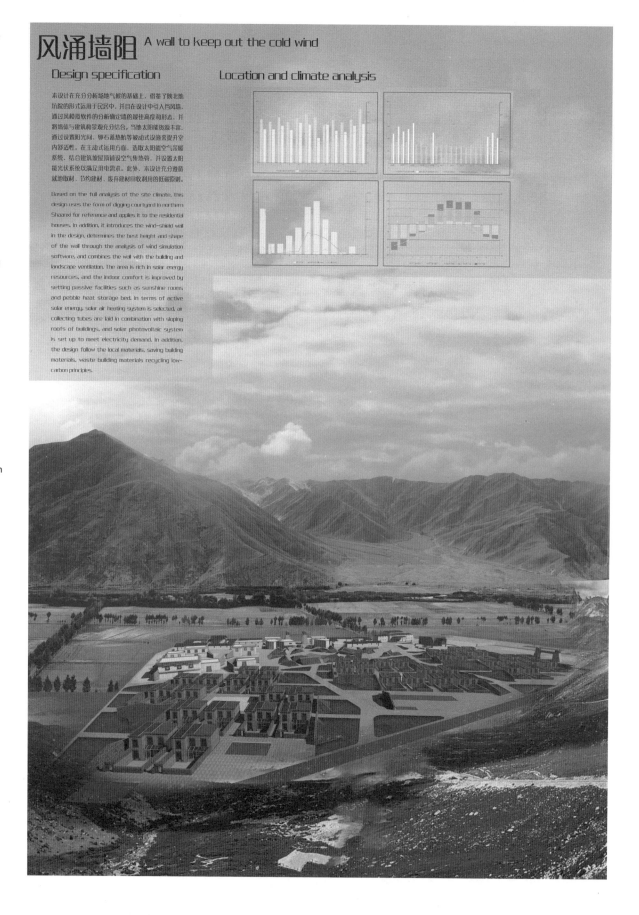

风涌墙阻 A wall to keep out the cold wind

1. 文化景观 Cultural landscape
2. 超市 The supermarket
3. 村委会 The village committee
4. 卫生院 Institutes of health
5. 警务处 Police force
6. 文化活动站 Cultural activity station
7. 垃圾中转站 Garbage transfer station
8. 景观步行带 Landscape walking belt
9. 特产售卖 Products sold
10. 民俗体验 Folk experience
11. 餐饮服务 Food and beverage service
12. 住宿服务 Accommodation services
13. 牧民定居住宅区 Herdsmen settled in residential areas

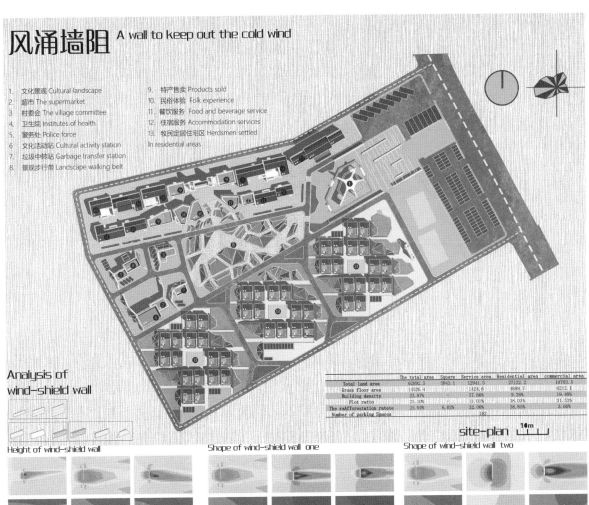

site-plan

Analysis of wind-shield wall

Height of wind-shield wall

H=2m   2H=4m   3H=6m

Shape of wind-shield wall one

Shape of wind-shield wall two

Material of wind-shield wall
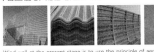

Wind wall at the present stage is to use the principle of aerodynamics, in accordance with the implementation of the environmental wind tunnel experimental results processed into a certain geometric shape, open hole ratio and different hole shape combination wind dust-controlling wall, make the circulation of air from the outside, through the wall, the wall on the inside of the form, the interference of air flow to achieve lateral winds, medial weak wind, small wind, the lateral inside without the effect of the wind.

Stone cage view wall: the cage is most often made of stainless steel wire mesh, and then bonded with adhesive or ring fasteners to connect them together, into the wool stone.

Simulated condition

Meteorological parameters of outdoor wind environment in Nagqu city

Season	Month	Average wind speed	Maximum wind direction
winter	12 2	9.03	W, WS
summer	6 8	7.2	W, WS
Transition season	9 11, 3 5	7.86	W, WS

Due to the great loss caused by cold wind in winter, the simulation parameters of wind environment in this scheme are all meteorological data in winter.

After adding the wind-shield wall

Site profile

Regional architecture

Regional architecture refers to the architecture that responds to the natural environment such as terrain, landform and climate of a certain region, as well as the cultural environment such as lifestyle, cultural customs and religious beliefs, and conforms to the local technical and economic conditions.

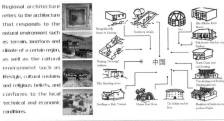

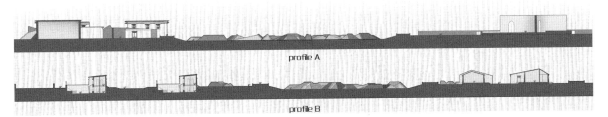

profile A

profile B

风涌墙阻 A wall to keep out the cold wind

Site planning

Functional system
In the commercial district, the hotel is located closest to the historic car park and plaza, and the rest of the functional buildings arranged alternately on both sides of the street. In the service area, the garbage station is downwind, and the village committee is in the shape of "L" facing the landscape belt.

Planning structure
From the entrance to the central landscape to form the main landscape corridor, attracting the main stream of people on the provincial road into the scenic area. The central landscape extends into commercial and residential areas. At the same time, a secondary axis is formed in the east-west direction of the business district to form a commercial pedestrian street, and the central activity venue of each group in the residential district also forms a secondary structure in space.

Walking system
We separates the pedestrian system of the landscape business district from that of the residents' activity area, prevents the circulation lines of foreign tourists and local residents from crossing and contradicting, and ensures the privacy of residents' life and the openness of tourists' sightseeing in the landscape business district, zigzagging and changeable circulation lines are adopted to increase the pleasure of tourists.

Garage system
The motor vehicle lane forms a ring road in the site, and is connected with some secondary road networks interspersed, so that the vehicle traffic in the region is smooth, and the transportation of loading and unloading goods in the commercial area and the transportation of residents in the residential area form a complete system.

Wind shield system
The windshield system is mainly based on the main landscape belt of the service area. Because of the need to resist the west wind, the wall tends to be north-south. In residential areas, there are windshields in the middle of each cluster. Residential courtyards are also arranged around the retaining wall.

Landscape system
Through the extension of the central landscape, the central landscape system and the commercial street and residential clusters on both sides form a landscape penetration, so that the different clusters form an organic whole. Cultural landscape is set up at the entrance of the site. Through the raised platform, a good viewing gallery is formed in the direction of the provincial road to attract tourists to enter.

Residential planning

Comparisn of alternative schemes site wind simulation

Scheme A   Scheme B   Scheme C   Scheme D

Residential houses are divided into three groups, which are equipped with wind shield system in the middle. The dwellings are lined with symbols.

Traffic square planning

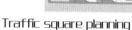

Set up characteristic cultural landscape in traffic square.

Iconography tianmen   abstract   The long end points in the direction of the heavenly gate.

The formation process of the pit courtyard

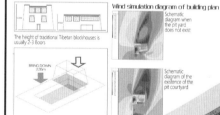

The height of traditional Tibetan blockhouses is usually 2-3 floors.

Wind simulation diagram of building plan
Schematic diagram when the pit yard does not exist
Schematic diagram of the existence of the pit courtyard

BRING DOWN 2.35m
Sink the ground 2.35 meters and place it in the blockhouse

Three energy-saving advantages of forming a pit yard

The natural soil layer becomes the insulation layer of the first floor enclosure structure of the blockhouse.

Set up a windshield around the yard. Prevent cold wind from entering of the blockhouse.

On the winter solstice. Ensure that the sun room has 4hs Unobstructed light

Wind simulation diagram of building facade
Schematic diagram when the pit yard does not exist
Schematic diagram of the existence of the pit courtyard

Active solar technology

1. Vacuum glass tube (collector)
2. Motor to pump air
3. Blow into the cold air duct
4-6. Pipes for conveying hot air
7. A container for exchanging heat from wind and water
8. Pipes for conveying hot air

Cold air enters from the duct under the eaves. Heated up by solar collectors. Then it is transported to the floor layer through the pipe embedded in the wall and blown into the room.

Passive solar technology

Open small windows during the day. Let the pebbles absorb solar heat radiation.
Close small windows at night. The pebbles release the heat absorbed during the day.

Perspective of Sunlight Room   Pebble Thermal Storage Kong

Prefabricated building technology

Dig the yard. Prefabricated lightweight aggregate concrete slabs in the factory. On-site assembly

1. Lightweight aggregate concrete
2. Bonding layer
3. Binder
4. Foam glass panel
5. Polymer anti-cracking mortar
6. Flexible water-resistant putty
7. Decorative surface

Assemble the prefab roof and enclosure structure. And set aside the sun room

1. Wall insulation material
2. Insulation material pipe
3. Dustproof fiber mesh
4. Wooden floor
5. Insulated ceiling

Paste insulation layer and facing material outside the enclosure structure

1. Vacuum tube
2. Insulation materials
3. Insulation board
4. Concrete

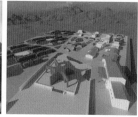

风涌墙阻 A wall to keep out the cold wind

Commercial and service area planning

Folk customs exhibition space: Tibetan opera and Tibetan costume performance.

There are parking lots and campsites on the north side, so the gate was built on the north side of the commercial street.

The main buildings of ethnic culture-themed hotels highlight the style and characteristics of Tibetan ethnic minority dwellings. The decoration and decoration of the hotel's lobbies, restaurants and guest rooms use colors and patterns that can highlight the characteristics of ethnic minority cultures.

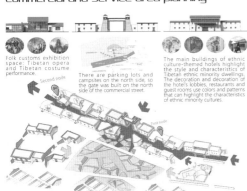

Assuming that the southwest wind is the prevailing wind direction in Nagqu

Wind blowing from the southwest. The direction of the wind wall is perpendicular to the wind direction

Two types of windshield walls are installed in the residential area. At the same time as windshield, it is also a landscape.

Layout the main vertical windshield landscape belt along the direction perpendicular to the wind direction

Schematic diagram of windshield wall in service area

Schematic diagram of windshield wall in residential area

Schematic diagram of windshield wall in landscape area

The use of regional materials

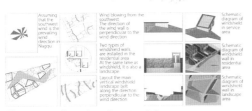

The decorative finishes of the houses are made of local gravel and stone binder. Highlight the original appearance of the material.

The windshield wall of the yard in the residential area is filled with crushed stones in an iron cage.

The concrete buffer and the excavated soil form a windshield wall in the landscape area.

Other low-carbon technologies

Solar pv on the landscape
Light strip detailed section

Electricity converted by solar panels
Supply LED strips on the windshield wall
As landscape lighting

Low carbon emissions throughout the cycle

Low carbon in the whole process

dismantle — Dig a hole — Prefabricated

Perspective

Tibetan opera performance

Commercial Street Entrance

Walk down the street

Buildings in service areas

Low carbon emissions throughout the cycle

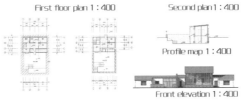

Technical drawing

Profile map 1:400

First floor plan 1:400   Second plan 1:400

First floor plan 1:400   Second plan 1:400

Profile map 1:400

Front elevation 1:400

Photovoltaic system installation area estimation

Shed roof   Shed roof   Double slope roof

Installable area = building base area ×r= land area × building density ×r

	Shed roof	Pitched roof (slope between 20 and 30)
r	0.7	0.6

	Service area	Residential area	commercial area
Total land area	10941.9	27122.2	19766.0
Building density	17.00%	9.20%	19.40%
r	0.7	0.7	0.6-0.7
The installation area	523.0	1746.7	2484.9

综合奖・优秀奖
General Prize Awarded・
Honorable Mention Prize

注 册 号：100514
Registration No：100514
项目名称：万物生
　　　　　Growth of All Things
作　　者：罗雪松、王　珍
Authors：Luo Xuesong, Wang Zhen
参赛单位：河南工业大学
Participating Unit：Henan University of Technology
指导教师：张　华
Instructor：Zhang Hua

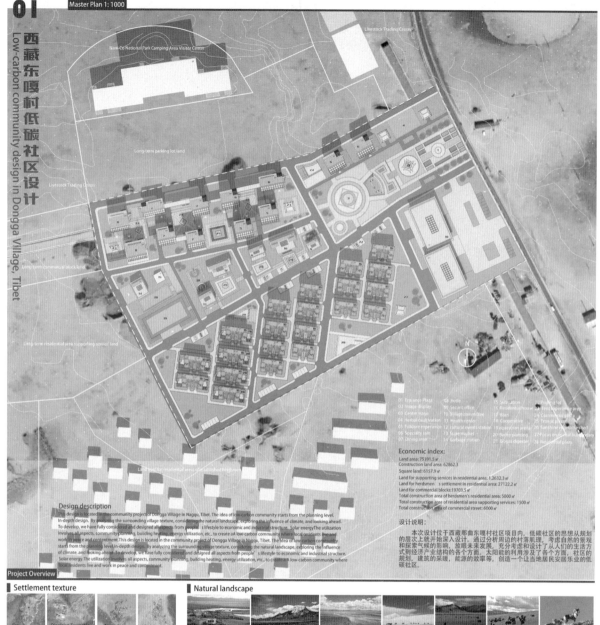

西藏东嘎村低碳社区设计
Low-carbon community design in Dongga Village, Tibet

01 Master Plan 1:1000

Design description
This design is located in the community project of Dongga Village in Nagqu, Tibet. The idea of low-carbon community starts from the planning level. In-depth design. By analyzing the surrounding village texture, considering the natural landscape, exploring the influence of climate, and looking ahead. To develop, we have fully considered and designed all aspects from people's lifestyle to economic and industrial structure. Solar energy The utilization involves all aspects, community planning, building heating, energy utilization, etc., to create aA low-carbon community where local residents live and work in peace and contentment. This design is located in the community project of Dongga Village in Nagqu, Tibet. The idea of low-carbon community starts from the planning level. In-depth design. By analyzing the surrounding village texture, considering the natural landscape, exploring the influence of climate, and looking ahead. To develop, we have fully considered and designed all aspects from people's lifestyle to economic and industrial structure. Solar energy. The utilization involves all aspects, community planning, building heating, energy utilization, etc., to create aA low-carbon community where local residents live and work in peace and contentment.

设计说明：
本次设计位于西藏那曲东嘎村社区项目内，低碳社区的思想从规划的层次上即开始深入设计，通过分析周边的村落肌理，考虑自然的景观和探索气候的影响，放眼未来发展，充分考虑和设计了从人们的生活方式到经济产业结构的各个方面。太阳能的利用涉及了各个方面。社区的规划，建筑的采暖，能源的效率等，创造一个让当地居民安居乐业的低碳社区。

Economic index:
Land area: 75391.5 ㎡
Construction land area: 62862.3
Square land: 6157.9 ㎡
Land for supporting services in residential area: 1,2632.3 ㎡
Land for herdumen's settlement in residential area: 27122.2 ㎡
Land for commercial blocks:19703.5 ㎡
Total construction area of herdsmen's residential area: 5000 ㎡
Total construction area of residential area supporting services: 7500 ㎡
Total construction area of commercial street: 6000 ㎡

Project Overview

Settlement texture
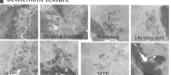

Natural landscape

Village　Mountain　River　Snow scene　Winter pasture　Summer pasture　Wild animal

Available energy analysis
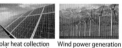
Solar heat collection and power generati　Wind power generation　Biomass

1.The average sunshine time in Nagqu is more than 2886/hour, with long sunshine time, high sunshine rate and strong solar radiation. The development of solar energy has unique advantages.

2.Tibet has annual wind energy reserves of 93 billion kWh, equivalent to 33.65 million tons of standard coal per year, ranking seventh in the country.

3.The vast agricultural and pastoral areas have always used biomass energy as their main energy source through direct combustion.

Climate analysis

Wind environment　Sun radiation

Relative humidity　Multiple enthalpy-humidity technology measures

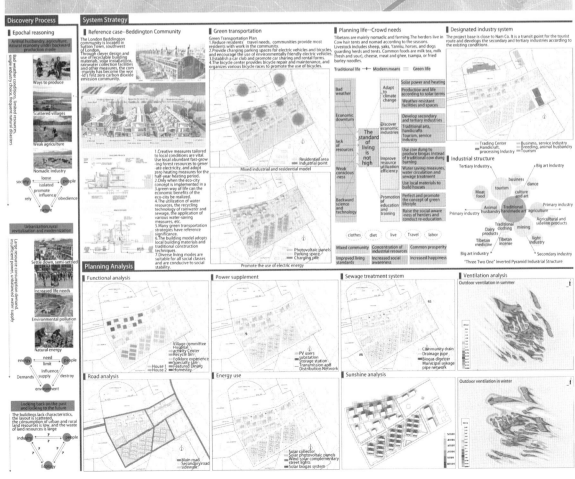

04 西藏东嘎村低碳社区设计——万物生
Low-carbon community design in Dongga Village, Tibet

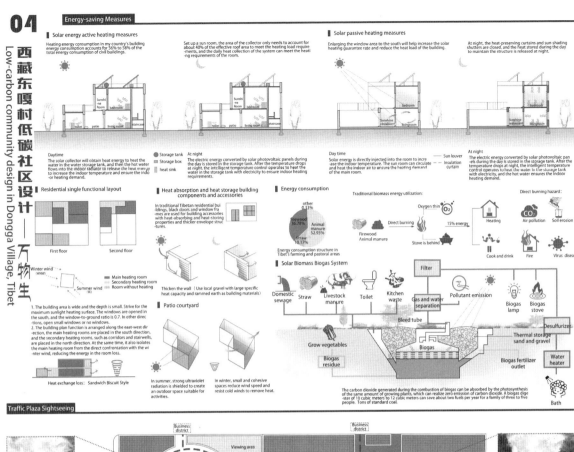

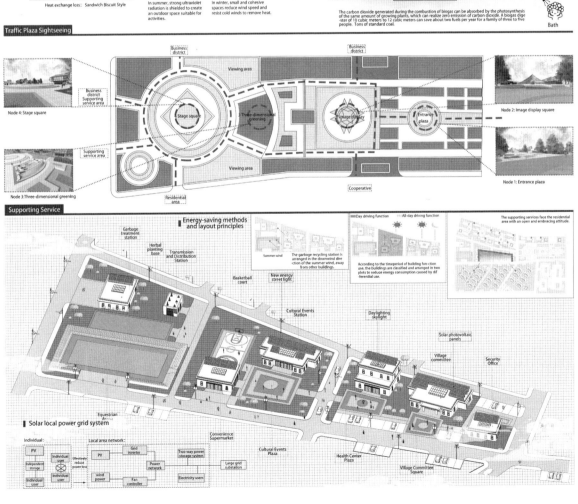

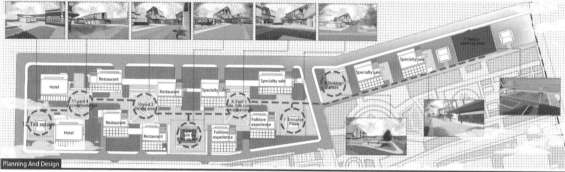
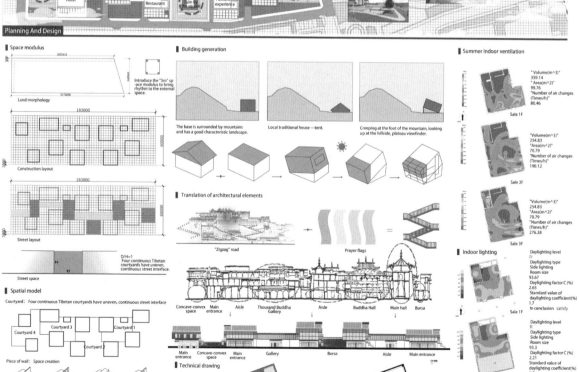

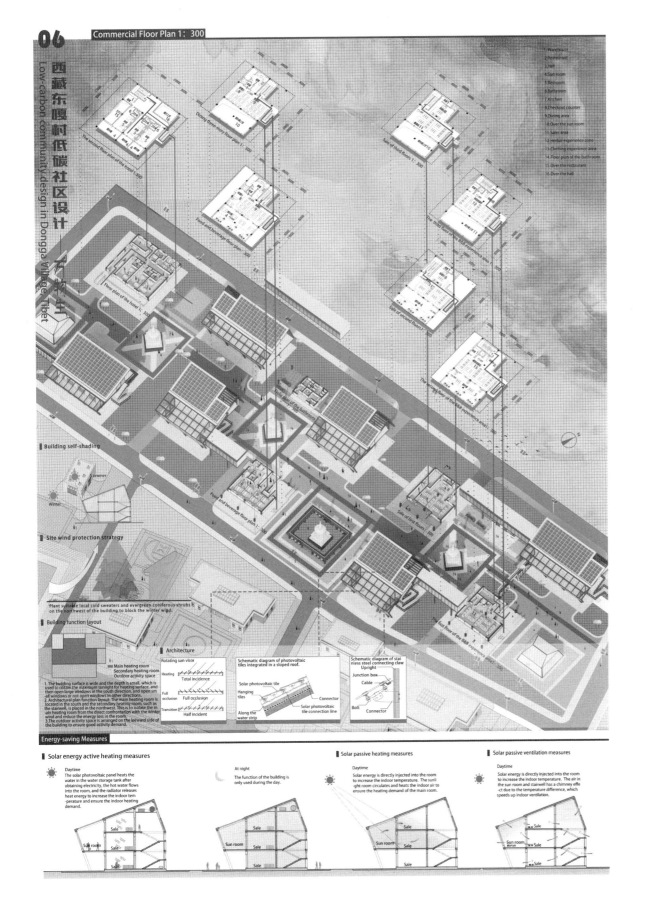

综合奖·优秀奖
General Prize Awarded·
Honorable Mention Prize

注 册 号：100569
Registration No：100569

项目名称：碳固·能转
Carbon Sequestration·
Energy Circulation

作　　者：高 博、朱可迪、罗 犇
Authors：Gao Bo, Zhu Kedi, Luo Ben

参赛单位：重庆大学
Participating Unit：Chongqing University

指导教师：张海滨、周铁军
Instructors：Zhang Haibin, Zhou Tiejun

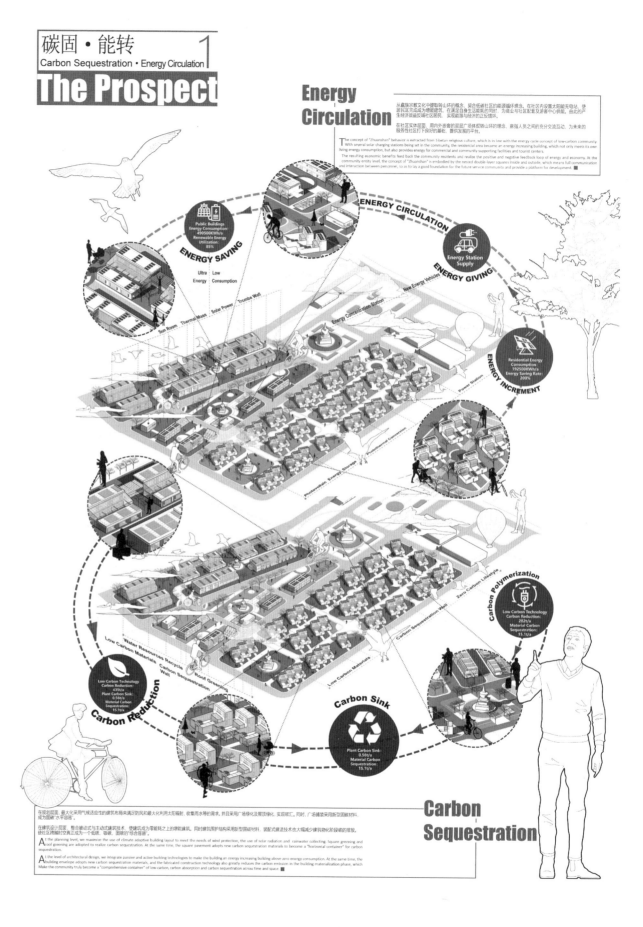

碳固·能转
Carbon Sequestration · Energy Circulation 2

The Investigation Phase

Location

Project land

The site is located in the southeast corner of bango county. Bango county is subordinate to Naqu City, with an average altitude of 4700m. It is located between Namucuo and bamucuo, the second largest lake in Tibet, only 30km away from both lakes. The project is a low-carbon community in Dongga village, Qinglong Township, bango County, Tibet.

Site

Integrated service center

Residential building

Cooperative industrial base

The site is only connected with the outside world through the East Township Road, which can lead to Lhasa, Naqu and Namucuo scenic spots. It is only 30km away from Namucuo and can be reached by driving for 30min. It is suitable for the development of tourism related industries and can be used as a backup service point for Namucuo tourism.

Resources

Water Resource Analysis
Abundant Lakes
There are Namucuo, bamucuo, rencuo, Jiangcuo, pengcuo, rutiaco, dongkacuo, bangoco, selinco, guomangcuo and other lakes around the base, which are rich in available water resources.

Mountain Resource Analysis
Continuous Hills
Around the base, there are mountains such as Anjun, Rida, Lagore, Garan, Digu, turi and sandurier, which are rich in geothermal energy.

Crowd

Residents are mainly Tibetans | Tourists are mainly middle-aged | Half male, Half female | Practitioners are mainly businessmen

Climate

Winter | Spring | Summer
Autumn | Whole Year

Mainly southeast wind
The average wind speed is between 6-8m/s, and the maximum wind speed can be more than 50, which will cause certain damage to production, life and buildings. It is necessary to do a good job in building wind protection.

Direct radiation | Scattered radiation | Total radiation
The amount of solar radiation is sufficient, and it is almost direct radiation throughout the year.

Annual average dry bulb temperature | Annual average precipitation | Annual average relative humidity
The days whose temperature is lower than 0 ℃ are about 250 days a year, and the maximum temperature is about 25 ℃. The precipitation is more in summer and autumn and less in winter and spring.

Activity

People Categories | Activity Demand | Activity Area

Practitioner — Dwelling — Residential Area
Builder — Eating — Supporting Service Land
Locals — Shopping / Entertaining / Interaction / Touring / Experiencing — Traffic Plaza Area
Tourists — Communication — Commercial Block Land

Botany

The Materialization Phase

Prefabricated Building

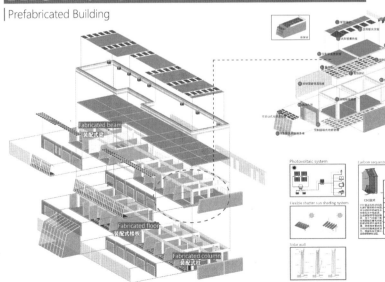

Fabricated beam 装配式梁
Fabricated floor 装配式楼板
Fabricated column 装配式柱

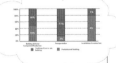

Photovoltaic system | Carbon sequestration solar wall
Flexible shutter sun shading system
Solar wall

The carbon emission in the materialization phase includes the carbon emission in the production stage of building materials components, the carbon emission in the transportation stage of building materials, and the carbon emission in the on-site installation and configuration stage. At this stage, we will try our best to improve the energy consumption structure of the factory and use renewable energy; we also use materials composed of components with low carbon emission factor; At the same time, we will eventually make more use of new energy during transportation and installation, recycle construction waste, shorten the construction period as far as possible and reduce per capita carbon emissions.

The proportion of carbon emission of traditional cast-in-situ buildings in the production and transportation stage is higher than that of fabricated buildings, mainly because the rough management of cast-in-situ method leads to large material consumption and waste, which increases the generation of carbon emission. The use of prefabricated components can reduce the carbon emission of buildings, improve the prefabrication rate of buildings, and promote the development of low-carbon buildings. ∎

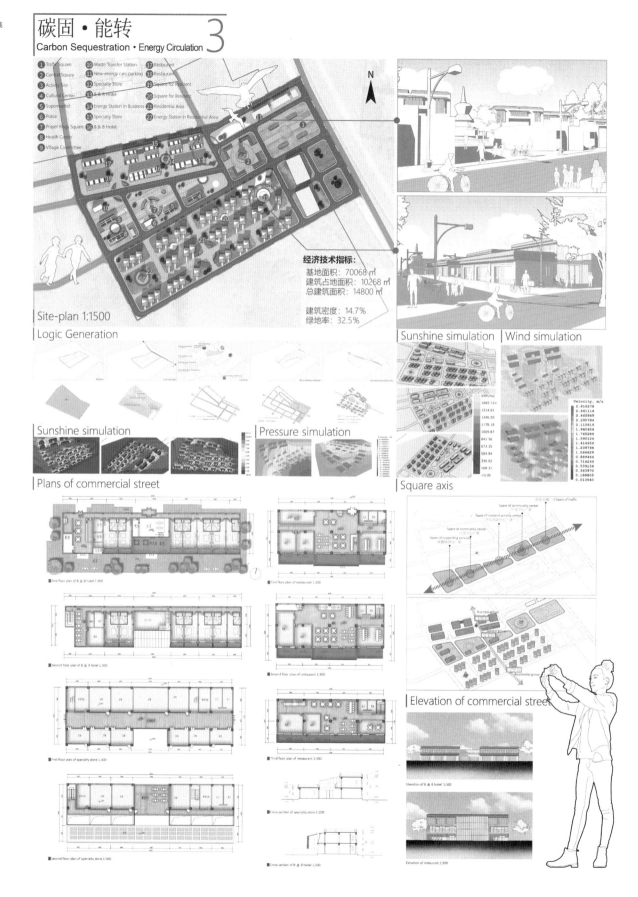

碳固·能转 4
Carbon Sequestration · Energy Circulation

The Operation Phase
Active technology

 Solar air heating system
 Heat recovery ventilation system
 Solar collector

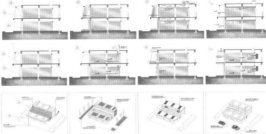

 Solar air heat collection and hot air heating system
 Solar water heating system
 Air source heat pump

Community Guidelines
1. It is advisable for all All community members to commute on foot within the community.
2. All community members must use an induction cooker for cooking.
3. When a car is needed, it must be electric.
4. Encourage community members to participate in planting trees.
5. Use passive heating as much as possible and reduce air conditioning.
6. Energy-efficient appliances must be used
7. Water saving appliances must be used and rainwater should be made good use of.
8. Rubbish sould be sorted and recycled.

The Recovery Phase
Life cycle

manufacture → transportation → construction → operation → recycling

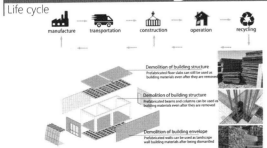

Demolition of building structure — Prefabricated floor slabs can still be used as building materials even after they are removed

Demolition of building structure — Prefabricated beams and columns can be used as building materials even after they are removed

Demolition of building envelope — Prefabricated walls can be used as landscape wall building materials after being dismantled

"Zero" Carbon Emission

Energy Consumption	
Item	Numerical Value(Kw·h/a)
Residential Area	192500
Public Building Area	490500
New Energy Vehicle	762120
New Energy Bus	883008
Domestic Waste Treatment	7402.2
Sewage disposal	12775
Carbon Emission	
Item	Numerical Value
Building Material Production Phase	8849.1t
Building Material Transportation Phase	1679.3t
Site Installation Phase	117.1t
Carbon Sink	
Plant	0.56t
Carbon Sequestration	
Material	15.1t
Solar Power	2500000Kw·h/a
Solar Heat	550000Kw·h/a

综合奖·优秀奖
General Prize Awarded · Honorable Mention Prize

注　册　号：100574
Registration No：100574

项目名称：融合社区·阳光市集
Convergence Community · Sunshine Town

作　　者：兰良建、梁今浒、张俊哲
Authors：Lan Liangjian, Liang Jinhu, Zhang Junzhe

参赛单位：华南理工大学
Participating Unit：South China University of Technology

指导教师：王　静
Instructor：Wang Jing

融合社区·阳光市集 1
Convergence Community · Sunshine Town

/ 设计说明 | Design Notes

本方案旨在打造一个以低碳为主题、西藏文化融合交流为骨架、社区各功能之相互渗透并显示独特地域文化的融汇社区。

其一，西藏文化与低碳设计结合。抓取西藏特色文化意象，赋予低碳方式新表达，充分利用当地丰富的绿色能源，延续住宅区肌理，展示社区地域特色。

其二，游客与当地居民和文化的交流融合。设立文化体验节点，串联并激活街区空间，建设文化交流的平台。

其三，立体和弹性公共空间与气候相适应，结合阳光房与院落设置立体流线与拓展场地，容纳多元活动。

This plan aims to create a fusion community with the theme of low carbon, the integration and exchange of Tibetan culture as the framework, the mutual penetration of various community functions, and the display of unique regional culture.

One is the combination of Tibetan culture and low-carbon design. Grasp the cultural imagery with Tibetan characteristics, give new expressions to low-carbon methods, and make full use of the local abundant green energy. Continuing the texture of the residential area, showing the regional characteristics of the community.

Second, the exchange and integration of tourists, local residents and culture. Set up cultural experience nodes, connect and activate the block space, and build a cultural exchange platform.

Third, the three-dimensional and flexible public space adapts to the climate. Combine the sun room and courtyard to set up a three-dimensional flow line and expand the venue to accommodate multiple activities.

/ Regional Features of Dwellings

 Flat roofs
 The color of the exterior wall
 Pastoral tents
 Bianma Wall
 Scripture / Colorful canopy

/ Orientation Analysis Based on Solar Radiation

Orientation based on average daily incident radiation on a vertical surface.

/ Climate Analysis

西藏2011-2020年平均气温（单位：℃）

The location of the product is severely cold in winter and cold in summer; building insulation and heating are prominent propositions of construction engineering.

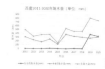
西藏2011-2020年风速（单位：km/h）

From the chart above, we can see that the daily maximum wind speed in Tibet is 7 on average. The wind resources are abundant and should be fully utilized and complemented by solar power generation.

西藏2011-2020年降水量（单位：mm）

Annual precipitation in Tibet is unstable, sometimes less than 200mm, sometimes 800mm; with little in winter and more in summer, and water collection and water supply facilities are needed.

/ Wind Analysis

Wind Direction / Wind Speed Distribution Frequency in Summer

Wind Direction / Wind Speed Distribution Frequency in Winter

Wind Direction / Wind Speed Distribution Frequency Monthly

/ Active Architectural Technology Analysis

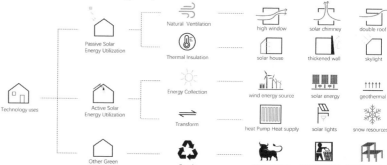

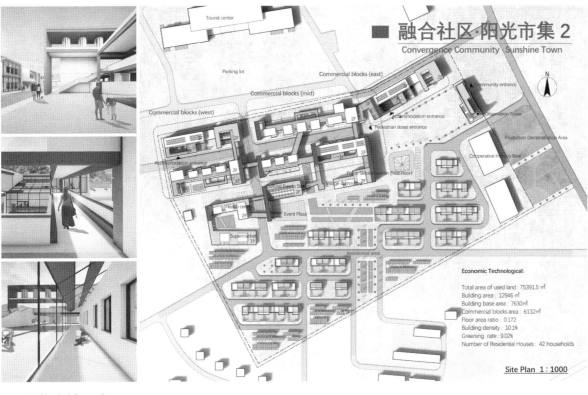

融合社区·阳光市集 2
Convergence Community · Sunshine Town

Economic Technological:

Total area of used land: 75391.5 ㎡
Building area: 12946 ㎡
Building base area: 7630㎡
Commercial blocks area: 6132㎡
Floor area ratio: 0.172
Building density: 10.1%
Greening rate: 9.02%
Number of Residential Houses: 42 households

Site Plan 1:1000

/ Logical Generation

Projection: Land area 75391.5㎡. Construction land area of 62892.3㎡, including Square land, herders residential land, commercial block land.

Axis: According to the plan for the layout of the building, the axis is radiant, the center converges in the square.

Route: Using nodes to connect their different commercial streets in series makes the flow line more flexible and interesting.

Orientation: Calculate and utilize the optimal orientation of the building in this area.

Texture: Adjust the architectural layout of residential areas to make them more integrated with the parts that have been built.

Site: To the north is the visitor center, south of the site is abuilt residential area for inhabitant.

Road network: The road is meshed and has the possibility of continuing to extend for the second phase of construction.

Node: Take advantage of traditional Tibetan architectural forms: Crossing street building.

Solar house & yard: The combination of sunroom and courtyard.

Solar cell: Solar cells are arranged on the south side, both to take advantage of sunlight resources and to keep out the wind.

/ Sunshine Duration

/ Sunshine Hours Throughout the Year   / Sunshine hours of Winter

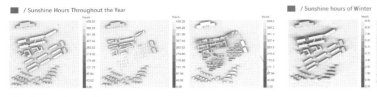

/ Cutaway Perspective

1-1 Section 1:300   **2-2 Section 1:300**

融合社区·阳光市集 3
Convergence Community · Sunshine Town

通风和蓄热分析 | Ventilation and Thermal Storge Analysis

餐饮单元 | Catering Unit

In winter, the building mainly uses solar houses to maintain the temperature within a certain range and form thermal pressure which promotes air circulation.

In summer, shading system attached on the daylighting roof protects indoor environment from excessive heating. Air vents are opening to facilitate air convection.

商业零售单元 | Commercial Retail Unit

采光 | Daylighting

At winter night, PCM and thermal roof release the heat collected in daytime. The solar house acts as a climate buffer and accommodates both indoor walking and commercial activities.

In summer, high roof windows provide a means for thermal ventilation.

通风和蓄热 Ventilation & Thermal

住宅单元 | Dwelling Unit

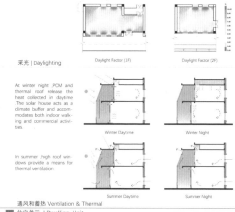

采光 | Daylighting

In winter, the building uses solar house to heating inoor enviroment. PV panel on the roof collect solar energy for household use.

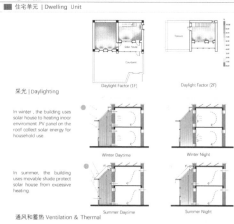

In summer, the building uses movable shade protect solar house from excessive heating.

通风和蓄热 Ventilation & Thermal

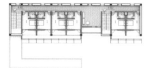

Commercial Blocks (West) 3F Plan 1:300

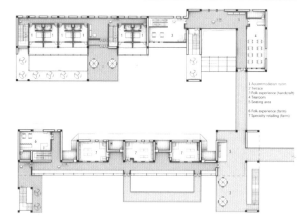

Commercial Blocks (West) 2F Plan 1:300

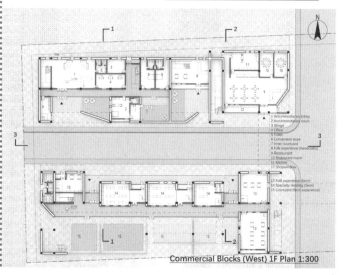

Commercial Blocks (West) 1F Plan 1:300

融合社区·阳光市集 4
Convergence Community · Sunshine Town

■ / Unite Analysis

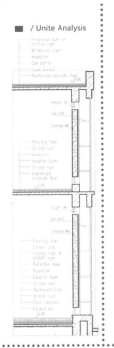

■ / Unite Analysis

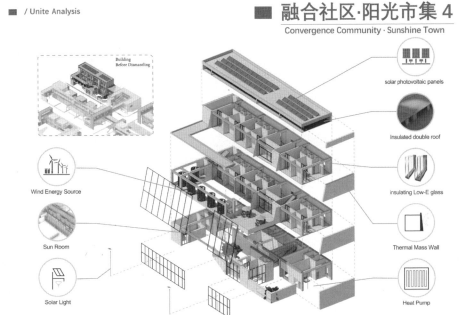

- solar photovoltaic panels
- Insulated double roof
- insulating Low-E glass
- Thermal Mass Wall
- Heat Pump
- Wind Energy Source
- Sun Room
- Solar Light

■ / Imagery Analysis

Entrance: The observation tower and the platform combine to form a landmark structure.

Steps: Places for cultural exchanges, activities and performances.

Decoration: Using Tibetan imagery to reflect local cultural characteristics.

Scriptures: reflect the local religious and cultural characteristics.

Equipment: Choose equipment close to local cultural objects and integrate into local culture.

Solar energy: make full use of the abundant local solar energy resources.

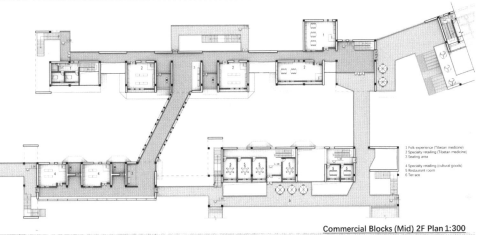
Commercial Blocks (Mid) 2F Plan 1:300

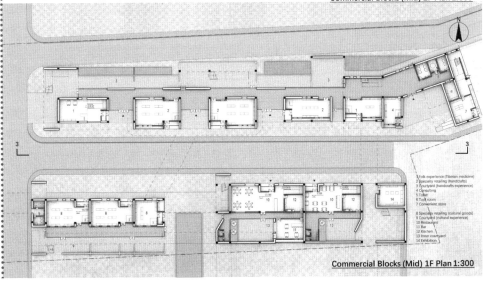
Commercial Blocks (Mid) 1F Plan 1:300

融合社区·阳光市集 5
Convergence Community · Sunshine Town

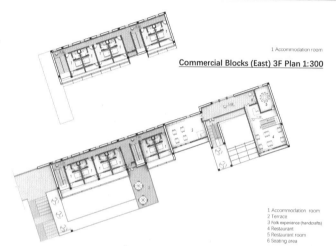

1 Accommodation room

Commercial Blocks (East) 3F Plan 1:300

1 Accommodation room
2 Terrace
3 Folk experience (handcrafts)
4 Restaurant
5 Restaurant room
6 Seating area

Commercial Blocks (East) 2F Plan 1:300

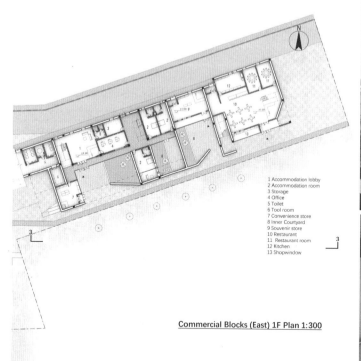

1 Accommodation lobby
2 Accommodation room
3 Storage
4 Office
5 Toilet
6 Tool room
7 Convenience store
8 Inner Courtyard
9 Souvenir store
10 Restaurant
11 Restaurant room
12 Kitchen
13 Shopwindow

Commercial Blocks (East) 1F Plan 1:300

融合社区·阳光市集 6
Convergence Community · Sunshine Town

■ / Functional Division Analysis

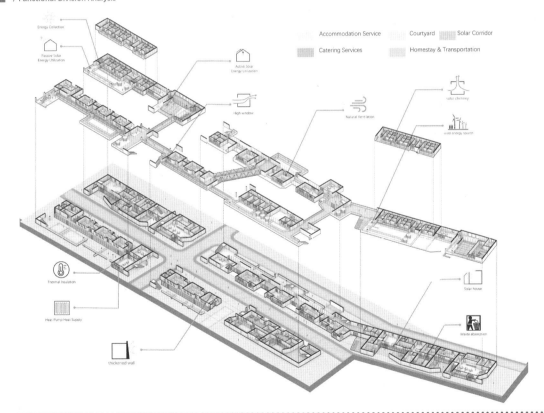

South facade 1:500

3-3 Section 1:500

综合奖·优秀奖
General Prize Awarded · Honorable Mention Prize

注 册 号：100586
Registration No：100586

项目名称：斯贝廓罗
Samsara

作　　者：刘奕辰、徐子牧、邓紫岚、
　　　　　康旭东、吴伯超、周　浩
Authors：Liu Yichen, Xu Zimu,
　　　　　Deng Zilan, Kang Xudong,
　　　　　Wu Bochao, Zhou Hao

参赛单位：湖南大学
Participating Unit：Hunan University

指导教师：徐　峰、严湘琦、彭晋卿、
　　　　　李洪强、周　晋、肖　坚
Instructors：Xu Feng, Yan Xiangqi,
　　　　　Peng Jinqing, Li Hongqiang,
　　　　　Zhou Jin, Xiao Jian

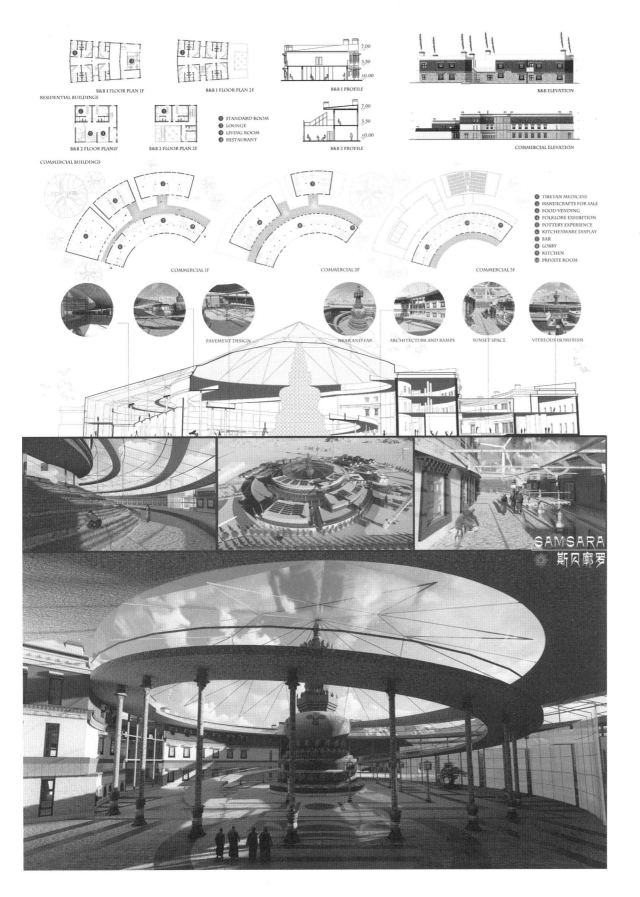

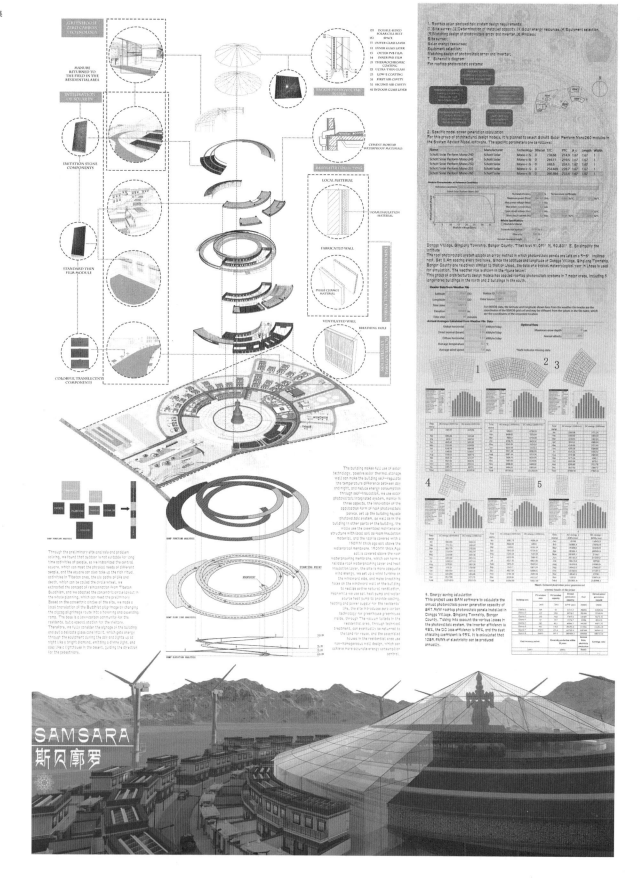

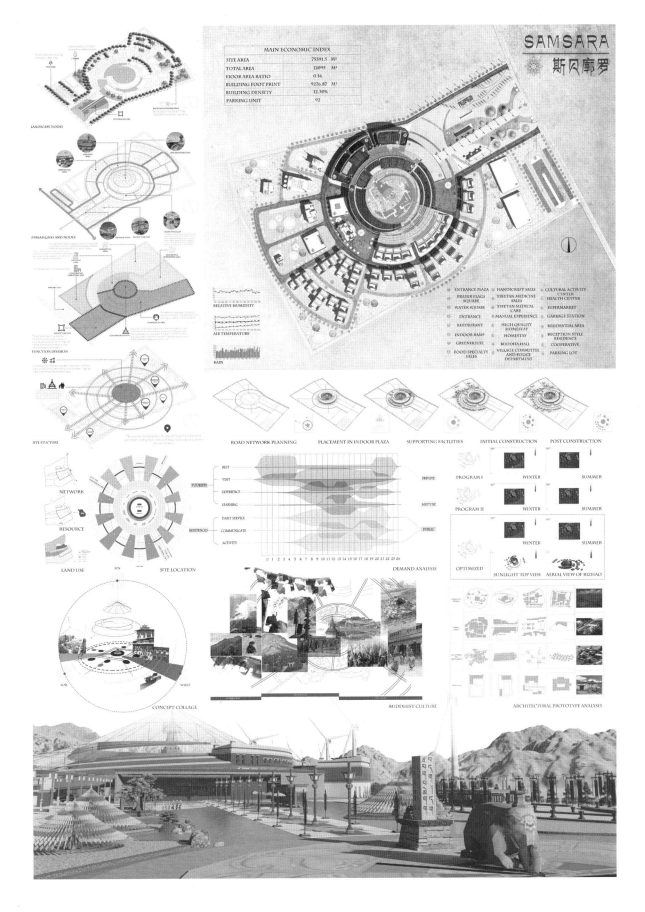

综合奖·优秀奖
General Prize Awarded · Honorable Mention Prize

注　册　号：100591
Registration No：100591

项目名称：轮回之径
　　　　　The Path of Reincarnation

作　　者：李小转、邵　红、王奇鹏
Authors：Li Xiaozhuan, Shao Hong, Wang Qipeng

参赛单位：西安建筑科技大学
Participating Unit：Xi'an University of Architecture and Technology

指导教师：李　涛
Instructor：Li Tao

轮回之径 The Path of Reincarnation 01
西藏班戈县低碳社区设计

Design Description
The project is designed as a low-carbon community integrating residence and tourism. The north side of the site is a commercial street under Tibetan culture, the south side is a residential area for herdsmen to settle down, and the east side is the concentration point of the entire site and the community. Image entrance, considering the unique regional culture of Tibet in architectural design, including climate, customs, crowd activities, etc., technically considering active and passive strategies, using solar photovoltaic power generation, solar collectors, rainwater collection, biogas treatment and other strategies Achieve low-carbon, green, and healthy communities.

Economic Index
Total land area:	62892㎡
Total building area:	59400㎡
Building base area:	23052㎡
Building density:	36.65%
Floor area ratio:	0.944
Greening rate:	24%
Parking spaces:	30个

Site Analysis
The project is located in Dongga Village, Qinglong Township, Bangor County, with a north latitude of 90.801 and an east longitude of 31.091. It is about 30 kilometers away from Namtso Scenic Area, 16 kilometers away from Qinglong Township and 82 kilometers away from Bangor County. The provincial highway S206 Banluo line runs through the east side of the base in the north-south direction. The project is planned to be built into a low-carbon community integrating residence and tourism reception, including commercial blocks, herdsmen's houses and supporting service facilities in residential areas.

Site

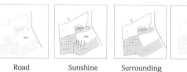
Road　Sunshine　Surrounding　Site

Regional Culture

 Symbolizes the harmony of heaven, earth, people and animals.

 Show devotion to the Buddha, you can get rid of the suffering of reincarnation.

 It embodies the rich experience created and accumulated by the Tibetan people in the long-term production and practice.

Current Situation and Problems
 The infrastructure is poor, and there are existing newly built houses nearby.

 The stability of power supply is poor, and the communication is unstable.

 With tap water system, no drainage measures.

 Large amount of solar radiation, long sunshine time.

Commercial Street Facade

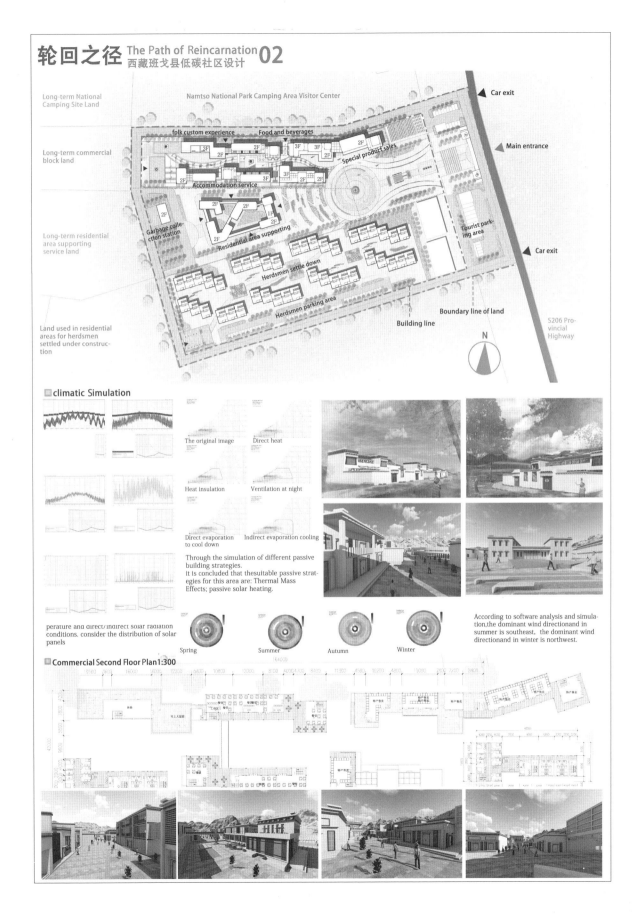

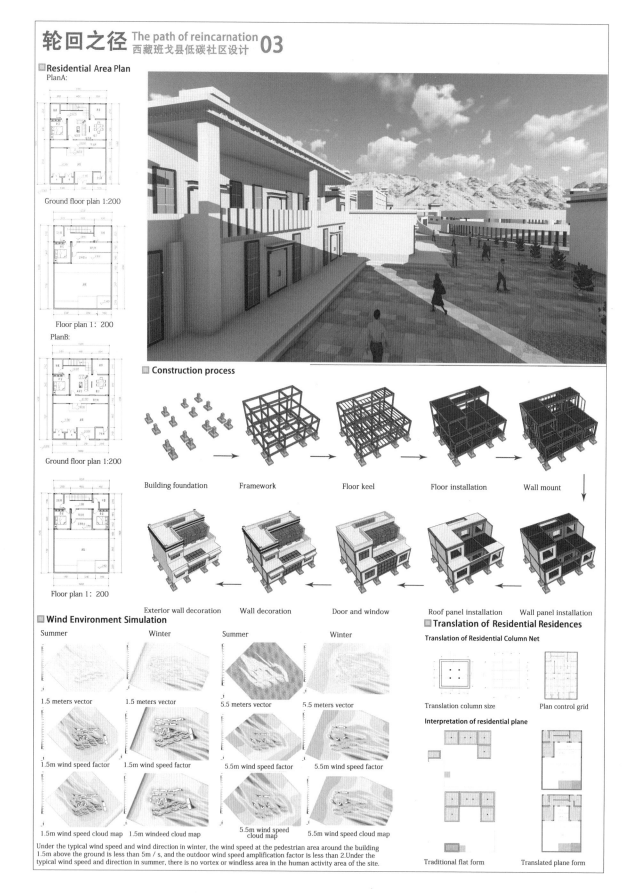

轮回之径 The Path of Reincarnation 04
西藏班戈县低碳社区设计

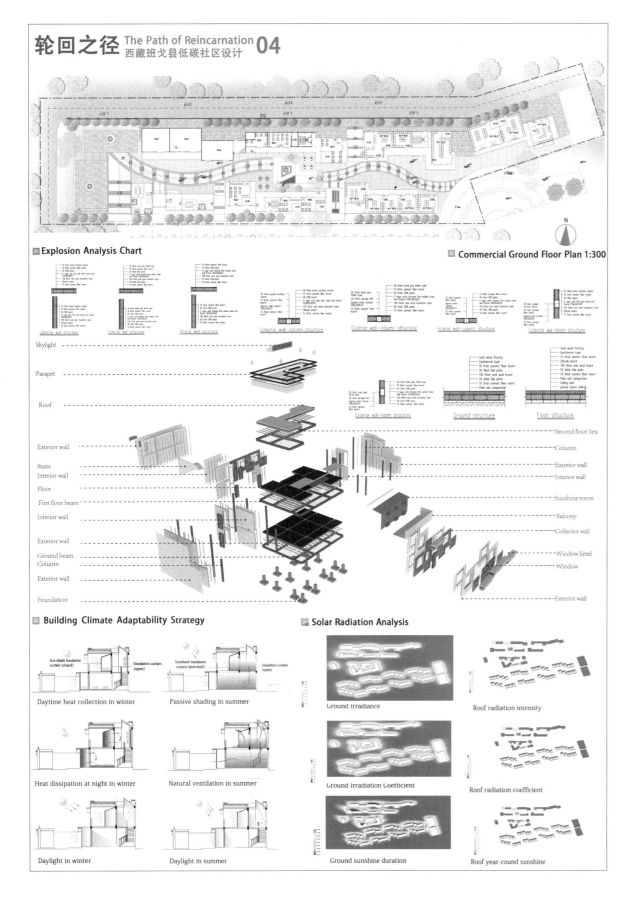

综合奖·优秀奖
General Prize Awarded · Honorable Mention Prize

注 册 号：100634
Registration No：100634

项目名称：向阳而生，逐光而行
Facing the Sun, Chasing the Sun

作　　者：雷珊珊、梁　欢、何秀敏、郑何山、雷宸骁
Authors：Lei Shanshan, Liang Huan, He Xiumin, Zheng Heshan, Lei Chenxiao

参赛单位：西安建筑科技大学
Participating Unit：Xi'an University of Architecture and Technology

指导教师：何　泉
Instructor：He Quan

向阳而生 逐光而行 Facing the Sun, Chasing the Sun
西藏班戈县青龙乡东嘎村低碳社区

chapter1: pre-program

1. Background Analysis
Location Analysis

设计说明：

西藏——最接近太阳的地方。本设计以"向阳、逐光"为主题，通过场地设计争取最佳日照。在建筑设计中采用直接受益式、特朗勃墙、阳光房等多种策略组合设计发挥太阳能的最大采暖效率，以改善室内热环境。在建筑设计安装光伏系统、雨雪水收集系统节约能源。

五彩经幡藏地的象征，哪里有经幡哪里就有善良、吉祥。在方案中我们以经幡为原型设计经幡广场。

2. Climate Analysis

Location Analysis

Prayer flags / the Potala Palace / Stone

Red top / Sutra Hall

最佳朝向　　哈湿图

According to the simulation, the best orientation is 5° south by East, so the best orientation of the scheme building is south

春季风玫瑰图　夏季风玫瑰图　秋季风玫瑰图

By simulating and analyzing the wind direction of each season, it is found that the dominant wind direction is southeast wind in summer and northwest wind in winter. Wind prevention is considered according to the wind direction in the scheme.

冬季风玫瑰图
全年风向

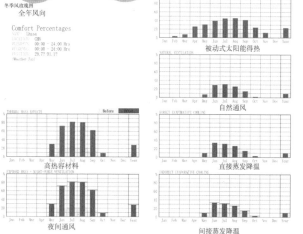

被动式太阳能得热

自然通风

高热容材料

直接蒸发降温

夜间通风

间接蒸发降温

向阳而生 逐光而行 Facing the Sun, Chasing the Sun
西藏班戈县青龙乡东嘎村低碳社区

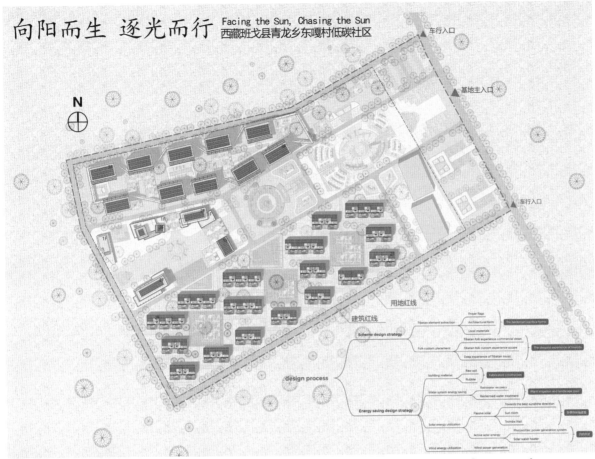

Landscape Analysis

Logical Generation

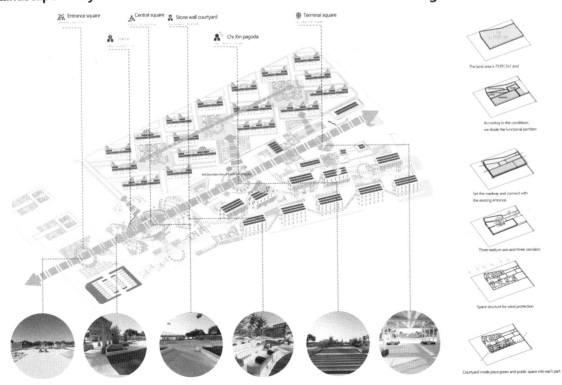

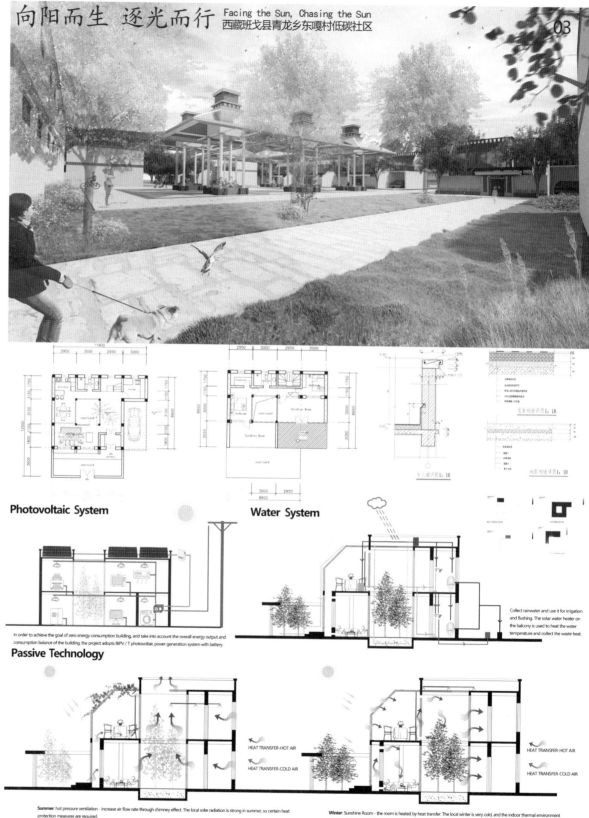

向阳而生 逐光而行
Facing the Sun, Chasing the Sun
西藏班戈县青龙乡东嘎村低碳社区

04

2021 台达杯国际太阳能建筑设计竞赛获奖作品集

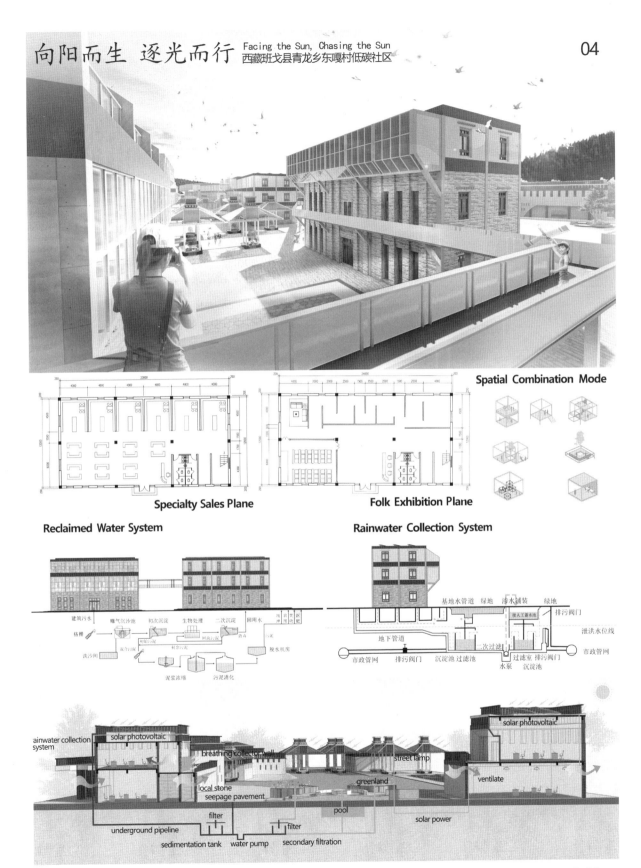

Specialty Sales Plane

Folk Exhibition Plane

Spatial Combination Mode

Reclaimed Water System

Rainwater Collection System

Low Carbon Strategy of Residential Supporting Buildings

向阳而生 逐光而行
Facing the Sun, Chasing the Sun
西藏班戈县青龙乡东嘎村低碳社区

Part 4: Tibetan folk commercial street

Design Ideas

In the functional design of commercial building design, the characteristics and area of four functional requirements are considered, and it is concluded that an independent business unit of 200 square meters is embedded in vertical traffic. The accommodation room is provided with a sunny room in the south to ensure that the room has appropriate temperature in summer night and winter; The sunshine room with outstanding catering has the function of viewing restaurant. The floors of four different functions are arranged and combined to form three types of 10 scattered commercial buildings.

We design a solar room, a heat collection and storage wall in the building, and a photovoltaic system on the roof to make use of solar energy resources. Prayer flags and colored roofs are placed between the buildings to form a semi outdoor space. The glass daylighting roof can be opened and closed freely according to the weather.

A commercial Floor Plan

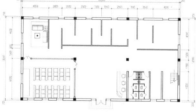
Ground floor plan (folk experience) 1:150

Second floor plan (catering service) 1:150

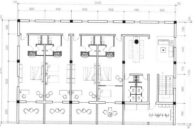
Third floor plan (accommodation service) 1:150

B commercial Floor Plan

Ground floor plan (specialty sale) 1:150

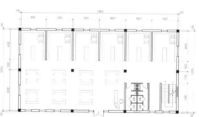
Second floor plan (specialty sale) 1:150

Third floor plan (catering service: 1:150

Tibetan Element Extraction

Traditional streets

Tibetan tent

Tibetan blockhouse

Prayer flags

Spatial Combination Mode

向阳而生 逐光而行 Facing the Sun, Chasing the Sun
西藏班戈县青龙乡东嘎村低碳社区

C Commercial Floor Plan

Ground floor plan (specialty sale) 1:150

Second floor plan (catering service) 1:150

Third floor plan (catering service) 1:150

Building Energy Saving Structure

Ground structure details 1:10
- 20 cement mortar
- concrete
- 60 polystyrene board
- concrete
- Plain soil compaction

Detailed drawing of roof structure 1:10
- 10 asphalt felt waterproof layer
- 20 cement mortar leveling course
- Polystyrene organic foam plastic board
- 100 cement coke slag slope making layer
- Reinforced concrete perforated slab

Parapet detail 1:10

Reclaimed Water System

Rainwater Collection System

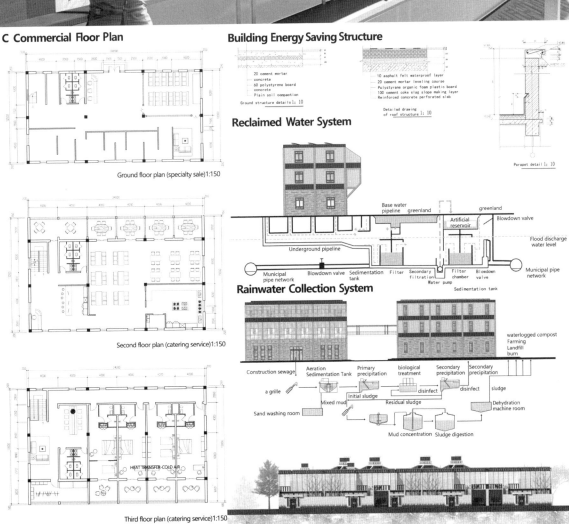

概念设计项目获奖名单
Concept Design Project

综合奖·优秀奖
General Prize Awarded · Honorable Mention Prize

注 册 号：100014
Registration No：100014

项目名称：田园"彝"居——以水为介质的太阳能和风能循环系统社区
Rural Yi Settlement-Community of Solar and Wind Energy Circulation Systems Using Water as the Medium

作　者：宋文雯、黎德伦、王昱晴、方钦
Authors：Song Wenwen, Li Delun, Wang Yuqing, Fang Qin

参赛单位：四川农业大学
Participating Unit：Sichuan Agricultural University

指导教师：董翔宇、陈川、张丽丽
Instructors：Dong Xiangyu, Chen Chuan, Zhang Lili

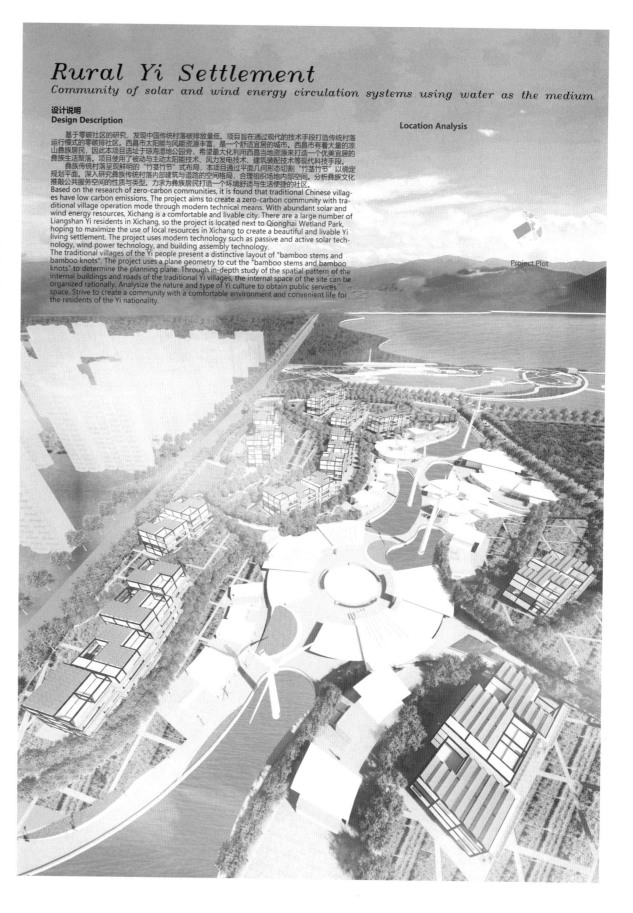

Rural Yi Settlement
Community of solar and wind energy circulation systems using water as the medium

设计说明
Design Description

　　基于零碳社区的研究，发现中国传统村落碳排放量低。项目旨在通过现代的技术手段打造传统村落运行模式的零碳排社区。西昌市太阳能与风能资源丰富，是一个舒适宜居的城市。西昌市有着大量的凉山彝族居民，因此本项目选址于琼海湿地公园旁，希望最大化利用西昌当地资源来打造一个优美宜居的彝族生活聚落。项目使用了被动与主动太阳能技术、风力发电技术、建筑装配技术等现代科技手段。
　　彝族传统村落呈现鲜明的"竹茎竹节"式布局，本项目通过平面几何形态切割"竹茎竹节"以确定规划平面。深入研究彝族传统村落内部建筑与道路的空间格局，合理组织场地内部空间。分析彝族文化推敲公共服务空间的性质与类型，力求为彝族居民打造一个环境舒适与生活便捷的社区。

Based on the research of zero-carbon communities, it is found that traditional Chinese villages have low carbon emissions. The project aims to create a zero-carbon community with traditional village operation mode through modern technical means. With abundant solar and wind energy resources, Xichang is a comfortable and livable city. There are a large number of Liangshan Yi residents in Xichang, so the project is located next to Qionghai Wetland Park, hoping to maximize the use of local resources in Xichang to create a beautiful and livable Yi living settlement. The project uses modern technology such as passive and active solar technology, wind power technology, and building assembly technology.
The traditional villages of the Yi people present a distinctive layout of "bamboo stems and bamboo knots". The project uses a plane geometry to cut the "bamboo stems and bamboo knots" to determine the planning plane. Through in-depth study of the spatial pattern of the internal buildings and roads of the traditional Yi villages, the internal space of the site can be organized rationally. Analyze the nature and type of Yi culture to obtain public services space. Strive to create a community with a comfortable environment and convenient life for the residents of the Yi nationality.

Location Analysis

Project Plot

Rural Yi Settlement

Climate Condition Analysis

Solar Radiation

Xichang annual solar radiation rich solar energy resources available.

Best Orientation
The best orientation of the building is southeast.

Temperature
People feel more uncomfortable in winter than in summer.

Rainfall
Rainfall throughout the year has dry and rainy seasons.

Green Building Technology Comparison

Passive Solar Heating

Natural Ventilation

High Specific Heat Capacity Material

Through comparison, it is found that passive solar heating technology has little improvement in indoor thermal comfort, while the use of heat storage materials with high specific heat capacity can greatly improve indoor thermal comfort. So we combine the two technologies. At the same time, we consider ventilation to improve indoor thermal comfort.

Site Analysis

Sea and Land Breeze

Surrounding Function

Surrounding People

Surrounding Traffic

Base Resource Analysis

- Available water source
- Site height difference

- Combing the terrain
- Enhancing site ventilation
- Uneven area division

Ensuring sunshine for residence

Needs Analysis

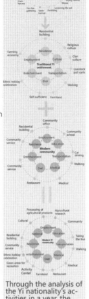

Through the analysis of the Yi nationality's activities in a year, the living needs of a Yi community are obtained, and then combined with the needs of the modern community, the needs of a modern Yi community are obtained.

The Layout of The Traditional Settlements of The Yi Nationality

Legend: Bamboo Stem, Bamboo Branches, Bamboo Joint, Residential settlement, Public event space, contour line, Folk house

The feature:
1. Take the road as the bamboo stem and the place where the crowd gathers as the bamboo joint.
2. Branches are drawn at the bamboo joints, and a settlement is formed on the bamboo branches.
3. Public venues are located on both sides of the road, and close to settlements.
4. The dwellings are arranged along the contour line.

Layout Generation

Cut out the shape of the bamboo stem with a circle.

Determine settlements and roads according to the shape of the bamboo.

Residential buildings are generated in accordance with the contour lines.

The public space is located at the bamboo joint, and the farmland is at the location closest to the settlement.

Economic and Technical Indicators

Planning land area: 91500㎡　work area: 2497㎡　Volume rate: 0.47
Total surface area: 43116㎡　Medical building area: 1270㎡　Building density: 17.10%
Residential area: 32690㎡　Educational building area: 1432㎡　Green area rate: 59.70%
Commercial area: 2184㎡　Community service building area: 3043㎡

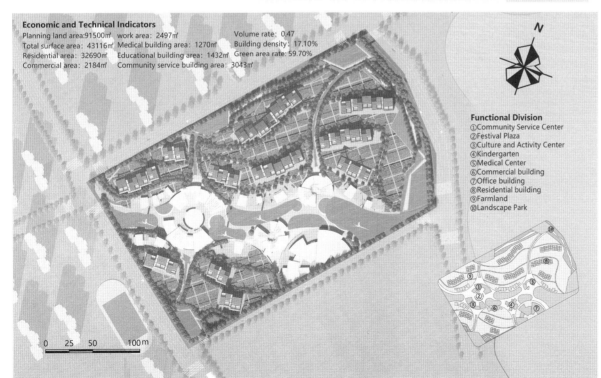

Functional Division
① Community Service Center
② Festival Plaza
③ Culture and Activity Center
④ Kindergarten
⑤ Medical Center
⑥ Commercial building
⑦ Office building
⑧ Residential building
⑨ Farmland
⑩ Landscape Park

Rural Yi Settlement

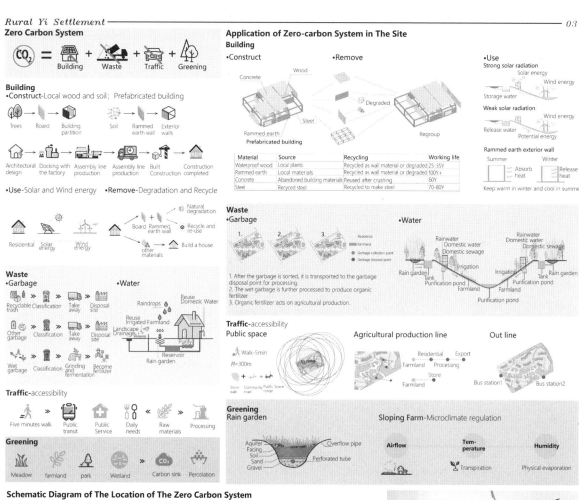

Zero Carbon System

CO_2 = Building + Waste + Traffic + Greening

Building
- **Construct**-Local wood and soil; Prefabricated building
- **Use**-Solar and Wind energy
- **Remove**-Degradation and Recycle

Waste
- Garbage
- Water

Traffic-accessibility

Greening

Application of Zero-carbon System in The Site

Building
- Construct
- Remove
- Use

Strong solar radiation / Weak solar radiation

Rammed earth exterior wall — Keep warm in winter and cool in summer

Material	Source	Recycling	Working life
Waterproof wood	Local plants	Recycled as wall material or degraded	25-35Y
Rammed earth	Local materials	Recycled as wall material or degraded	100Y+
Concrete	Abandoned building materials	Reused after crushing	60Y
Steel	Recyced steel	Recycled to make steel	70-80Y

Waste
- Garbage
- Water

1. After the garbage is sorted, it is transported to the garbage disposal point for processing.
2. The wet garbage is further processed to produce organic fertilizer.
3. Organic fertilizer acts on agricultural production.

Traffic-accessibility
- Public space (Walk-5min, R=300m)
- Agricultural production line
- Out line

Greening
- Rain garden
- Sloping Farm-Microclimate regulation (Airflow, Temperature, Humidity)

Schematic Diagram of The Location of The Zero Carbon System

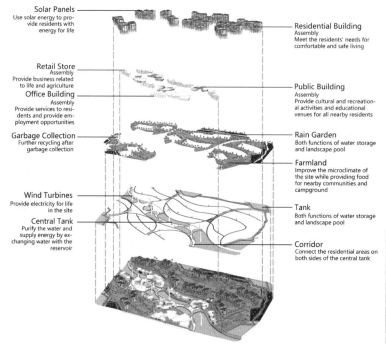

- **Solar Panels** — Use solar energy to provide residents with energy for life
- **Retail Store** — Assembly — Provide business related to life and agriculture
- **Office Building** — Assembly — Provide services to residents and provide employment opportunities
- **Garbage Collection** — Further recycling after garbage collection
- **Wind Turbines** — Provide electricity for life in the site
- **Central Tank** — Purify the water and supply energy by exchanging water with the reservoir
- **Residential Building** — Assembly — Meet the residents' needs for comfortable and safe living
- **Public Building** — Assembly — Provide cultural and recreational activities and educational venues for all nearby residents
- **Rain Garden** — Both functions of water storage and landscape pool
- **Farmland** — Improve the microclimate of the site while providing food for nearby communities and campground
- **Tank** — Both functions of water storage and landscape pool
- **Corridor** — Connect the residential areas on both sides of the central tank

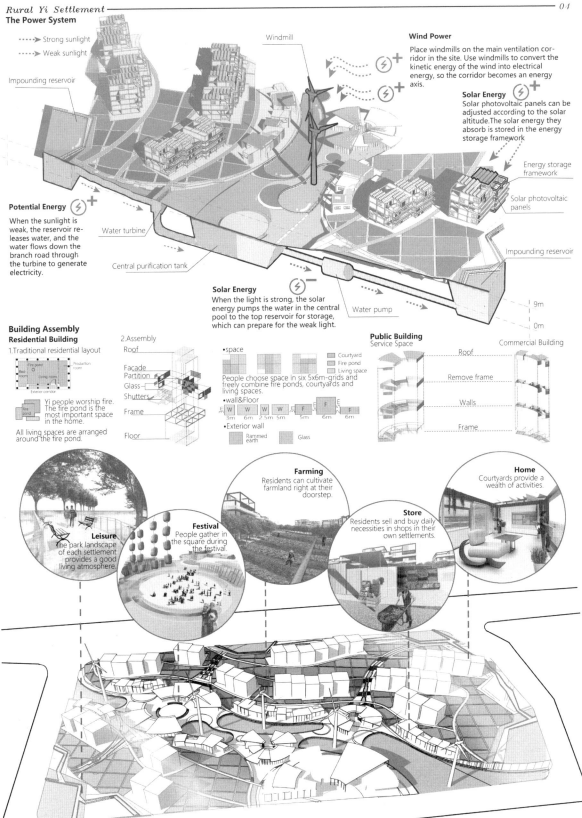

1 风曦·满居
WIND AND SUNSHINE · MAN RESIDENCE

综合奖·优秀奖
General Prize Awarded · Honorable Mention Prize

注册号：100031
Registration No:100031

项目名称：风曦 · 满居
Wind and Sunshine · MAN Design

作　者：沈　鑫、石　广、万金波、朱梦龙、王雅萱
Authors：Shen Xin, Shi Guang, Wan Jinbo, Zhu Menglong, Wang Yaxuan

参赛单位：沈阳大学
Participating Unit：Shenyang University

指导教师：赵雪峰、孟晓雷、滕　凌
Instructors：Zhao Xuefeng, Meng Xiaolei, Teng Ling

Climatic Analysis

Description of Design

本方案重点研究在沈阳特定的气候和地理条件下，以"空间"和"建造"为先导，统筹各专业目标、方法和流程，运用零耗能技术，形成"零碳"的物质循环体系，创造健康、舒适、便捷、充满地域文化特色的宜居社区。具体在规划、建筑、能源、文化四个层面开展：

1. 综合利用气候特征、场地地形以及河流特点。引入"绿轴"和"银带"概念，形成"一带三轴"的空间格局。合理组织低碳交通，配置垃圾回收和污水处理设施，组团内部以恢复满族院落生活为目标，对"院"的空间逻辑进行重新梳理。

2. 基于开放建筑理念，采用装配式结构体系，模块化建造，增加室内空间弹性。运用被动式太阳能技术，诱导通风技术，保温隔热构造降低能耗。使用乡土等生态型材料，探索满居的光伏一体化方法。

3. 采用光伏发电为主、风力发电为辅的再生能源策略，重点研究柔性用电方式。将地源热泵、太阳能热水、雨水回收、沼气系统联动，并采用智能化控制。

4. 采用御寒技术，打造室内外公共空间，延续邻里交往互助概念。恢复满族地域文化，倡导绿色时尚生活。

Under the specific climate and geographical conditions of Shenyang, this scheme focuses on the "space" and "construction" as the guide, coordinating the objectives, methods and processes of various disciplines, using zero energy consumption technology to form a "zero carbon" material circulation system, and creating a healthy, comfortable, convenient and livable community full of regional cultural characteristics. Specifically, it is carried out at the four levels of planning, architecture, energy and culture:

1. Comprehensive utilization of climate characteristics, site topography and river characteristics. The concepts of "green corridor" and "silver belt" are introduced to form a spatial pattern of "one belt, two corridors". Reasonably organize low-carbon transportation and configure waste recycling and sewage treatment facilities. The group recompiled the spatial logic of the "courtyard" with the goal of restoring the life of the Manchu courtyard.

2. Based on the concept of open architecture, the fabricated structural system is adopted and modular construction is adopted to increase the elasticity of indoor space. Passive solar energy technology, induced ventilation technology and thermal insulation structure are used to reduce energy consumption. Ecological materials such as rammed earth are used to explore the photovoltaic integration method of Manchu houses.

3. Adopt the renewable energy strategy of photovoltaic power generation supplemented by wind power generation, focusing on the flexible power consumption mode. The ground source heat pump, solar hot water, rainwater recovery and biogas systems are linked, and intelligent control is adopted.

4. Adopt cold protection technology to create indoor and outdoor public space and continue the concept of neighborhood communication and mutual assistance. Restore Manchu regional culture and advocate green and fashionable life.

Location

Peripheral Myoaric Analysis

Major surrounding public places.

Wetlands, mountains, and rivers.

Air flow rate analysis to the site.

Flow vector analysis to the site.

The mountains block the winter northwest wind.

Water vapor in the southeast wind zone in summer.

The green axis has good views along the river.

The mountain landscape penetrates into the site.

Regional Characteristics

装饰性五花山墙 Decorative five flower gable
防潮月台 Damp proof platform
排烟跨海烟囱 Smoke exhaust cross sea chimney
私有庭院 Private courtyard
保温阳光房 Insulated sunshine room
传统满族窗户 Traditional Manchu windows
硬山顶 Flush gable roof
传统青砖夯土材料 Traditional green brick and rammed earth material

Technological Roadmap

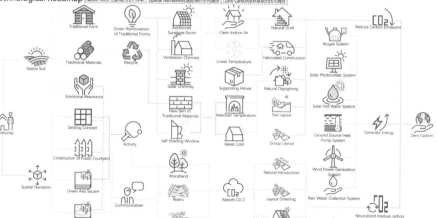

Native Soil 乡土建构的技艺传承 | Spatial Narration 院落叙事的空间编译 | Zero Carbon 绿色零碳的技术耦合

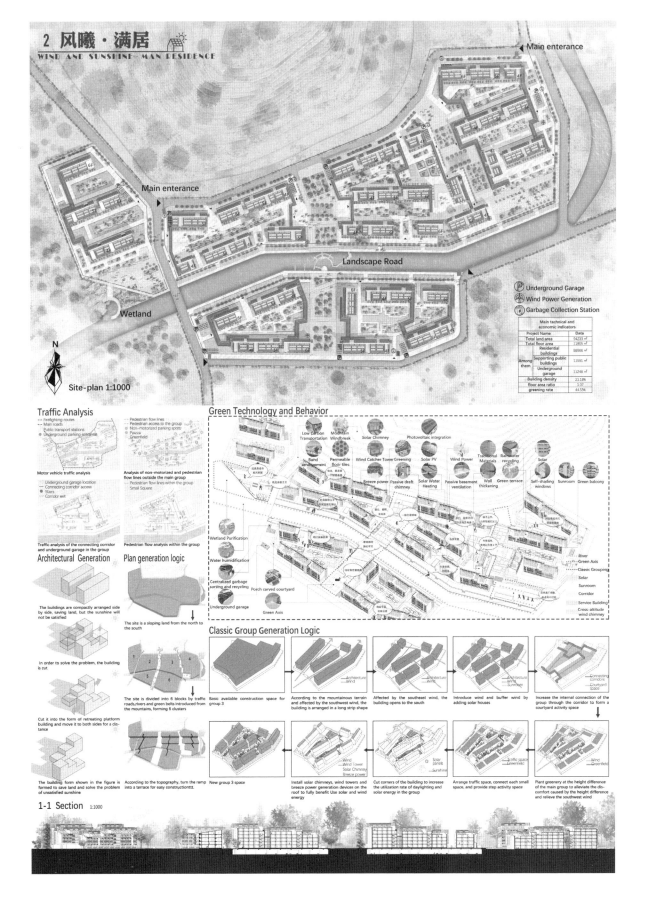

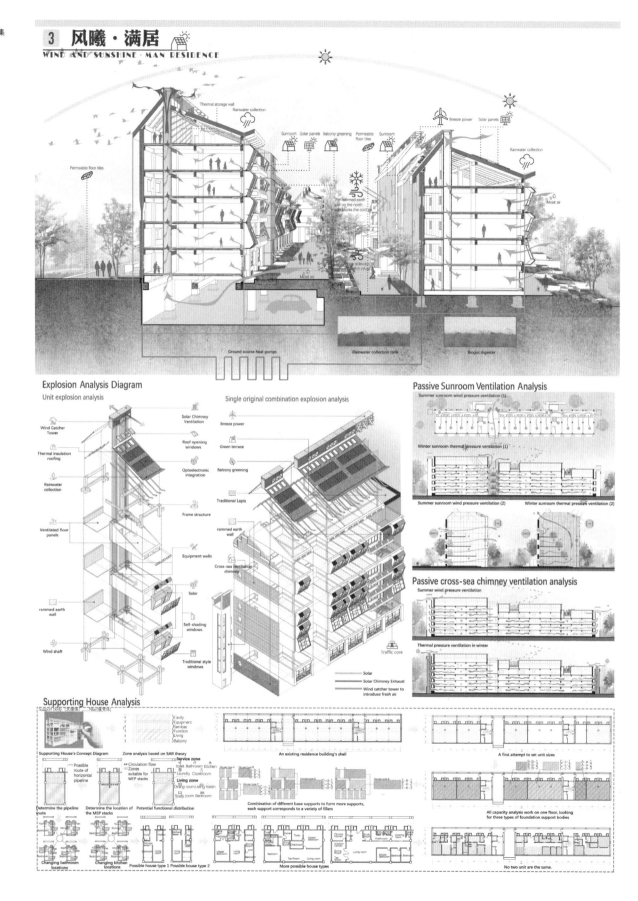

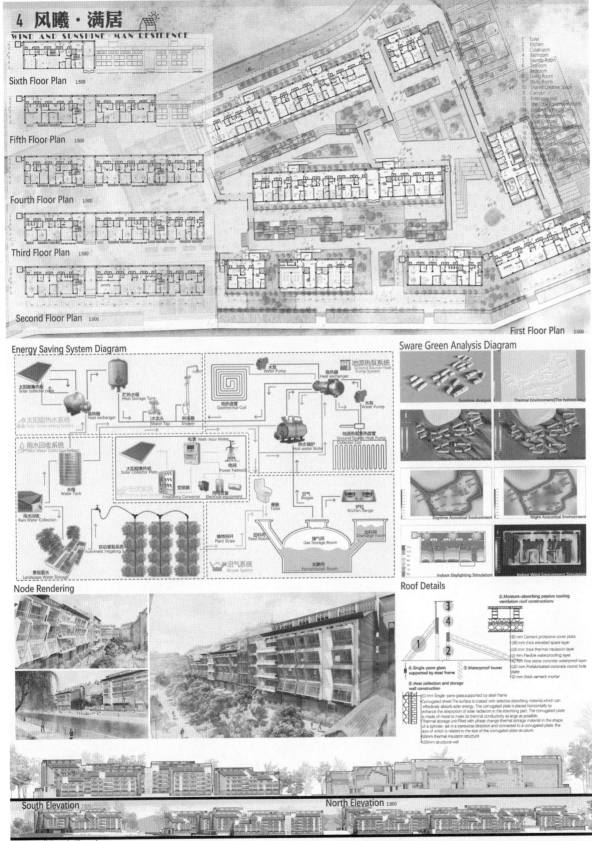

综合奖 · 优秀奖
General Prize Awarded · Honorable Mention Prize

注 册 号：100038
Registration No：100038

项目名称：归园·田居
Farming Community

作　　者：赵嘉诚、张尚书、桑弘翼、
　　　　　王嘉宇、李吉福、陈雨鑫、
　　　　　宋 怡
Authors：Zhao Jiacheng, Zhang Shangshu,
　　　　　Sang Hongyi, Wang Jiayu,
　　　　　Li Jifu, Chen Yuxin, Song Yi

参赛单位：中国矿业大学（北京）
Participating Unit：China University of
　　　　　　　　　Mining and Technology
　　　　　　　　　(Beijing)

指导教师：李晓丹
Instructor：Li Xiaodan

01. Architectural Design Description

苏州是一座园林城市，也是一座拥有千万级人口的城市，人们需要寻找一个舒适的居住空间，暂时远离闹市，安逸的享受生活。为此，我们在居住区设计中植入园林气息，同时又引入有农社区的概念，在居住区内设置农田，帮助人们去体验田园生活，感知从付出到收获的完整过程，享受大自然的馈赠。农业会产生许多秸秆和植物线粒，通过相关技术可充分利用其丰富的生物质能进行发电，以达到节能环保、废物利用的效果，同时，该地区具有丰富的太阳能资源和水资源，充分利用这些能源与技术可实现社区内部能源的自给自足和资源的循环供给与利用，营造一个绿色、开放、包容、共享的和谐社区，为实现我国"碳达峰"与"碳中和"目标贡献对策与方案。

Suzhou is a garden city, but also a city with a population of tens of millions, people need to find a comfortable living space, temporarily away from the downtown, comfortable to enjoy life.To this end, we implant garden flavor in residential area design, and at the same time introduce the concept of "farming community", set farmland in residential area, help people to experience the idyllic life, perceive the complete process from giving to receiving, and enjoy the gift from nature.Agriculture produces many straw and plant residues, through relevant technology can make full use of its abundant biomass energy to generate power, in order to achieve the effect of energy conservation and environmental protection, recycling, at the same time, the region has abundant solar energy resources and water resources, make full use of the energy and the technology can realize community internal energy self-sufficiency and the cycle of supply of natural resources,To build a green, open, inclusive and sharing harmonious community, and contribute countermeasures and plans to achieve China's "carbon peak" and "carbon neutral" goals.

02. Site Analysis

The base is located in Xiangcheng District, Suzhou City, Jiangsu Province, near a trapezoidal plot of Yangcheng Lake, the surrounding environment of the base is beautiful, pleasant scenery, suitable cli-mate, convenient transportation, the surrounding is relatively empty and no tall buildings block, suitable for residential design.

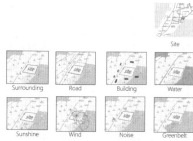

03. Conceptual Design

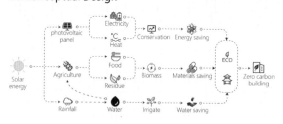

04. Climate Environmental Analysis

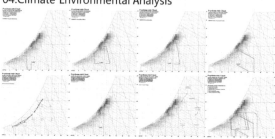

The analysis chart of base climate environment is obtained by software simulation.The region has four distinct seasons and a mild climate. Suitable building energy saving technologies include natural ventilation and evaporative heat transfer, and other building technologies include passive solar heating + high heat capacity materials.The corresponding technology can be used for design and construction to achieve the purpose of building energy saving and zero emissions.

Through software analysis and simulation, it can be seen that southeast wind and north wind prevail in this region, and the average temperature and average humidity of the wind are high,and the area has abundant rainfall. Therefore, the potential of natural ventilation can be considered to design buildings with good natural ventilation to achieve the purpose of cooling and dehumidification.

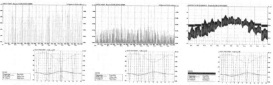

Through software simulation analysis, the region's cloud cover the sky is more, direct radiation and scattering radiation is balanced, at the same time, the average temperature is higher, in addition, according to the China Meteorological Administration, the region belongs to the four types of area of solar energy utilization, qiu dong season has abundant solar energy resources, therefore, can consider to use the solar energy technology, photovoltaic and solar heating, in order to save energy,Reduce energy consumption.

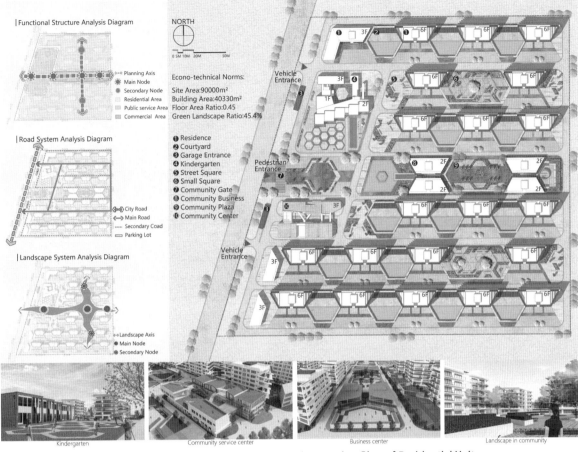

Functional Structure Analysis Diagram

- Planning Axis
- Main Node
- Secondary Node
- Residential Area
- Public service Area
- Commercial Area

Road System Analysis Diagram

- City Road
- Main Road
- Secondary Road
- Parking Lot

Landscape System Analysis Diagram

- Landscape Axis
- Main Node
- Secondary Node

Econo-technical Norms:
Site Area: 90000m²
Building Area: 40330m²
Floor Area Ratio: 0.45
Green Landscape Ratio: 45.4%

1. Residence
2. Courtyard
3. Garage Entrance
4. Kindergarten
5. Street Square
6. Small Square
7. Community Gate
8. Community Business
9. Community Plaza
10. Community Center

Kindergarten | Community service center | Business center | Landscape in community

05. Simulation Analysis of Wind Environment

Summer wind simulation diagram | Winter wind simulation diagram | Spring/Autumn wind simulation diagram

The working condition in summer and transition season reaches the requirement of "there is no vortex or no wind in the human activity area of the site"; In summer and transition season, the working condition reaches the requirement of "more than 50% of the outdoor Windows can be opened, and the wind pressure difference between the indoor and outdoor surfaces is greater than 0.5Pa".

In winter, the working conditions meet the requirements of "wind speed is less than 5m/s at 1.5m above the ground in the pedestrian area around the building, wind speed is less than 2m/s in the outdoor rest area and children's entertainment area, and outdoor wind speed amplification coefficient is less than 2".

06. Residential Unit Design

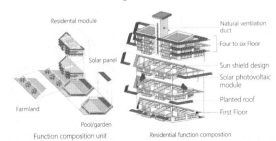

- Residential module
- Solar panel
- Farmland
- Pool/garden
- Function composition unit
- Residential function composition
- Natural ventilation duct
- Four to six Floor
- Sun shield design
- Solar photovoltaic module
- Planted roof
- First Floor

The design of the housing unit is closely tied to the design theme and concept, integrating the farm into the landscape-scented community and increasing the community's capacity for collaborative labor and organic renewal.

07. Function Plan of Residential Unit

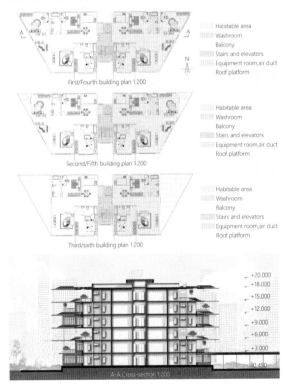

First/Fourth building plan 1:200

- Habitable area
- Washroom
- Balcony
- Stairs and elevators
- Equipment room, air duct
- Roof platform

Second/Fifth building plan 1:200

Third/Sixth building plan 1:200

A-A Cross-section 1:200

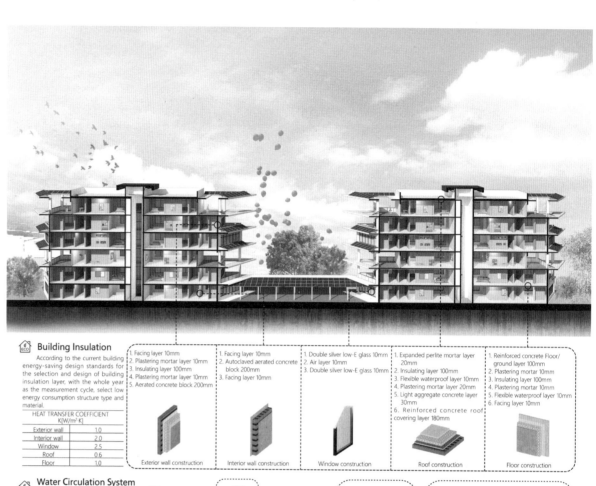

Building Insulation

According to the current building energy-saving design standards for the selection and design of building insulation layer, with the whole year as the measurement cycle, select low energy consumption structure type and material.

HEAT TRANSFER COEFFICIENT K[W/m²·K]	
Exterior wall	1.0
Interior wall	2.0
Window	2.5
Roof	0.6
Floor	1.0

Exterior wall construction
1. Facing layer 10mm
2. Plastering mortar layer 10mm
3. Insulating layer 100mm
4. Plastering mortar layer 10mm
5. Aerated concrete block 200mm

Interior wall construction
1. Facing layer 10mm
2. Autoclaved aerated concrete block 200mm
3. Facing layer 10mm

Window construction
1. Double silver low-E glass 10mm
2. Air layer 10mm
3. Double silver low-E glass 10mm

Roof construction
1. Expanded perlite mortar layer 20mm
2. Insulating layer 100mm
3. Flexible waterproof layer 10mm
4. Plastering mortar layer 20mm
5. Light aggregate concrete layer 30mm
6. Reinforced concrete roof covering layer 180mm

Floor construction
1. Reinforced concrete Floor/ground layer 100mm
2. Plastering mortar 10mm
3. Insulating layer 100mm
4. Plastering mortar 10mm
5. Flexible waterproof layer 10mm
6. Facing layer 10mm

Water Circulation System / Sponge City System

Rainwater recycling system including direct recycling of roof rainwater, the use of sponge city technology, ecosystem services as the goal, to improve community building low impact development of rainwater system, fully considering the rainwater storage, infiltration, water purification, recycling part of the surface runoff of rain water, will be relatively clean rainwater recycling by the filtration and purification process for flushing water in the rain.

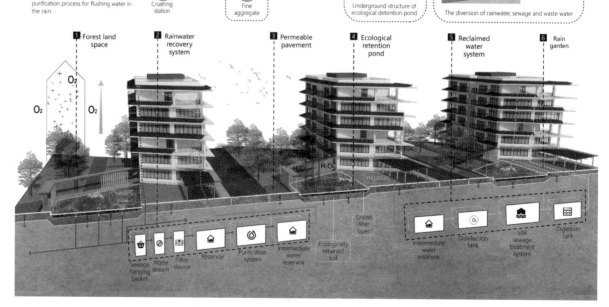

08. Building Energy-saving Technology

Solar Photovoltaic System

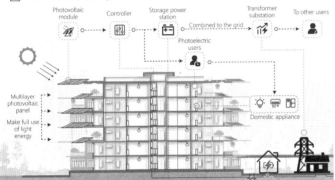

Biomass Power Generation System

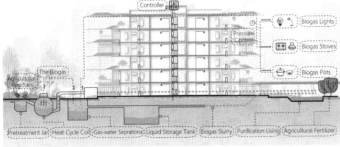

Natural Ventilation System

Direct blasting | Lateral blasting | Upside blasting | Passing blasting

Ventilating shaft | Household air duct | Positive pressure ventilation | negative pressure ventilation

The specially designed backstepping building form can allow the wind from different directions to easily pass through the building and form natural ventilation. At the same time, because the building involves air shafts and indoor air ducts, two forms of natural ventilation, positive pressure ventilation and negative pressure ventilation, can be formed to cool and dehumidify the building in the area.

Construction of ventilating shaft/ air duct

Vertical Planting System

According to the living habits of green plants and the depth of the root system, the building platform space is used to plant green plants in three dimensions to cool down and purify the air, increase the oxygen content in the building micro-circulation, promote the organic renewal of the building micro-ecology, so that the building can cope with the special climate environment of the region.

Roof planting | Roof construction
Soil planting | Soil construction

South Facade 1:200 | North Facade 1:200 | East Facade 1:200

综合奖·优秀奖
General Prize Awarded · Honorable Mention Prize

注 册 号：100041
Registration No：100041

项目名称：三"R"零碳太阳能社区
Three "R" Community-Zero Carbon Community

作 者：曾婧
Author：Zeng Jing

参赛单位：天津大学
Participating Unit：Tianjin University

指导教师：刘刚
Instructor：Liu Gang

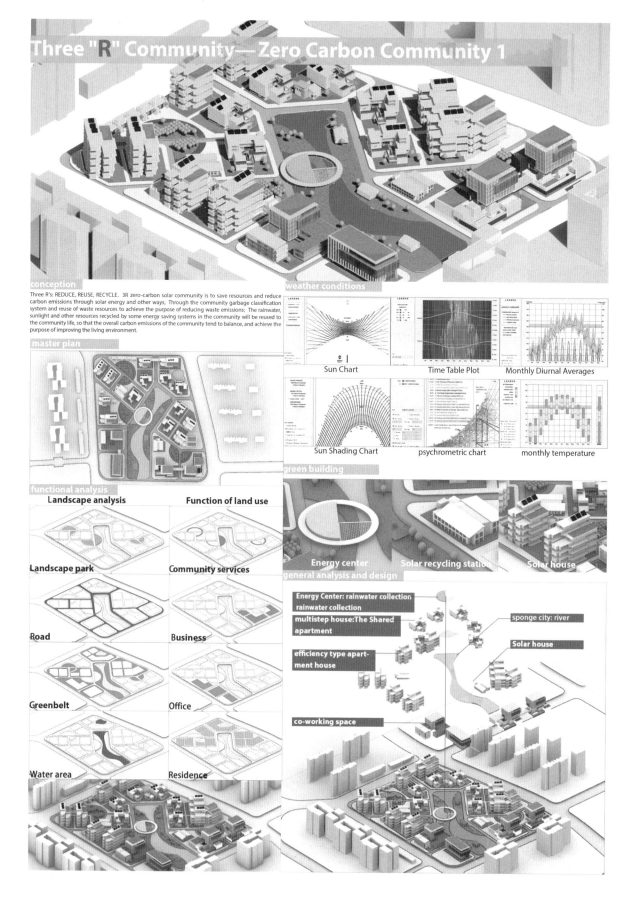

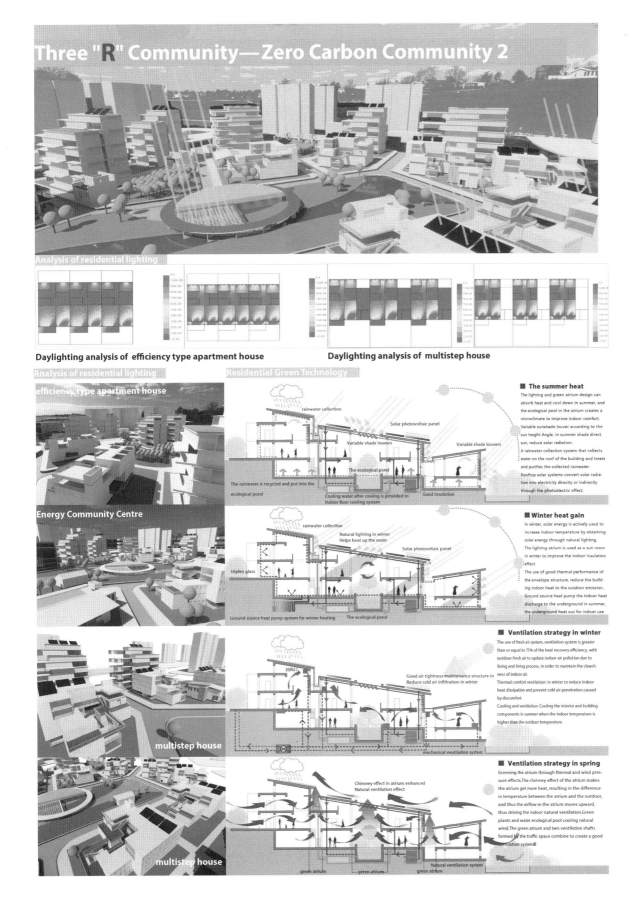

综合奖·优秀奖
General Prize Awarded · Honorable Mention Prize

注　册　号：100108
Registration No：100108
项目名称：零碳社区
　　　　　Zero Carbon Community
作　　　者：谭福利、葛云凡、王俊杰、
　　　　　孙欣欣
Authors：Tan Fuli, Ge Yunfan,
　　　　　Wang Junjie, Sun Xinxin
参赛单位：沈阳建筑大学
Participating Unit：Shenyang Jianzhu
　　　　　　　　　University
指导教师：赵　钧、高德战
Instructors：Zhao Jun, Gao Dezhan

Zero Carbon Community

项目位于辽宁省沈阳市大东区，社区总占地面积90000㎡，社区在设计的时候充分考虑了沈阳市的气候环境和人文特点，力求打造健康友好型社区，社区中采用了各种绿色建筑技术，如光伏、光热、装配式建筑等，最终实现零碳社区的目标。
The project is located in Dadong District, Shenyang City, Liaoning Province. The total area of the community is 90000 square meters. In the design of the community, the climate, environment and cultural characteristics of Shenyang City are fully considered, and the community is striving to build a healthy and friendly community. Various green building technologies are adopted in the community, such as photovoltaic, light and heat, prefabricated buildings, etc., so as to achieve the goal of zero carbon community.

Site analysis

Traffic analysis: the north side of the site is the East West expressway, and the south side is the subway. There are many bus stops around, so the traffic is convenient.

Analysis of surrounding plots: the north, East and west sides of the site are residential areas, and the south side of the site is the main commercial plot.

Analysis of surrounding facilities: there are good medical and educational resources around the site, and there is a large shopping center near the south side of the district government.

Location analysis

Site selection

Site status

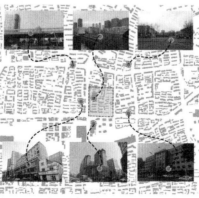

There is a reserved building inside the site. There is a shopping mall in the south, a viaduct in the north, a primary school in the East, and a government in the East. There are not many high-rise buildings around, which block the sun.

Wind frequency diagram

Solar radiation map

Enthalpy humidity diagram

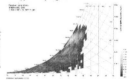

Temperature pattern

Relative humidity

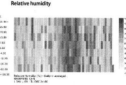

Zero Carbon Community

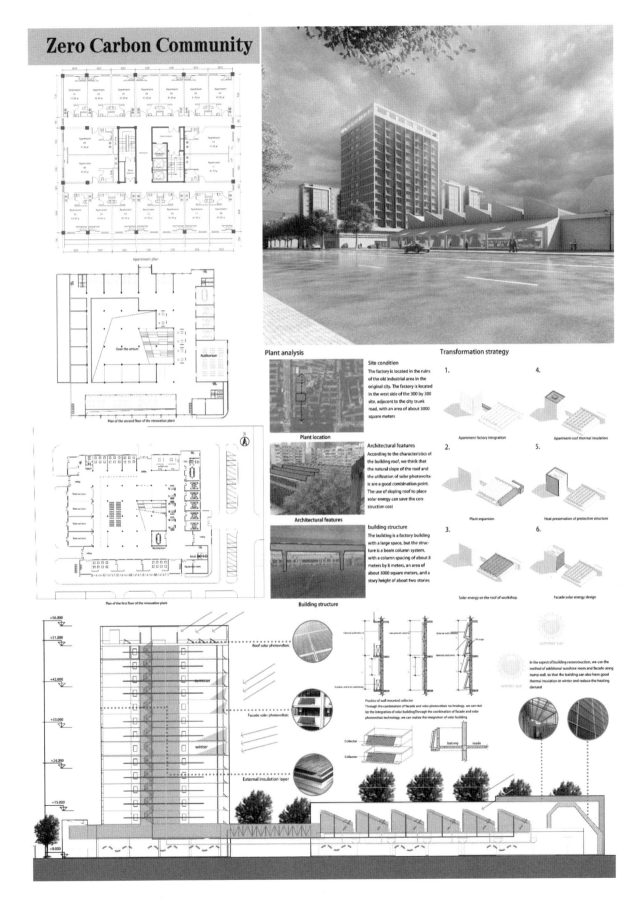

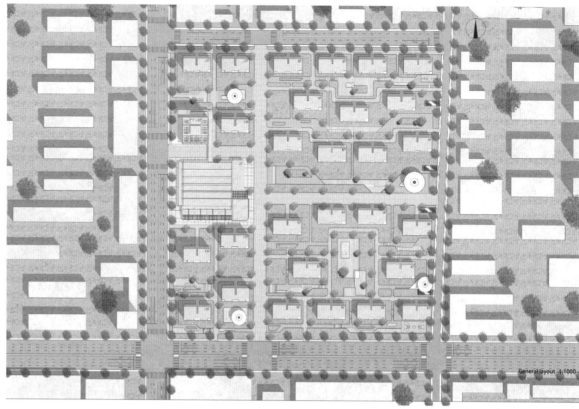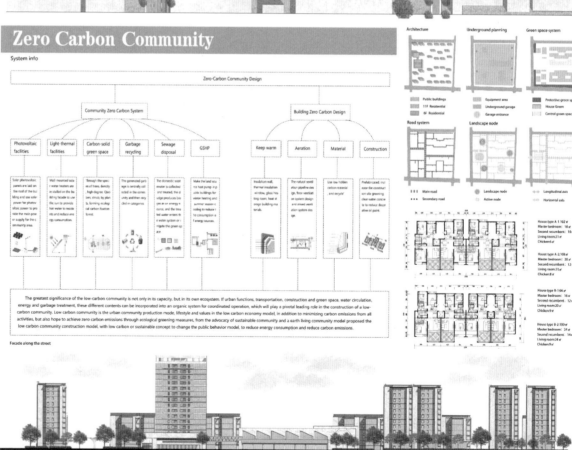

Zero Carbon Community

综合奖·优秀奖
General Prize Awarded · Honorable Mention Prize

注 册 号：100197
Registration No：100197

项目名称：零碳·侗寨
Zero Carbon · DONG Village

作　　者：吴起霖、莫静梅、郭孔洋
Authors：Wu Qilin, Mo Jingmei, Guo Kongyang

参赛单位：广西科技大学
Participating Unit：Guangxi University of Science and Technology

指导教师：叶雁冰
Instructor：Ye Yanbing

Zero Carbon · DONG Village II

Source of ideas

The base is located at the boundary of the Liujiang River, with the characteristics of traditional folk dwellings born from water. Dig deep into the Liujiang culture and trace back to the culture of "Landscape Longfield" in Liuzhou and the Dong culture living near the water. The layout of the folk dwellings is centered on the drum tower, surrounded by the arrangement of the folk dwellings, the wind and rain bridge across the village river, and the village gate as the edge point of the whole village. This kind of traditional village formed spontaneously due to the form of family settlement has become the source of design concept. On this basis, we integrate the zero-carbon strategy in the early stage into the design principle and adopt assembly technology to meet the traditional stilted building form of Dong village to form a characteristic zero-carbon Dong village and create more possibilities for future residential houses.

Free Dispersion of Residential Buildings · Drum tower in place

Wind and Rain Bridge double eaves, buckets structural characteristics · Built against the mountain, it has rich layers

Dong village life mode

Public Drum Tower Ping in front of a hundred banquet → Translated into public open leisure space → Stormy bridge talks and retail → Translated into the ground floor overhead space → rally

Public Drum Tower Ping in front of festival activities → Translated into public commercial office core building → Stage performance → Translate free landscape nodes → Cultivation of Characteristic Fields

General drawing energy saving method

Solar photovoltaic power generation — Roof solar panels are used to collect heat and generate electricity for residential buildings and centerbodies, which are arranged in different areas and supplied in different areas to reduce transportation loss.

Wind power generation — Wind power generation system is arranged against the river wind to collect power and directly serve the commercial center building.

Biogas power generation — Garbage and waste from residential and commercial centers will be collected through underground pipelines for centralized fermentation, and the generated biogas will be processed and transported back to them for use.

Rainwater harvesting — The rainwater of the site is collected by partition. After passing through the rainwater purification system, part of it reaches the building and part of it is used for landscape greening.

Water reuse — Each building carries out a segmenting medium water recycling cycle, with the upper part serving the upper part and the lower part serving the lower part to reduce transportation energy consumption.

Water circulation system — The combination of rainwater purification system and reclaimed water reuse system can achieve 80% of water recycling and meet the energy saving requirements of water in zero-carbon building energy consumption.

Active solar heating — The roof collector collects heat and transfers it to the house, so as to reduce the energy consumption caused by the heat demand and achieve the purpose of energy conservation and emission reduction.

Passive Solar Heating (Trumbo Wall) — The walls of the main and sub-centers are designed with unique structure. There is no mechanical power, no traditional energy consumption, and only rely on passive solar energy collection to provide heating for the building.

Ground source heat pump heating — The ground source heat pump system is arranged underground in the commercial center to achieve the effect of indoor suitable temperature difference and reduce the building energy consumption.

Site analysis

1. Site Functional Zoning
2. streamline
3. Equipment layout
4. Site energy acquisition
5. Landscape use
6. The field inlet
7. Rainwater collection
8. Site height difference processing
9. Building towards

General drawing planning and design

Put in the centrosomes · Surrounding layout of residence · The entrance concedes along the river · The wind and rain corridors connect the centrosome And the surrounding residences

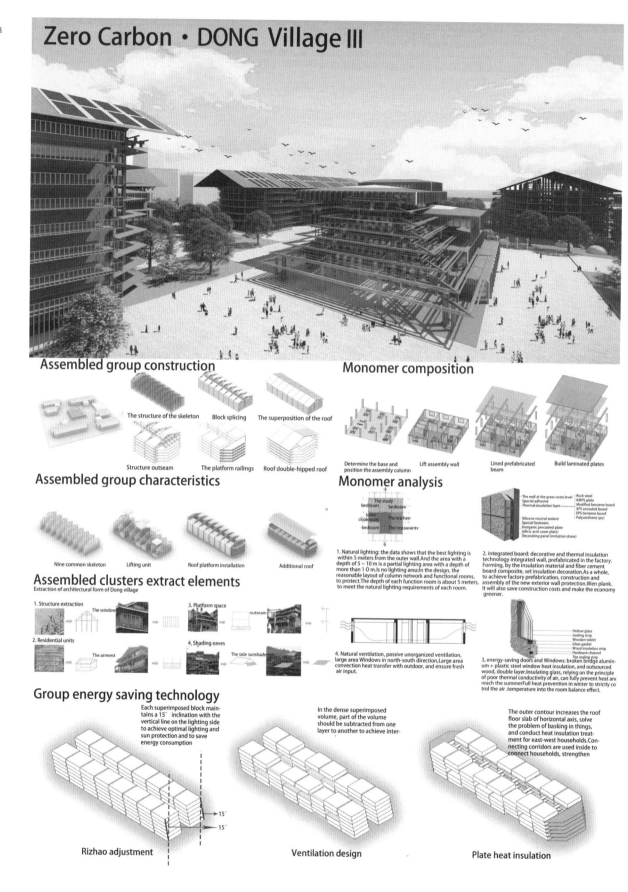

Zero Carbon · DONG Village IV

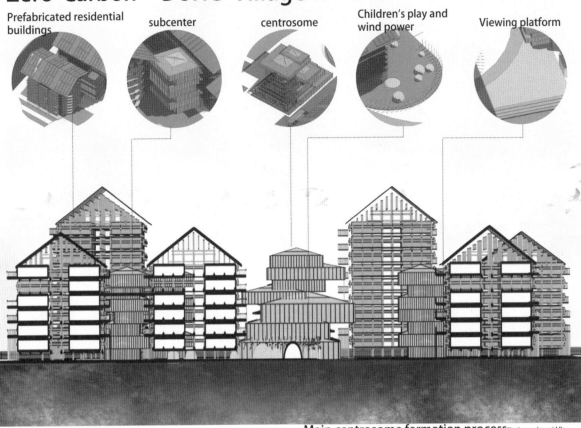

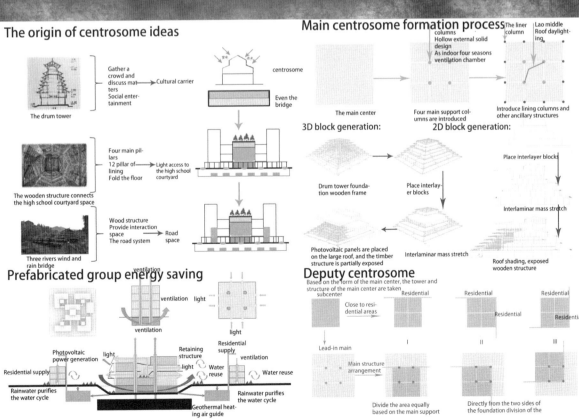

综合奖·优秀奖
General Prize Awarded · Honorable Mention Prize

注 册 号：100220
Registration No：100220

项目名称：浅草托岳
The Marsh Prop Up Huge Mountain

作　者：陈俊安、舒宏达、刘　旭、杨豪广、张佳峰
Authors：Chen Jun'an, Shu Hongda, Liu Xu, Yang Haoguang, Zhang Jiafeng

参赛单位：三江学院
Participating Unit：Sanjiang College

指导教师：焦自云、金　方
Instructors：Jiao Ziyun, Jin Fang

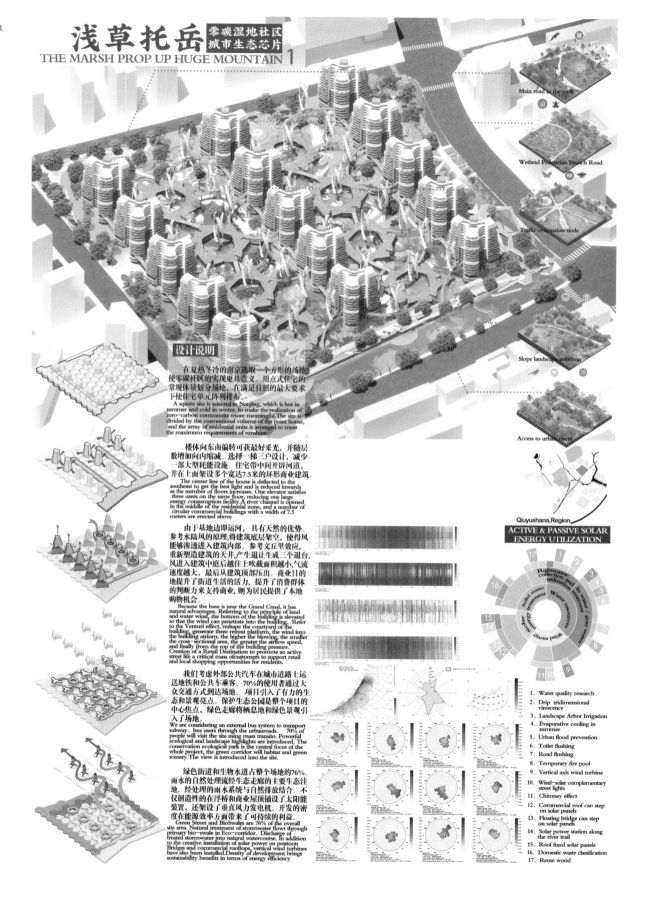

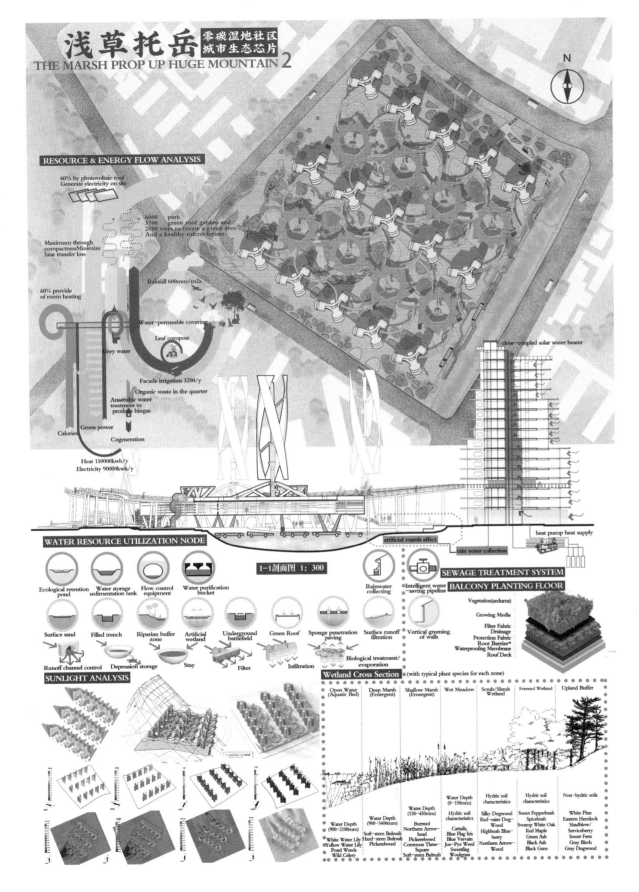

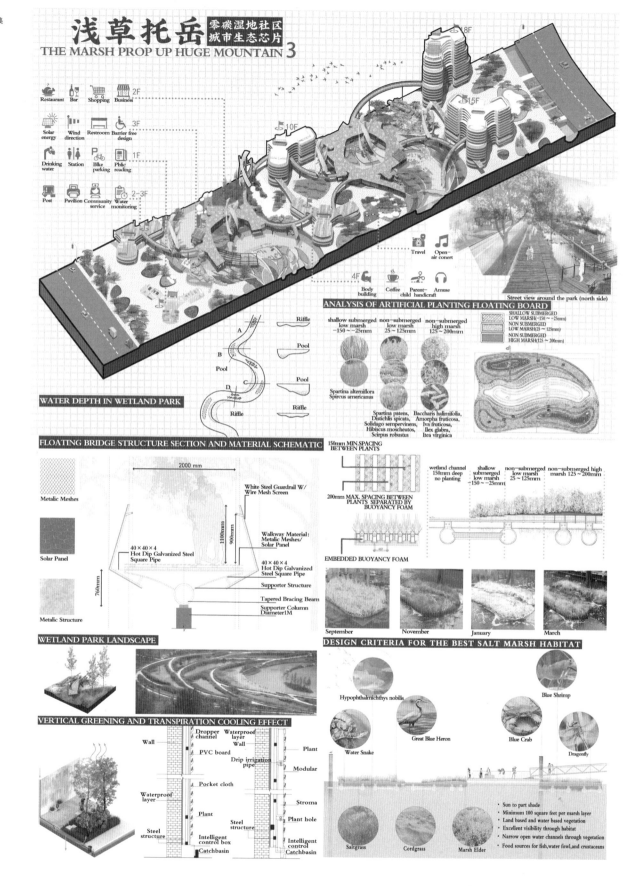

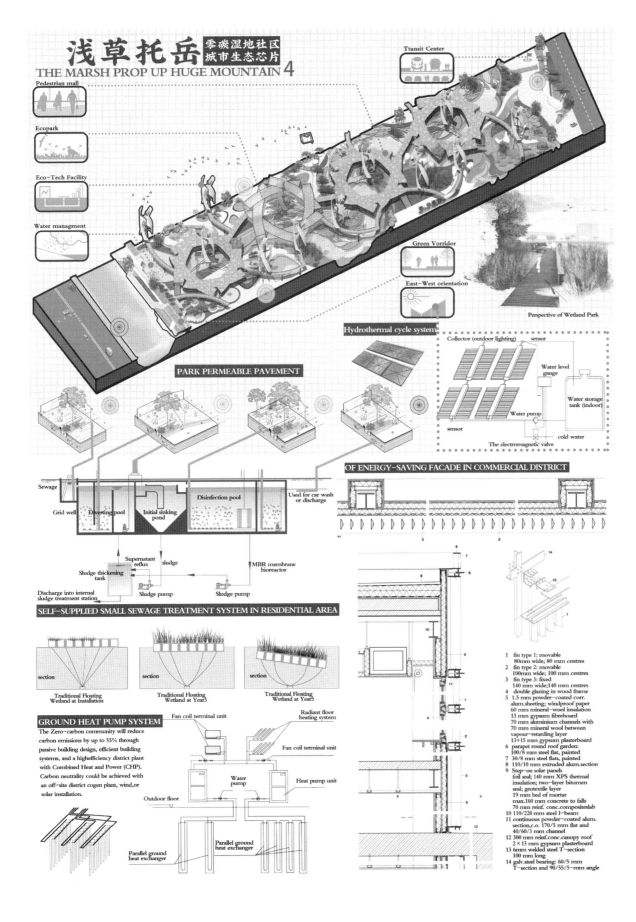

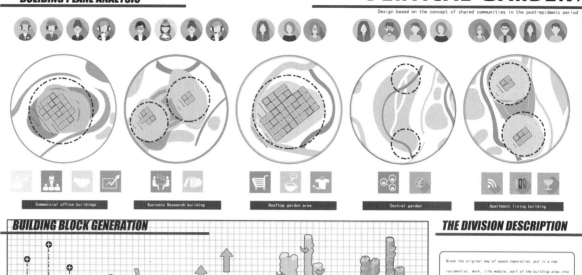

综合奖·优秀奖
General Prize Awarded · Honorable Mention Prize

注 册 号：100245
Registration No：100245

项目名称：后疫情垂直社区
　　　　　Vertical Garden

作　　者：曹屿
Author：Cao Yu

参赛单位：郑州轻工业大学
Participating Unit：Zhengzhou University of Light Industry

指导教师：李晓阳
Instructor：Li Xiaoyang

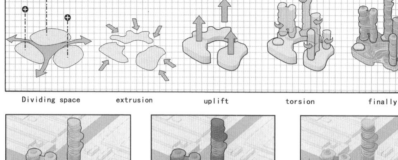

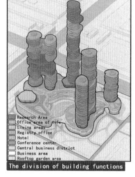

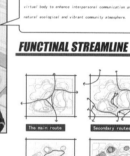

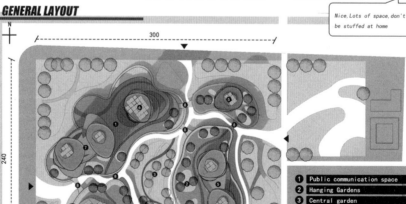

VERTICAL GARDEN 2

Design based on the concept of shared communities in the post-epidemic period

The effect diagram

High-altitude park

High-level space

Inside the building

Outer Circle Park

Central Park

Low-level space

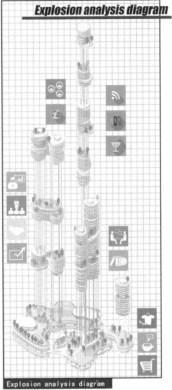
Explosion analysis diagram

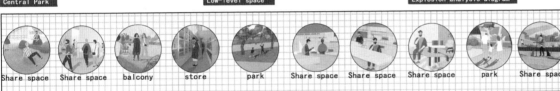
Share space | Share space | balcony | store | park | Share space | Share space | Share space | park | Share space

The scenario — What's wrong

Would you like to go to the sky garden for a spin, and go out and relax?

Well, I'll be right there. Do you want me to find you now?

Oh, you've arrived

I'm waiting for you

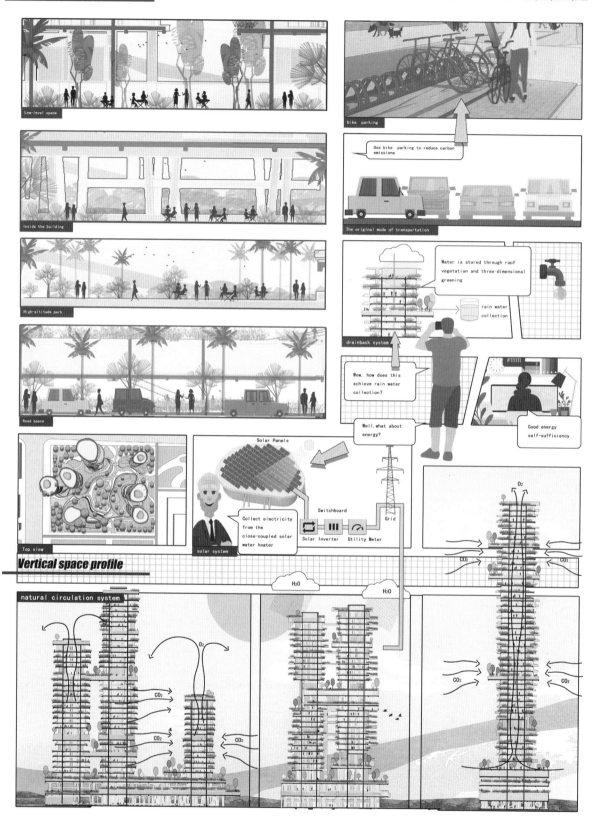

VERTICAL GARDEN₃

Design based on the concept of shared communities in the post-epidemic period

2021 台达杯国际太阳能建筑设计竞赛获奖作品集

Feature diagram2

Public communication space

PV façade

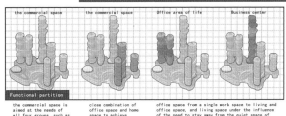

drainback system

analysis of body mass function

Off-duty hours

working hours

The tridimensional virescence effect diagram

Rainwater environment

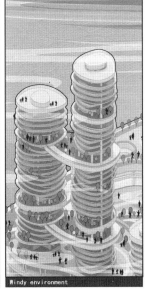

Windy environment

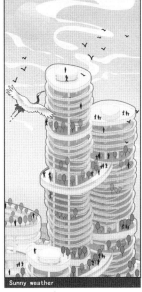

Sunny weather

Night environment

综合奖·优秀奖
General Prize Awarded · Honorable Mention Prize

注 册 号：100274
Registration No：100274

项目名称：城间一绿
Green Live Hood in the City

作　者：韦海璐、宋郭睿、张　锟、
　　　　袁家硕、曹　瑞、段恒玮
Authors：Wei Hailu, Song Guorui,
　　　　 Zhang Kun, Yuan Jiashuo,
　　　　 Cao Rui, Duan Hengwei

参赛单位：西北工业大学
Participating Unit：Northwestern Polytechnical University

指导教师：邵　腾、王　晋
Instructors：Shao Teng, Wang Jin

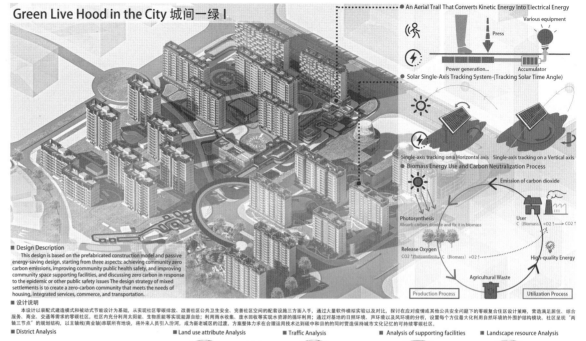

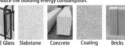

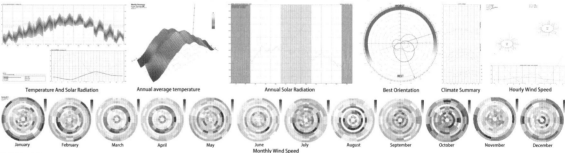

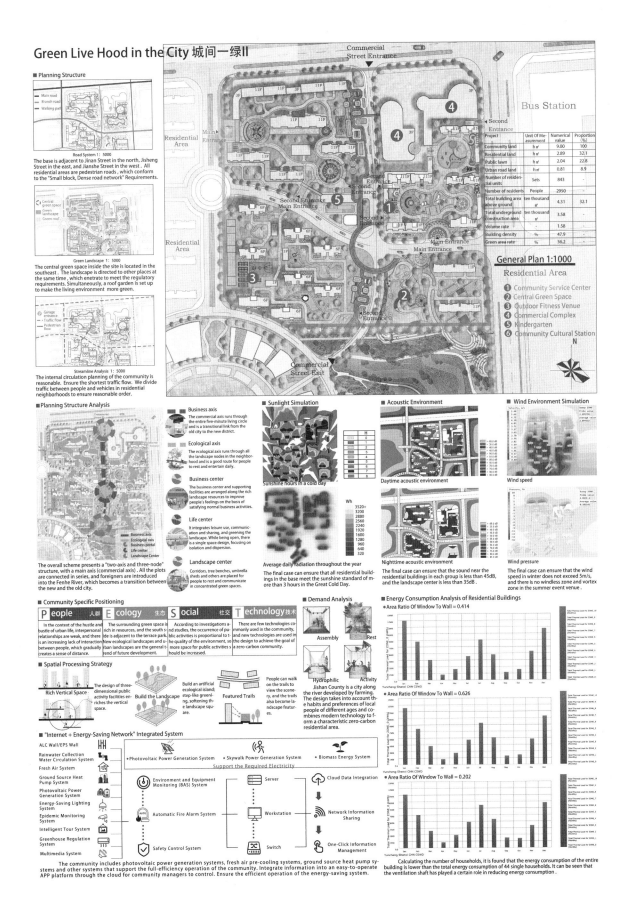

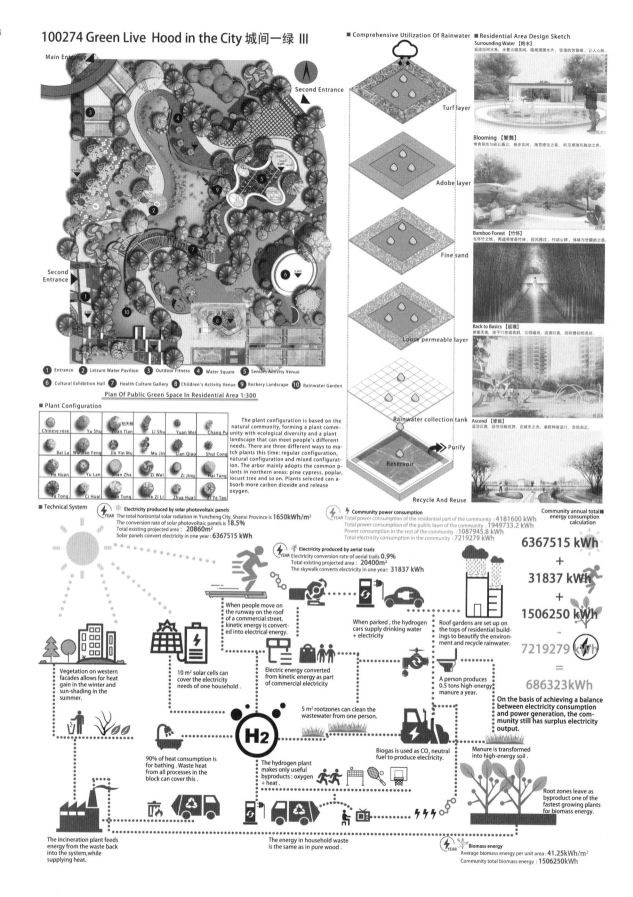

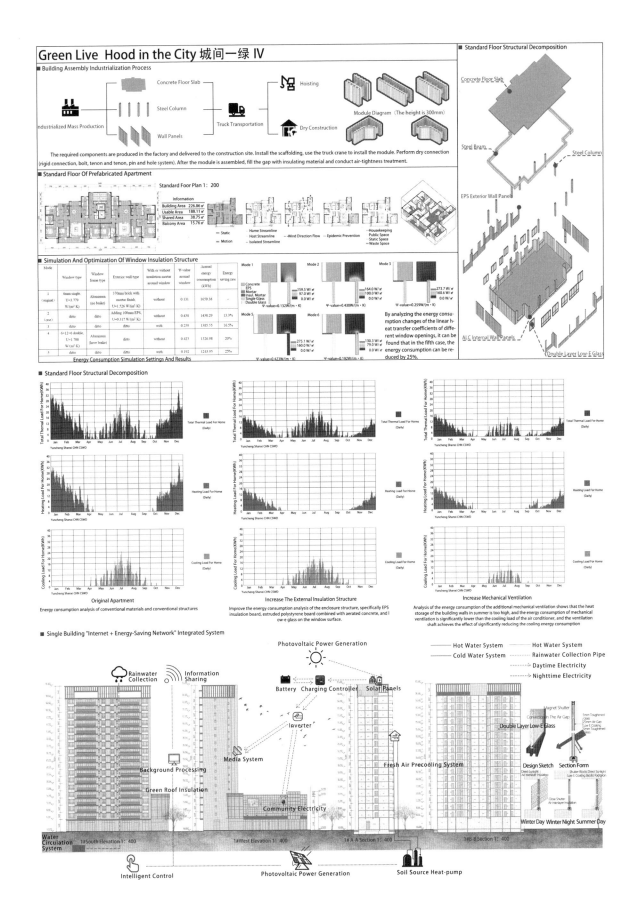

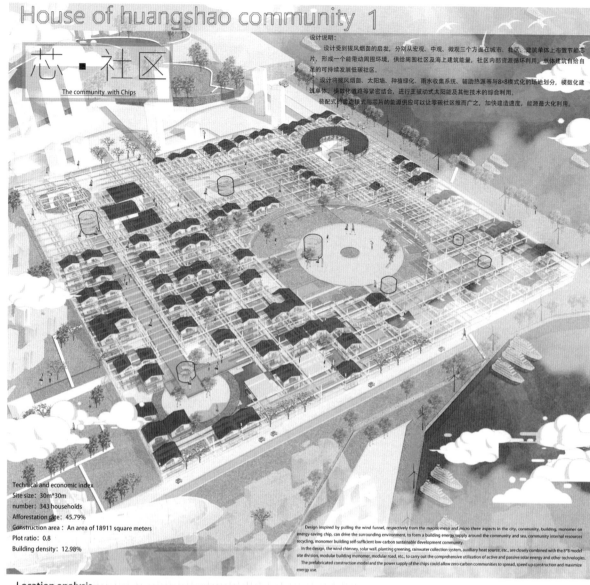

House of huangshao community 2

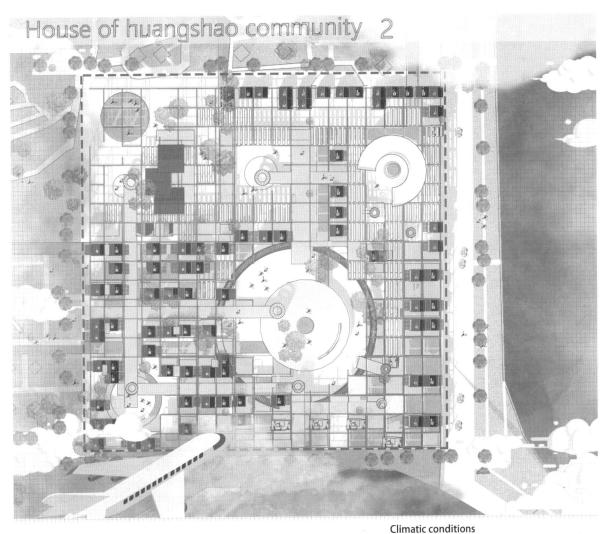

Mind mapping

- Passive Technology Utilization
  - Macro (Urban) — Centralized Heating
  - Zhong Guan (Residential Community)
    - Overhead Ventilation
    - Prefabricated Building
  - Micro (Single Building)
    - Draught
    - Windpipe — Harvest Energy
    - Window-To-Wall Area Ratio
    - Sun Wall
    - Heat Insulation
- Active Technology Utiliation
  - Macro (Urban)
    - Energy Storage And Recovery System — Photovoltaic Power Generation System
    - Reflux System
    - Heat Pump Heating
    - Water Cycle
  - Zhong Guan (Residential Community)
    - Solar Phase Change Heat Storage System
    - Air Collector
    - Rainwater Collecting
    - Heat Storage
  - Micro (Single Building)
    - Photovoltaic Curtain Wall
    - Low-Temperature Hot Water Floor Radiant Heating
- Other Green Technology
  - Zhong Guan (Residential Community)
    - Household Heat Metering
    - Tracking Collector
    - Heat Consumption Index
  - Micro (Single Building)
    - Shutters
    - Insulated Windows

Climatic conditions

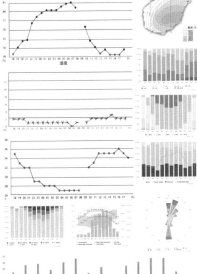

The trajectory state of human activity in it

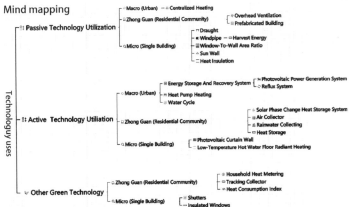

Crowd Description:
The crowd of the site is mainly tourists from outside, businessmen from the nearby community who make a living by selling seafood and local residents who live in the site for a long time. Foreign tourists can not only have food, entertainment, rest, experience local culture and local life here, but also promote the development and dissemination of local culture, promote the construction of BBB 0 spirit, and promote the economic development of the local and its surrounding areas. Due to the large area facing the sea in the south of the site, seafood and other agricultural products are abundant. Merchants living in the surrounding area can purchase and trade in large quantities here, which promotes the development of local economy and the surrounding economy. The local people who live here do not have to leave the community to carry out work, such as fishing, farming, catering, navigation, service, agriculture, etc.

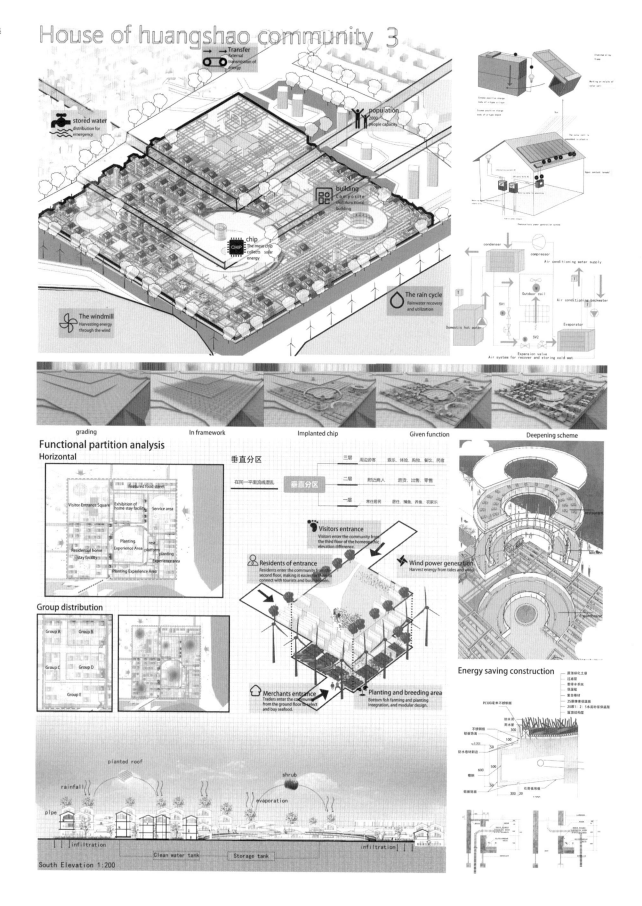

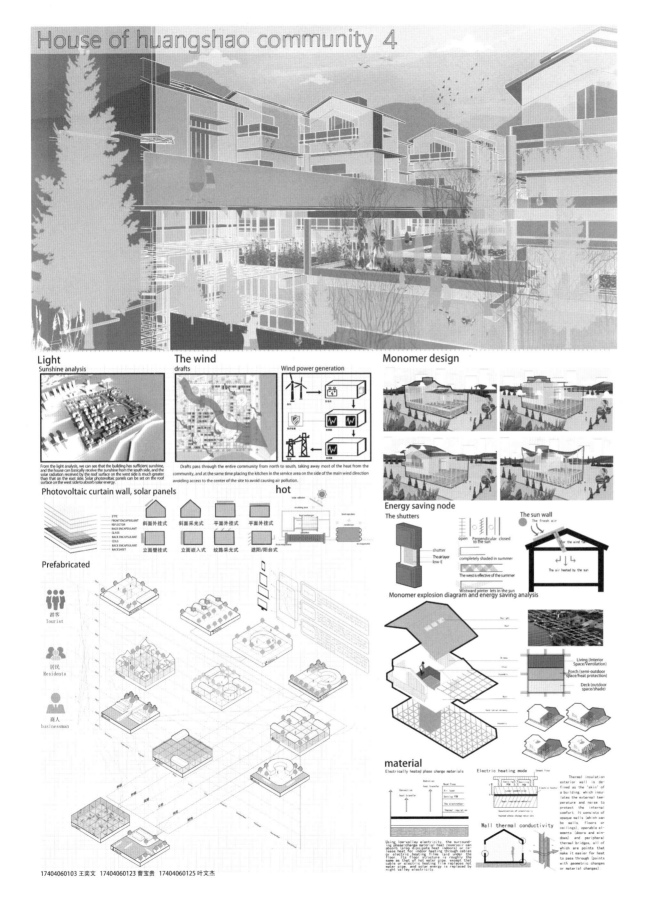

综合奖·优秀奖
General Prize Awarded · Honorable Mention Prize

注 册 号：100293
Registration No：100293

项目名称：零荷博益
        Zero-Carbon Lotus Community

作　　者：徐珂晨、沈奕辰、吴浩麒、
        姚双越、孙竞超、王子煜
Authors：Xu Kechen，Shen Yichen，
        Wu Haoqi，Yao Shuangyue，
        Sun Jingchao，Wang Ziyu

参赛单位：浙江大学
Participating Unit：Zhejiang University

指导教师：吴津东
Instructor：Wu Jindong

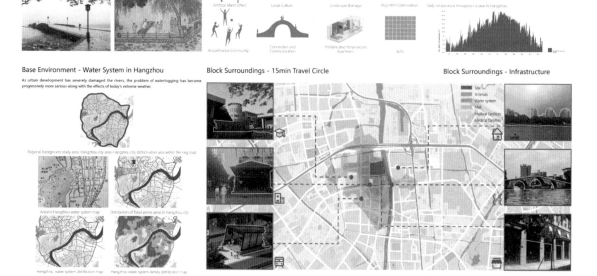

Site Analysis

Concept Deepening - Community Layout

Residential Module - House Type Combination

Wool algorithm - Site Road Planning

Wool Algorithm-Site road planning

The Primary Road

The Secondary Road

Genetic Algorithm - Residential Layout Optimization

Number=10 Optimal Solution

Number=11 Optimal Solution

Number=12 Optimal Solution

Optimal Solution
Number=10
Area=44000 m²
totalSunlightHours=175420.46h

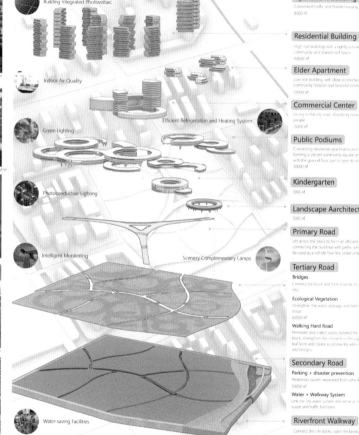

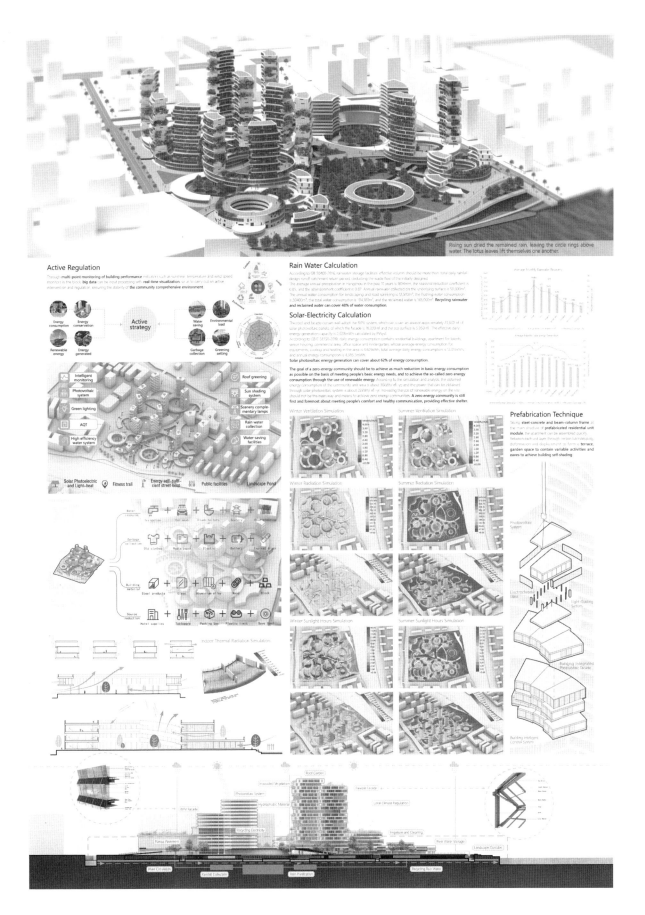

综合奖·优秀奖
General Prize Awarded·
Honorable Mention Prize

注 册 号：100299
Registration No：100299

项目名称：零社模方
M-Cube Zero Carbon Community

作　　者：马成俊、曹梦莹、杜照怡、纪雨辰
Authors：Ma Chengjun, Cao Mengying, Du Zhaoyi, Ji Yuchen

参赛单位：天津大学
Participating Unit：Tianjin University

指导教师：朱　丽、霍玉佼
Instructors：Zhu Li, Huo Yujiao

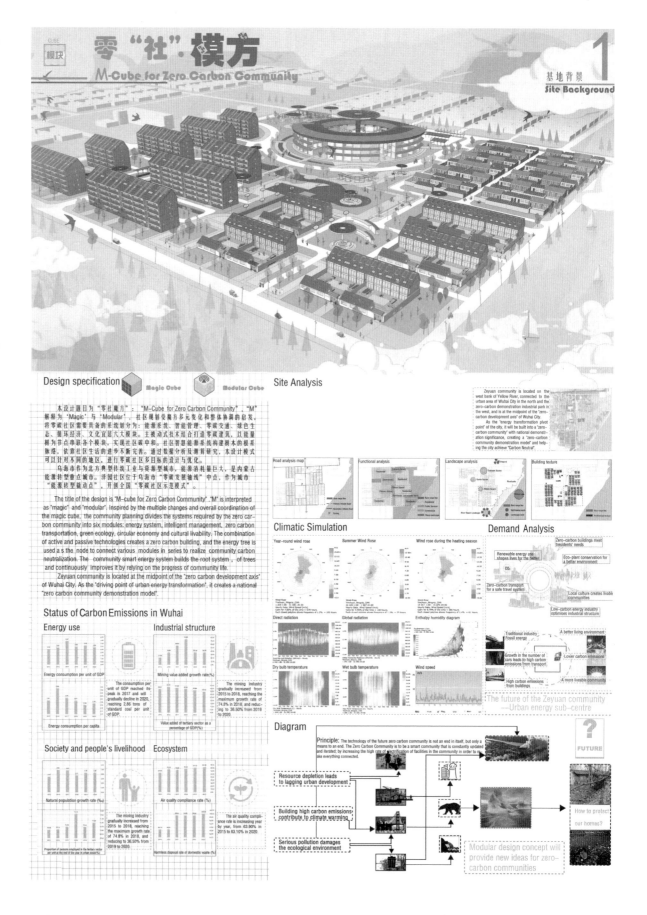

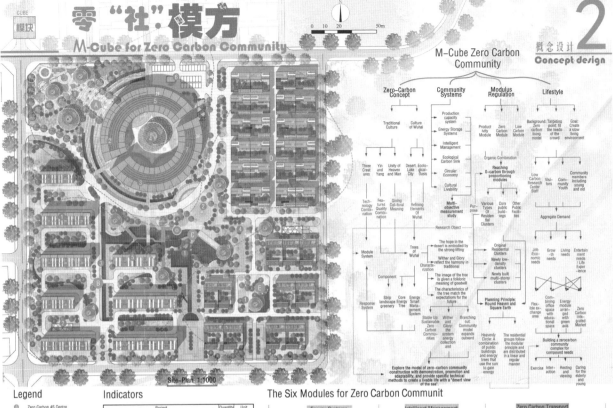

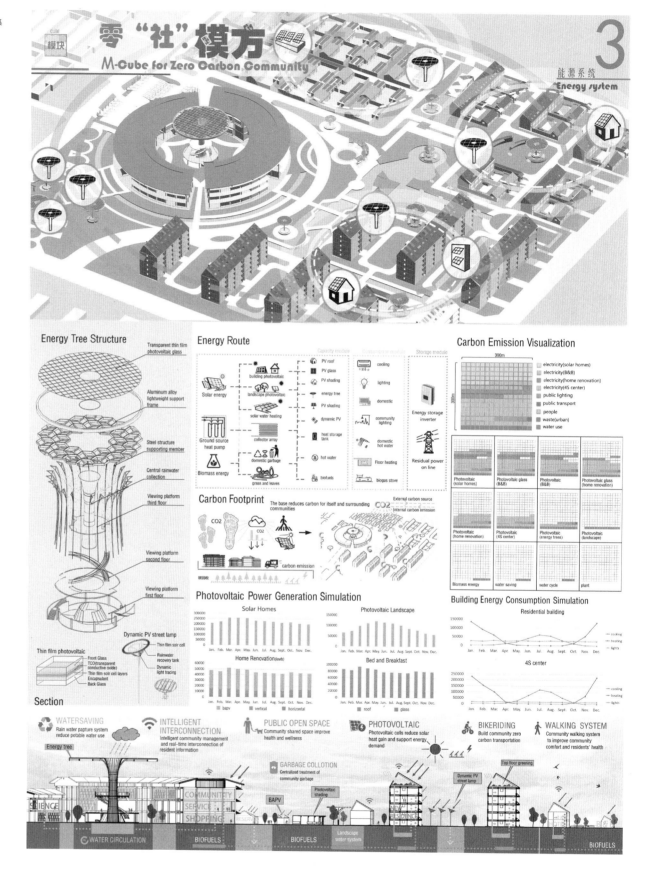

零 "社" 模方
M-Cube for Zero Carbon Community

4 零碳生活 Zero-carbon lifestyle

2021 台达杯国际太阳能建筑设计竞赛获奖作品集

Community principle

People-oriented principle. Low-carbon community construction is closely related to the lives of community residents, and the development of low-carbon behavior is ultimately the voluntary behavior of residents. It is a prerequisite for the success of the pilot construction of low-carbon communities. Therefore, the basic requirement for promoting the construction of low-carbon communities should be "people-oriented and enhancing residents' sense of well-being and access".

1,2: Waking up early and going first to the community smart store to purchase fresh goods. During the process, exchange information with fellow villagers on the development of low-carbon industry.
3: return to the B&B and make local specialties for the guests.
4,5: Take the guests to visit the community for low-carbon science popularization and to appreciate the beauty landscape
6: After sending off the guests, return to the zero-carbon residence and rest.

Single Building Generation
- Analysis of building roof form generation
- Integrated solar thermal design
- Zero carbon courtyard generation for B&B buildings

Grouping Functional Unit

90㎡ Unit plan 1:200
- Family of two
- Family of one child
- Family of two children
- Family of three children

125㎡ Unit plan 1:200
- Add the grandparents

Building Energy System
- sloping roof
- photovoltaic system
- solar cell
- natural ventilation
- small outside window area ratio
- heat storage wall
- solar collector
- modular spatial system
- standardization design
- prefabricated construction
- PV facade
- sun-shield

Solar house section 1:200

Solar PV Panels Solar Water Heater Air source heat pump Natural Ventilation

The community building design takes into account the daily maintenance and easy replacement of PV panels, and the size of PV panels in different locations are designed according to a uniform module at the beginning of the design. This ensures that standard PV modules can be replaced at any time during the entire life cycle of the community building.

Community Crowd Behavior Analysis

Community Service Workers | Scientific Researchers | Local adults | Local children | Local elderly | Out-of-town visitors

- Working Needs: Commuting / Outdoor work / Experiment
- Growing-up needs: B&B Reception / Plantation / Logistics
- Living Needs: Discussion / Exhibition / Education
- Entertainment And Leisure: Course Training / Retail / Cafeteria
- Life Experiences: Sports / Board Games / Outdoor Play
  Performance / Healing Care / Waterfront Events

Circular Community Cultural and Service Center Building

1. The core public building is located in the middle of the site to the north, surrounded by residential buildings and multiple landscape strips. In response to the planning concept and to create a community atmosphere with a sense of belonging, the complex functions required for a zero-carbon community are integrated into the modularizing building.

2. Each space module is networked with the intelligent community management system, and the system performs big data analysis to obtain real-time data on the interaction between residents and the community environment and residents' health data, establishing information interconnection between people and the community environment, thus making it easier to understand the operation and capacity information of each space module.

3. In order to achieve the functional index of zero carbon building, the roof is covered with pv panels. The photovoltaic panels are different from the previous blue-black color and instead use red dye-sensitized solar cells to highlight the centrality of the core building while enlivening the community. The outer circle of the building is equipped with a ring-shaped According to the analysis of sunlight radiation, the placement type of the eaves is determined, and the northern side of the eaves plays the role of wind shelter and wind guide. The northern side of the girdle plays the role of wind shelter and guide, while the southern side of the girdle is arranged with solar photovoltaic panels to fully receive light and help the zero-carbon building.

4. In accordance with the modular design concept of sustainability, the core public building is designed as a fully assembled building, which can be added or subtracted according to the future changing needs of each functional space. The inner circle of the building is equipped with adjustable sunshade louvers to facilitate shading or blocking the reflection of solar panels.

5. According to the public space needs of the crowd obtained from the research, more than 10 kinds of indoor public spaces are grouped into four categories: community services (green), supporting commercial (purple), exhibition and research (red) and comprehensive management office (blue), and the spaces are divided according to the needs and directions of people in the site.

6. The core is shaped so that it becomes the center of the community and becomes the residents' own public courtyard. The composite public buildings allow residents and visitors to walk the streets and stroll leisurely through various living scenes with a sense of urban life in the landscape. The energy tree in the center provides an observation deck with a high view of The view of Gander Mountain and Wuhai Lake in the distance, and the view of Wuhai can be seen everywhere. The community is an ideal living environment in the context of a zero-carbon economy.

综合奖·优秀奖
General Prize Awarded·Honorable Mention Prize

注　册　号：100304
Registration No: 100304

项目名称：城市地衣·零碳社区
　　　　　Urban Lichen·Zero Carbon Community

作　　　者：陈文博、昂媛媛、叶一帆
Authors：Chen Wenbo, Ang Yuanyuan, Ye Yifan

参赛单位：中央美术学院、中国矿业大学（北京）
Participating Units：Central Academy of Fine Arts, China University of Mining and Technology (Beijing)

指导教师：郑利军
Instructor：Zheng Lijun

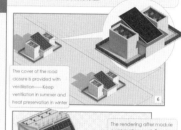
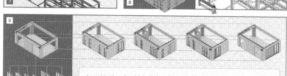
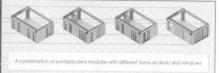
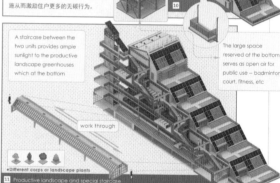
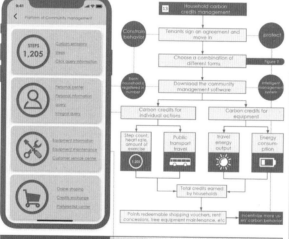
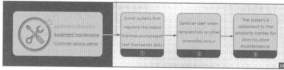

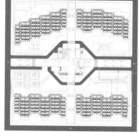

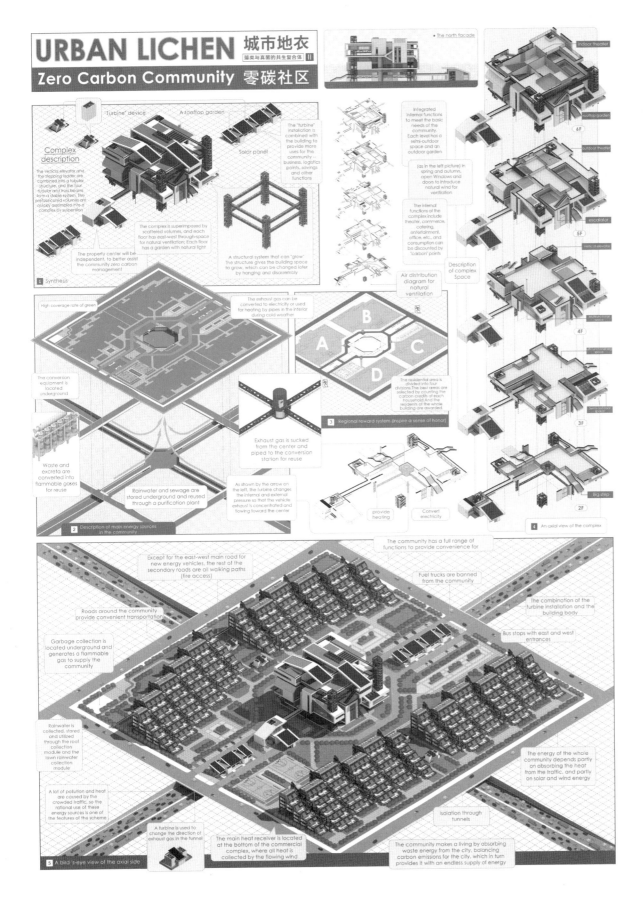

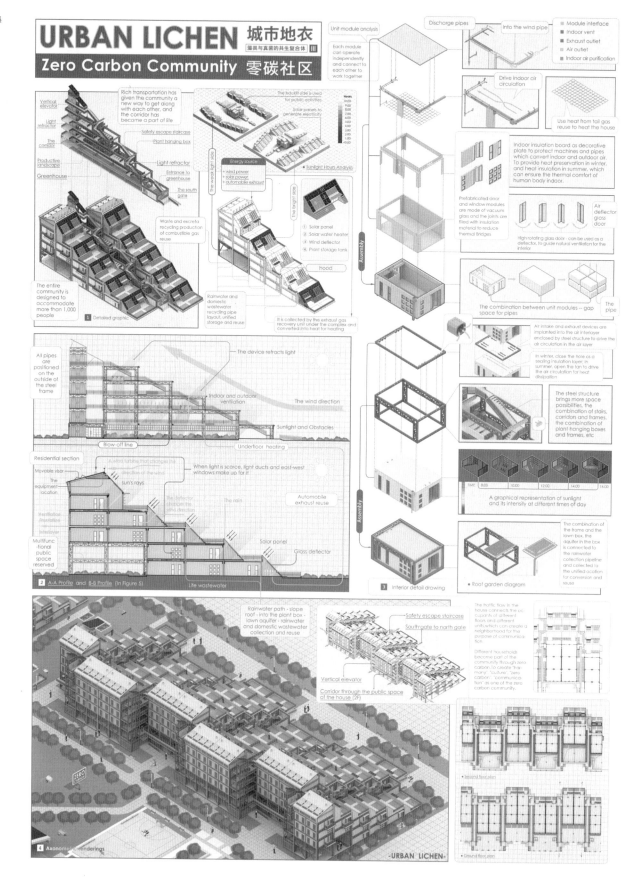

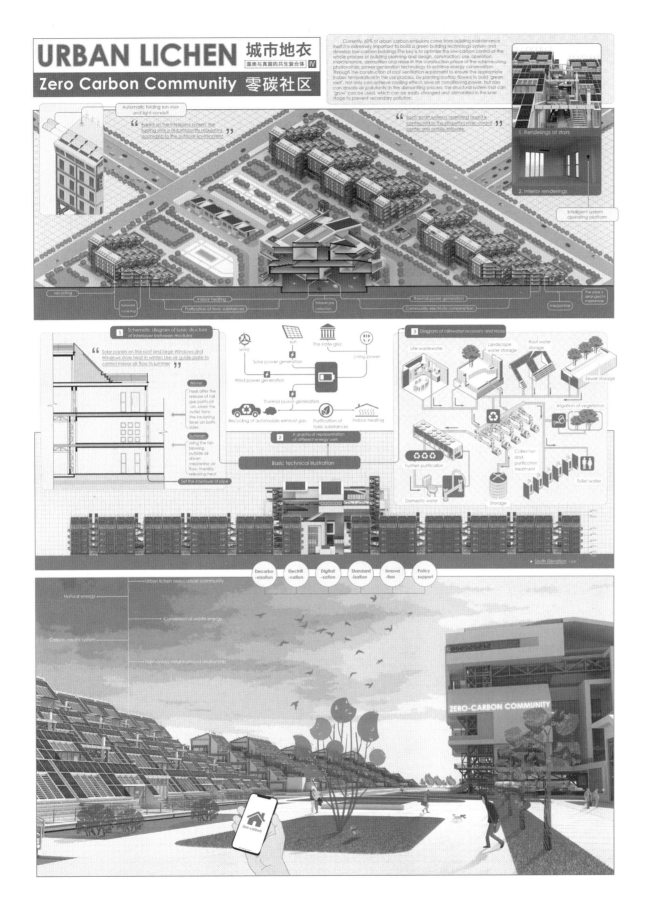

综合奖·优秀奖
General Prize Awarded · Honorable Mention Prize

注 册 号：100317
Registration No: 100317

项目名称：流动之"环"——基于智能化平台的意识型低碳社区设计
Mobile Loop

作　　者：高樱嘉、郑钰杭、杨雨欣
Authors：Gao Yingjia, Zheng Yuhang, Yang Yuxin

参赛单位：郑州大学
Participating Unit：Zhengzhou University

指导教师：付孟泽、吴　迪
Instructors：Fu Mengze, Wu Di

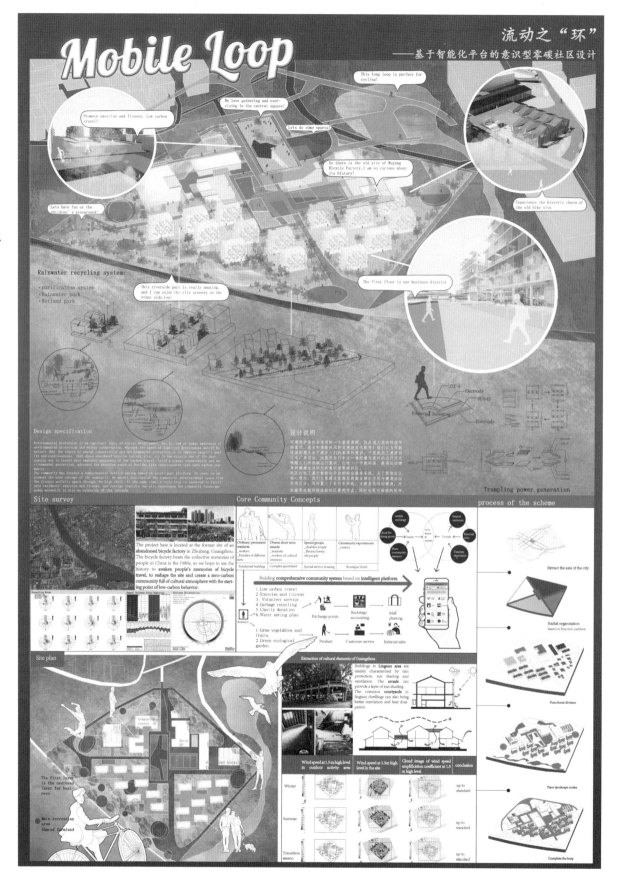

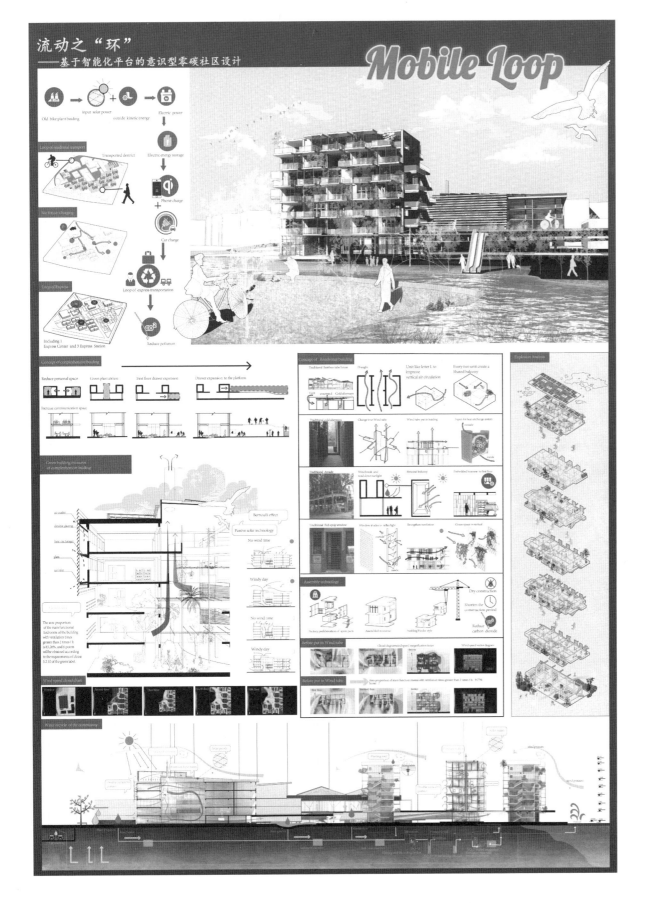

综合奖·优秀奖
General Prize Awarded · Honorable Mention Prize

注 册 号：100322
Registration No：100322

项目名称：缝合·社区——大理低碳社区
Urban Farming Community Stitching City

作　　者：刘潇天、李佳骏、陈曦玥、韩志琛
Authors：Liu Xiaotian, Li Jiajun, Chen Xiyue, Han Zhichen

参赛单位：代尔夫特理工大学、东南大学、天津大学
Participating Units：Delft University of Technology, Southeast University, Tianjin University

指导教师：李伟、刘魁星、刘婺
Instructors：Li Wei, Liu Kuixing, Liu Wu

缝合·社区　Dali Zero Carbon Community
Urban Farming Community Stitching City

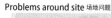

缝合·社区 Dali Zero Carbon Community

02

2021 台达杯国际太阳能建筑设计竞赛获奖作品集

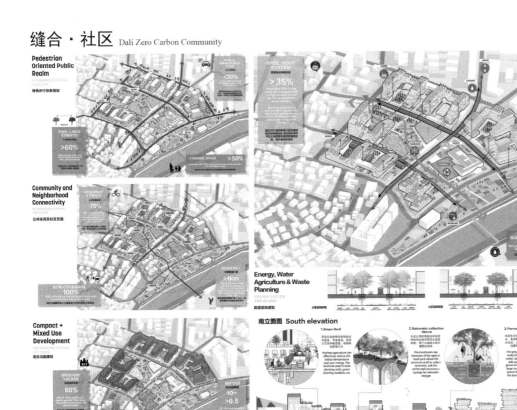

Site Plan 总平面图

缝合·社区 Dali Zero Carbon Community

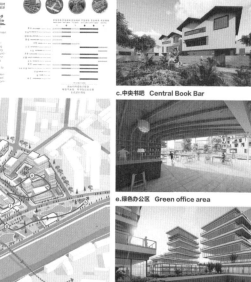

f. 农业展览馆 Agricultural Exhibition Hall

b. 农贸集市 Vegetable market

m. 居住片区 Residential area

i. 田园民宿 Idyllic Homestay

c. 中央书吧 Central Book Bar

e. 绿色办公区 Green office area

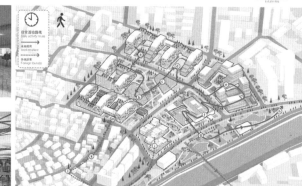

低碳社区的一天：本地居民
A day of local residents

低碳社区的一天：外来游客
A day of Foreign tourists

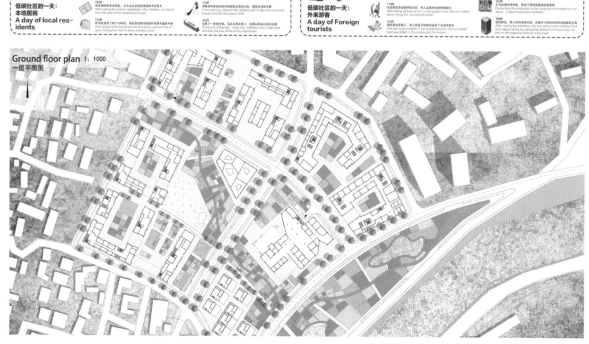

Ground floor plan 1:1000
一层平面图

东立面图 East elevation

缝合·社区
Urban Farming Community Stitching City

Clean Energy Supply System
能源供应系统图

OFFICE DISTRICT 办公区块   RESIDENTIAL DISTRICT 居住区块   B&B DISTRICT 民宿区块

Building Intelligent Control System
智能建筑控制系统

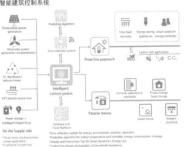

Carbon offset in tridimensional virescence
立体绿化空间

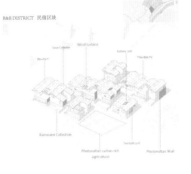

Material Recycling
材料循环

Inflow Category / Outflow Category

Ecological Landscape 生态景观

Carbon Emissions Calculation 碳排放计算

综合奖·优秀奖
General Prize Awarded · Honorable Mention Prize

注 册 号：100324
Registration No: 100324

项目名称：森之谷
　　　　　City Valley

作　　者：郭　龙、岳文茹、郑潇莹、王球锋
Authors: Guo Long, Yue Wenru, Zheng Xiaoying, Wang Qiufeng

参赛单位：西安交通大学
Participating Unit: Xi'an Jiaotong University

指导教师：王宇鹏
Instructor: Wang Yupeng

森之谷 CITY VALLEY 1
2021 台达杯国际太阳能建筑设计竞赛
CONCEPTUAL DIAGRAMS

作品从TOD开发模式导入，选用开放街区、绿色社区的设计理念。规划上采用小街区、密路网的棋盘式规划，结合风模拟羊毛算法结果，进行规划布局，并使用空中廊桥将各个区域相串联。建筑设计上吸纳中国传统院落的建筑形式，以垂直方形院落控制居住功能，商办建筑采用流线型设计语言。设计采用垂直绿化、雨水回收系统、太阳能板技术的运用，实现碳中和的目标，形成一个与地铁无缝接驳、步行友好、绿色生态的综合性住区。

This project is based on TOD development mode, and adopts the design concept of open block and green community. In the planning, the checkerboard planning of small blocks and dense road network is adopted. Combined with the results of wind simulation wool algorithm, the planning layout is carried out, and the air corridor bridge is used to connect each area in series. In terms of architectural design, the architectural form of traditional Chinese courtyard is absorbed, and the residential function is controlled by vertical square courtyard. The commercial office buildings adopt streamline design language. The design uses vertical greening, rainwater recycling systems and solar panel technology to achieve carbon neutrality. To form a seamless connection with the subway, pedestrian-friendly, green and ecological comprehensive residential area.

SITE ANALYSIS

CITY SCALE

NEIGHBORHOOD SCALE

CONCEPTUAL DIAGRAM
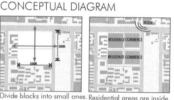
Divide blocks into small ones. Residential areas are inside.

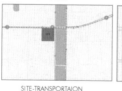
SITE-TRANSPORTATION

SITE-FUNCTION

Smaller blocks are divided in residential areas while the commercial ones are connected.

According to the simulation of the wind blowing trace, the wind passage is formulated by the defination of buildings.

HISTORY ANALYSIS

| Railway Line Opened | Military Factories Were Built | Established Staff Dormitory Area | High-level Mechanization | All-age Living Community |
| 20C 50S | 1954-1960 | 20C 60S | 21C | 2008-Now |

SUNLIGHT HOURS ANALYSIS
SPRING EQUINOX

SUMMER SOLSTICE

AUTUMN EQUINOX

WINTER SOLSTICE

WIND SIMULATION
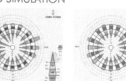
Wind Wheel of Whole Year　Wind Wheel of Summer　Wind Wheel of Winter

TEMPERATURE ANALYSIS
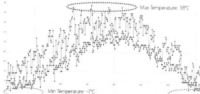

Xi'an, as a cold region where northeast wind prevails all year round, should focus on the use of wind direction for cooling in summer, and pay attention to the use of the sun's shadows.

森之谷 CITY VALLEY 2
2021 台达杯国际太阳能建筑设计竞赛

PROBLEM ILLUMINATION

HIGH-RISE PEDESTRIAN SYSTEM

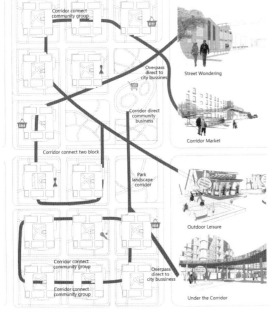

SITE PLAN

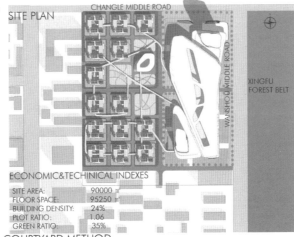

ECONOMIC&TECHINICAL INDEXES

SITE AREA:	90000 m²
FLOOR SPACE:	95250 m²
BUILDING DENSITY:	24%
PLOT RATIO:	1.06
GREEN RATIO:	35%

COURTYARD METHOD

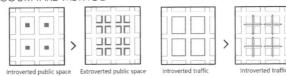

Introverted public space / Extroverted public space / Introverted traffic / Introverted traffic

The concept of "large open, small closed" is conducive to the formation of a vibrant community form, to ensure the convenience of management and residents' sense of belonging to the community

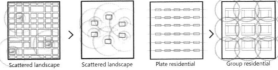

Scattered landscape / Scattered landscape / Plate residential / Group residential

Courtyard construction is conducive to building a close neighborhood relationship, enhance the sense of belonging.

PRIVACY HIERARCHY

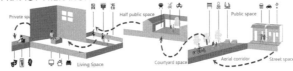

PUBLIC MODULAR FACILITY

SPORT PEDESTRIAN RESTING PARKING
POCKET PARK POOL POOL WATER PERMEABLE PATH

Public functional modulars are listed which can be added into the residential blocks aiming to inhance the relationship between residents.

TOD STRATEGY

The commercial areas and residential blocks are highly connected with the metro station which are aiming to recommend residents and clerks working in offices taking the public transportation rather than their cars.

SITE SECTION 1:1000

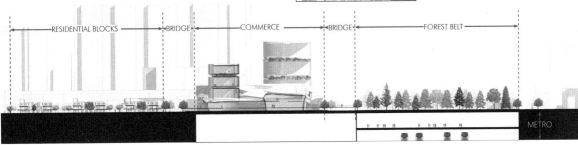

RESIDENTIAL BLOCKS — BRIDGE — COMMERCE — BRIDGE — FOREST BELT

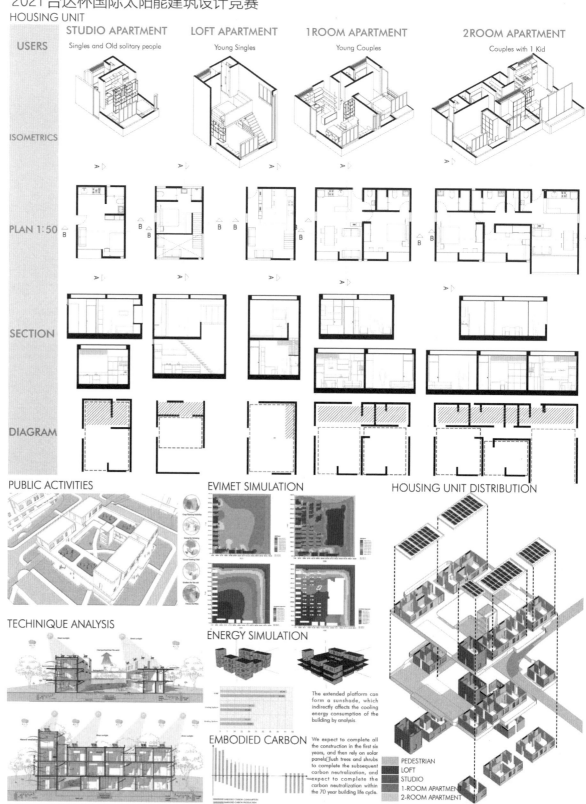

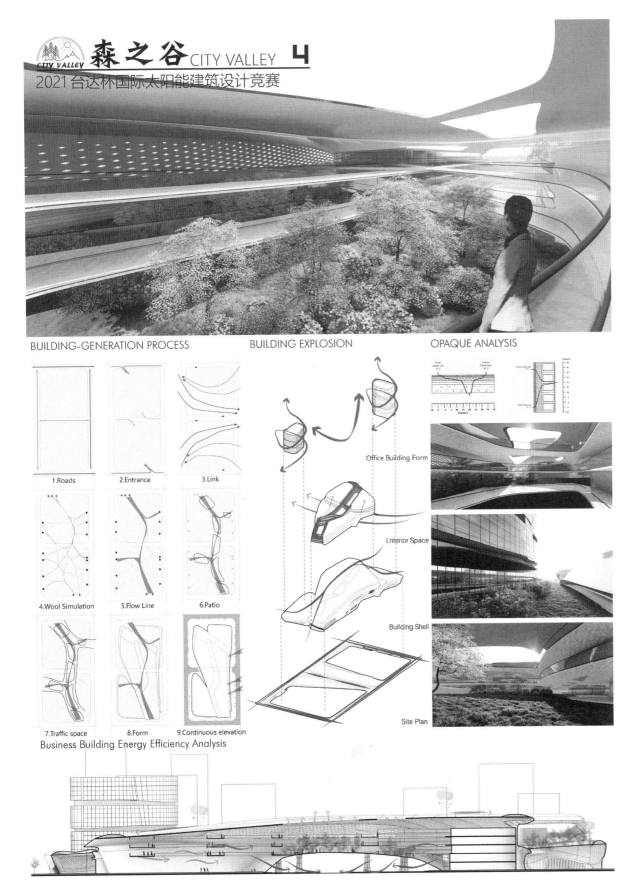

综合奖·优秀奖
General Prize Awarded · Honorable Mention Prize

注 册 号：100326
Registration No：100326

项目名称：生土楼"埂与生"
The Rebirth of Fujian Earth Building

作　者：朱占威、陈　果、沈曦煜、方阳开、王　涛
Authors：Zhu Zhanwei, Chen Guo, Shen Xiyu, Fang Yangkai, Wang Tao

参赛单位：合肥工业大学
Participating Unit：Hefei University of Technology

指导教师：王　旭
Instructor：Wang Xu

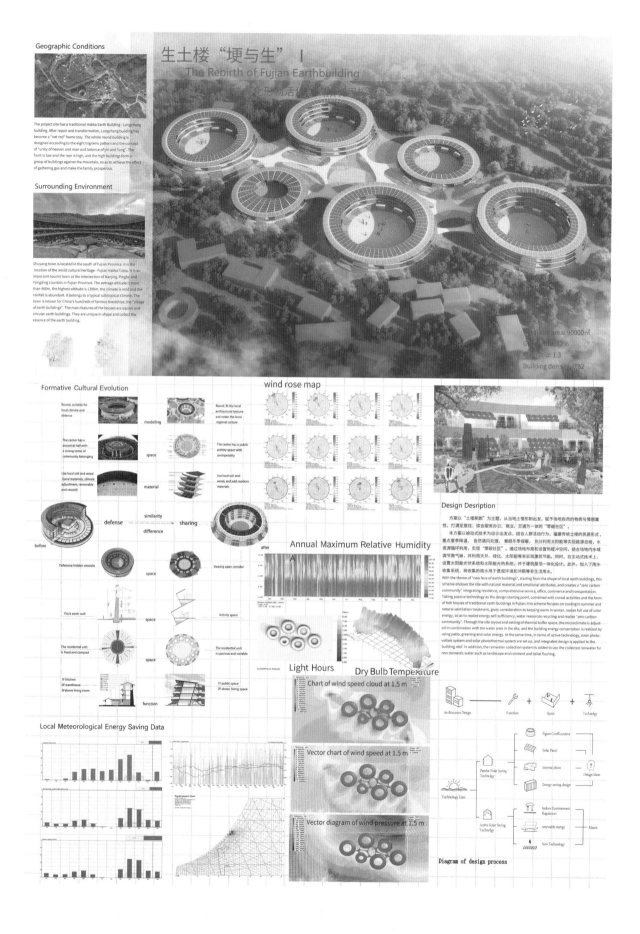

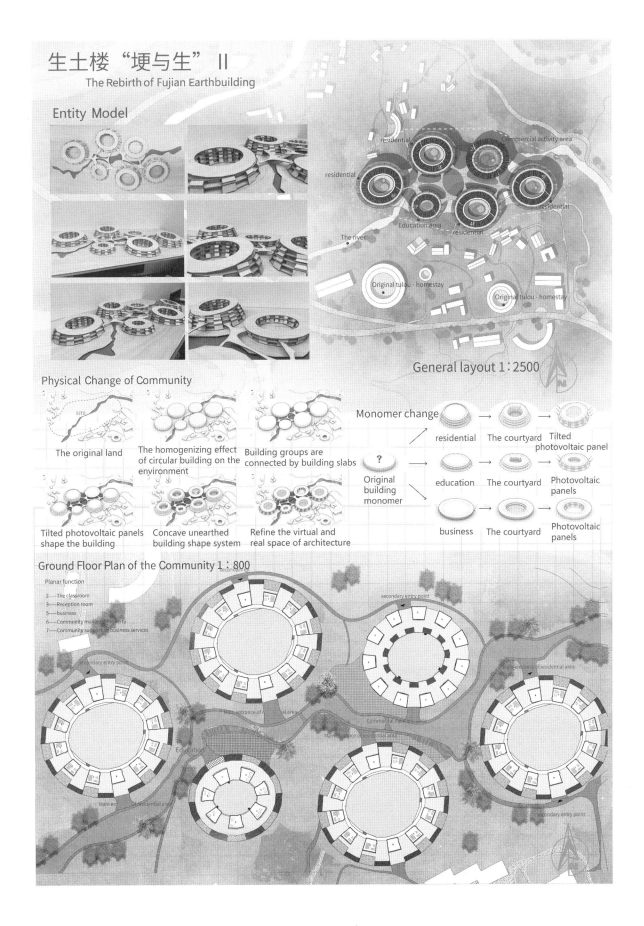

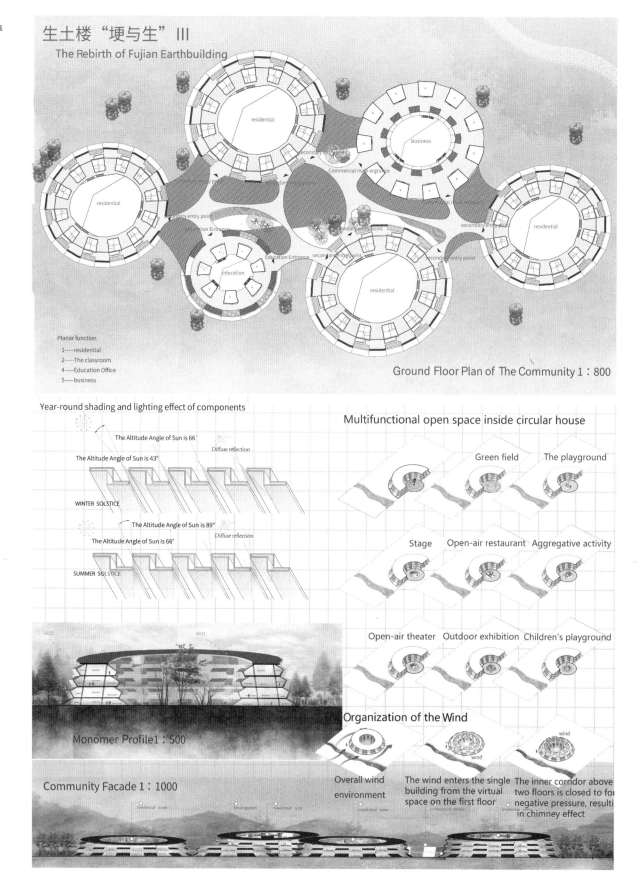

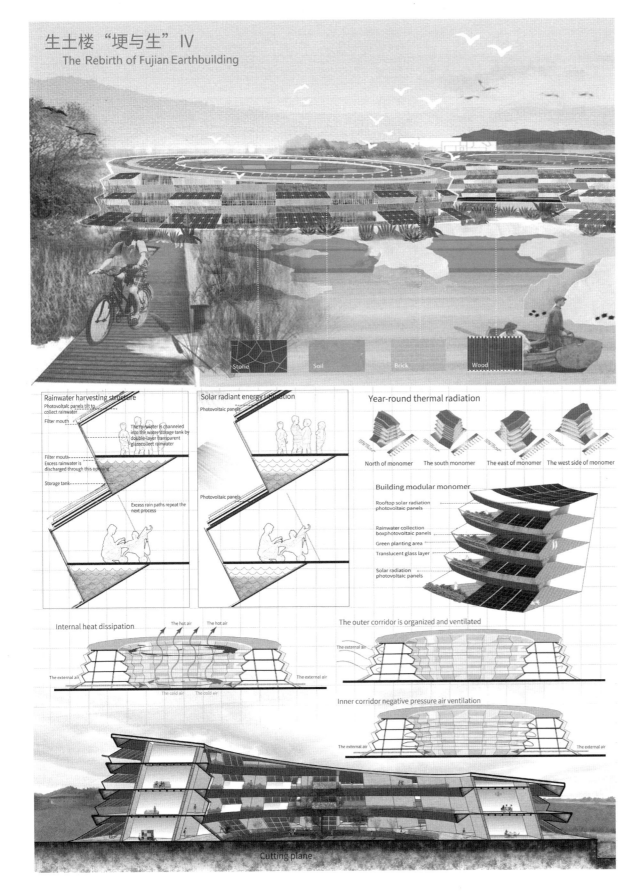

综合奖·优秀奖
General Prize Awarded · Honorable Mention Prize

注 册 号：100353
Registration No．100353
项目名称：共享办公
　　　　　Office Community
作　　者：陈悦妍、谢燕婷、华蓓蕾
Authors：Chen Yueyan, Xie Yanting, Hua Beilei
参赛单位：南京工业大学
Participating Unit：Nanjing University of Technology
指导教师：郭　兰
Instructor：Guo Lan

Office Community I

Concept Analysis

Found Problem

Commuting Energy Consumption

The commuting modes of office workers are mostly divided into three types: public transportation, private car, and bicycle. Among them, public transportation and private cars consume a lot of electricity, water, fossil energy, and also produce a lot of carbon dioxide.

Commuting Time Consumption

According to the survey, most office workers' homes and workplaces are separated. There are many problems in commuting, such as bus congestion and tidal traffic congestion, which will consume a lot of time and energy and bring great pressure to office workers.

Closeness of Person-to-person Communication

Interpersonal relationships are indispensable at all times, but modern cities have closed the communication between people. High-rise housing reduces the opportunities for people to communicate face-to-face. Most residents prefer to stay at home, neighbouring relations become cold.

Closedness of Communication Between Man and Nature

With the development of the city, modern living space is seriously derailed from nature. Enjoying the natural landscape has become a very luxurious way of life for modern urban people. People can only plant potted plants in a small balcony space to sustenance their yearning for nature.

Solve Problem

Sub-regional office

The design adopts the form of office points arranged by the company in different areas, so that the work and residence of employees are solved within the community. This makes the company's office more flexible and reduces the commuting distance of employees.

Prefabricated Apartment With Living and Office

The design modularizes the living space and office space and puts them into the frame to form a prefabricated apartment with mixed living and office functions.

Create Communication Space

Set up intermediate spaces and public spaces on residents' travel routes to create opportunities for communication.

Integrated Green Route

Arrange natural spaces on the travel routes of residents and within the sight range.

Design Ideas

本方案以"共享办公"为主题，赋予场地人文和自然的零碳环保属性。
本方案以装配式技术和集成化手段为设计出发点，结合社区人群活动行为和传统社区居住形式，重点加强居民对社区和自然的归属感，利用光伏板围护结构、口袋公园、社区客厅、蓝色广场等实现社区零碳与居民互动。
在建筑的立面和屋面以及停车场设置了不同的光伏板材料，并将建筑屋顶、口袋公园、社区客厅做了集成化设计，高效地完成对生活污水的处理净化和雨水的收集再利用。
With the theme of "shared office", this plan gives the site a zero-carbon environmental protection attribute of humanity and nature.
This plan takes prefabricated technology and integrated methods as the design starting point, combines community activities and traditional community living forms, focuses on strengthening residents' sense of belonging to the community and nature, and uses photovoltaic panels to protect the structure, pocket parks, community living rooms, and the blue square and others realize zero-carbon interaction with residents in the community.
Different photovoltaic panel materials are installed on the building's facade and roof and parking lot, and the building roof, pocket park, and community living room are integrated design to efficiently complete the treatment and purification of domestic sewage and the collection and reuse of rainwater.

Base Location

Gulou District

Base Analysis

District Analysis

Time consumption an hour | Power consumption 38 kWh
Time consumption an hour and a half | Gasoline consumption 4 liters
Time consumption 40 minutes | Gasoline consumption 1.1 liters

The base is far away from the urban area, and people's commuting time takes 0.6 to 1.5 hours, which is very inconvenient. And it needs to consume a lot of energy and produce a lot of carbon dioxide at the same time.

Crowd analysis

Residents in this community are mainly young people living alone and middle-aged people with a family. Their commuting methods are mainly by taking the subway and driving a private car. And through the analysis of behavioral activities, the community lacks neighborhood communication, outdoor activities are not active, and the communication between people and nature is closed.

Office Community II

General Layout 1:1000

Waterfront Design

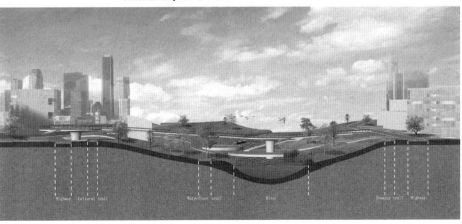

Office Community III

Modularity · Residential Area

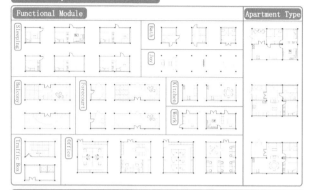

Prefabricated · Residential Area

According to relevant project calculations, prefabricated buildings have the advantages of saving resources. Therefore, the building is assembled from prefabricated components, and a large number of building parts are produced and processed in the workshop. Design standardization and management informationization have made component standards, high production efficiency, and corresponding component costs reduced. With the digital management of the factory, the entire prefabricated building is cost-effective.

Folding Wall
When the house is idle or wants to change the indoor layout, the wall can be folded and used for many times, so as to save the building space, reduce the construction waste and reduce the energy consumption of building materials in the community.

Prefabricated Frame
The demolition of steel structure buildings hardly produces construction waste, which can be recycled and reused. This greatly reduces the energy consumption of building materials in the community.

Breathing Orifice Wall
The wall is equipped with a large number of opening and closing holes, which will be intelligently adjusted according to the air temperature, humidity and air velocity detected by the sensor connected to it.
In summer, the holes are wide open to ensure air circulation, and at the same time reduce the internal temperature of the building with the help of climbing plants attached to it; most of the holes will be closed during heat preservation in winter, and only a small part of the holes will be left as air inlets and outlets.

Magnetic levitation mobile wall
The wall facing the corridor can be moved freely using magnetic levitation technology, which can meet the needs of community residents to hold large-scale events.

Solar Glass Window

Corridor Exchange Seat
Neighbor exchange and nature dialogue.

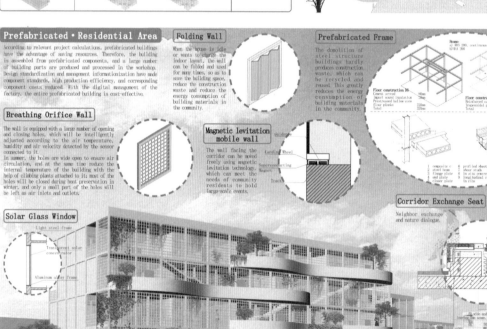

Office Community IV

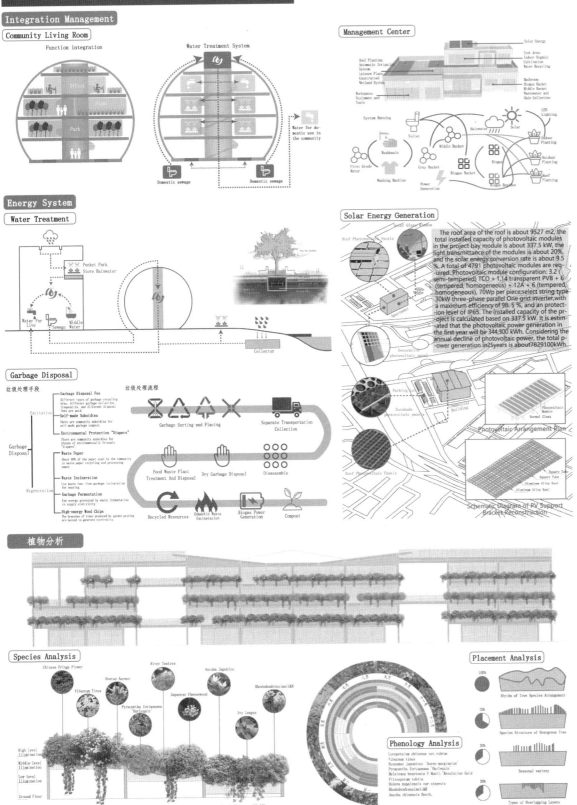

不期而"寓"
零碳社区概念设计
Zero-Carbon Community Design

综合奖·优秀奖
General Prize Awarded · Honorable Mention Prize

注 册 号：100372
Registration No：100372
项目名称：不期而"寓"
　　　　　Zero Carbon Community
　　　　　Design
作　　者：于　洋、顾他一、吴一舟、
　　　　　滕苏彦
Authors：Yu Yang, Gu Tayi,
　　　　　Wu Yizhou, Teng Suyan
参赛单位：南京工业大学
Participating Unit：Nanjing University of
　　　　　Technology
指导教师：薛春霖、罗　靖
Instructors：Xue Chunlin, Luo Jing

■ Question

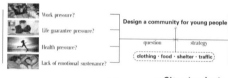

■ Site Selection

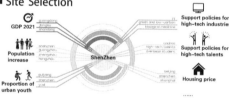

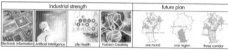

Longhua District is located in the geographic and geometric center of Shenzhen and the central axis of urban development in the Guangdong–Hong Kong–Macao Greater Bay Area. It is adjacent to six districts and one city. Owning Shenzhen North Railway Station, a large comprehensive transportation hub in the Guangdong–Hong Kong–Macao Greater Bay Area, it only takes 15 minutes to reach Hong Kong West Kowloon, forming a half-hour living circle with Hong Kong and Guangzhou, and a four-hour living circle with Wuhan and Xiamen.

■ Site Analysis

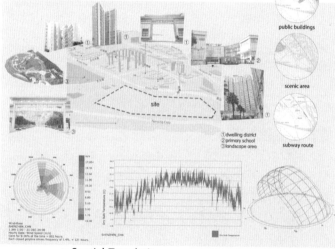

■ Humanistic Analysis

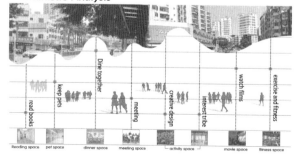

■ Spatial Translation

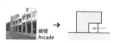
骑楼 Arcade
①Protect from wind and rain, prevent sun exposure and create a cool environment
②Become a shared space for customers, good for interpersonal relationships

外廊 Veranda
①Have good ventilation effect
②Become a shared space for customers, good for interpersonal relationships

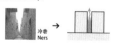
冷巷 Ners
①Keep it long-term free from solar radiation, smooth air circulation
②Vertical temperature difference can form convection with the hot air in the buildings on both sides

庭院 Courtyard
①Have good ventilation and lighting effect
②Play a role in air purification
③People get the beauty of nature and art in architecture

不期而"寓" II
零碳社区概念设计
Zero-Carbon Community Design

2021 台达杯国际太阳能建筑设计竞赛获奖作品集

Technical and Economic Index
- Total land area: 91230 ㎡
- Construction area: 59472 ㎡
- Green rate: 42%
- Volume rate: 3.2
- Building density: 65%
- Building height: 49.3m

General layout 1:1500   N

■ **Concept Generation**

1. Site is wide and long, the depth is short

2. The depth direction is divided into 3 sections. Two residential and one commercial

3. Blocks rise with the boundary

4. Operate the blocks and create different spaces

5. Refine the block and form the final community

■ **Climate Simulation**

■ **Axis and Node**

■ **Solar Energy Circulation**

■ **Site Planning Explode**

① Residential Area
② Roof Runway
③ Connection Space
④ Business Area
⑤ Footpath System
⑥ Greening System

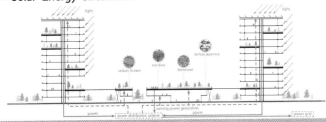

■ **Water Resource Circulation**

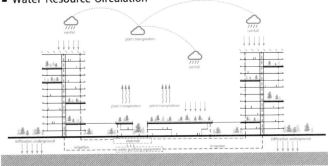

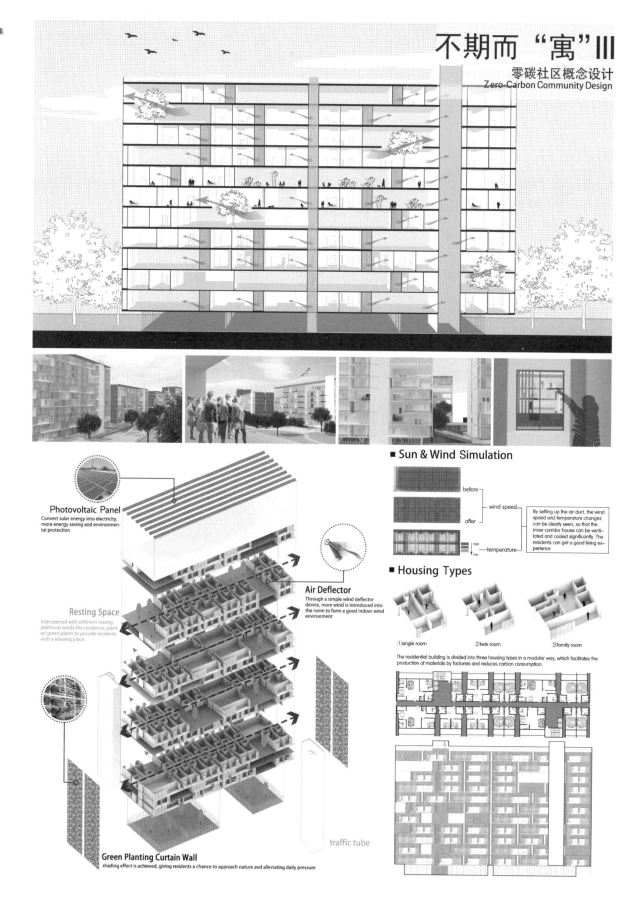

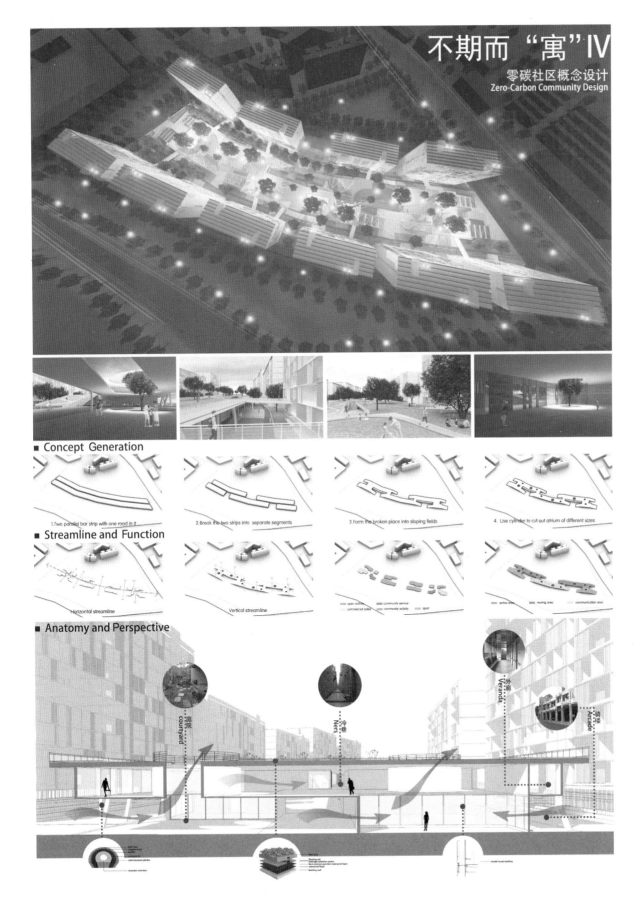

综合奖·优秀奖
General Prize Awarded·
Honorable Mention Prize

注 册 号：100380
Registration No：100380
项目名称：江风摇巷
　　　　　River Wind Shaking Alley
作　　者：吕诗祺
Author：Lü Shiqi
参赛单位：三峡大学
Participating Unit：Three Gorges University

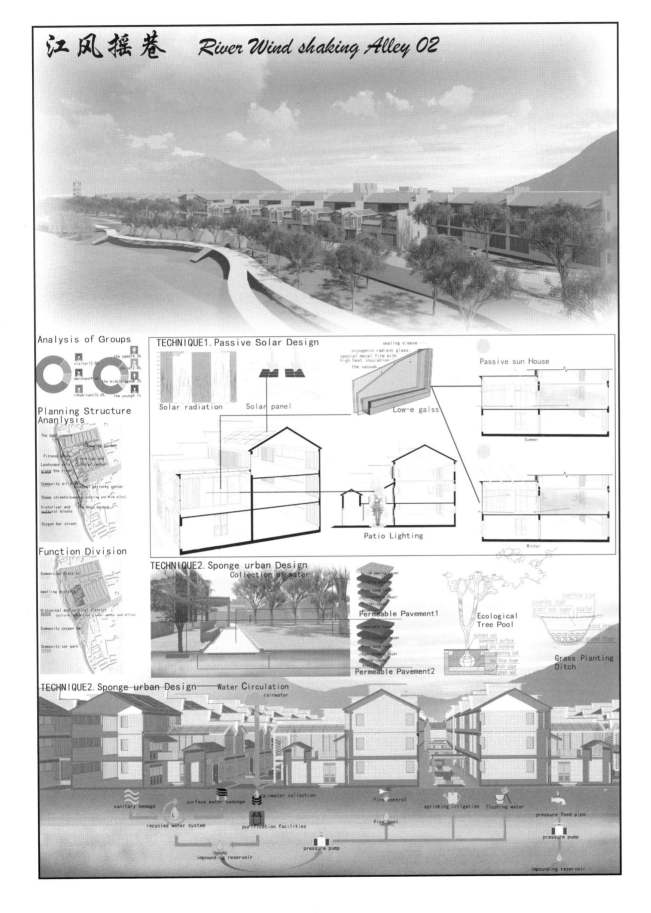

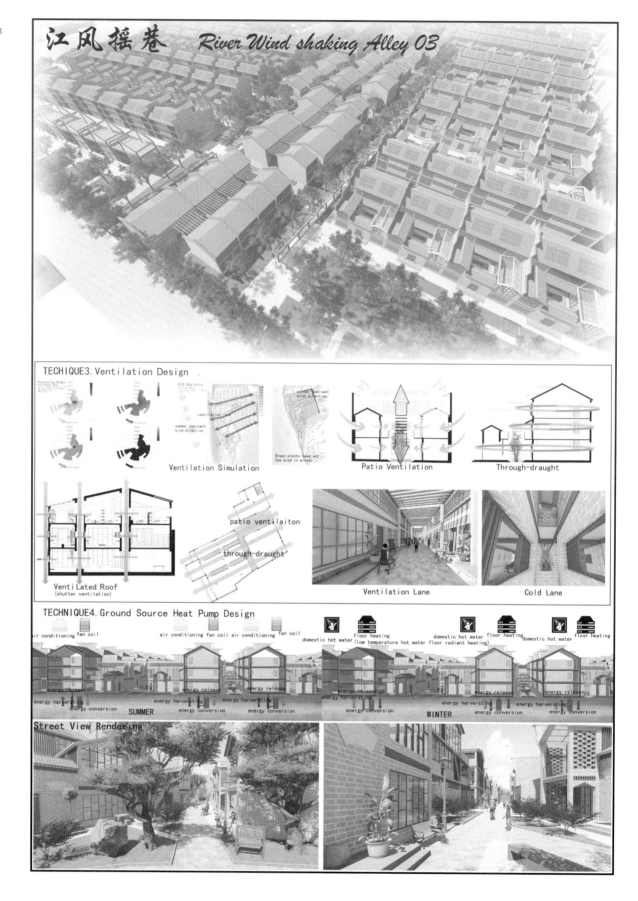

综合奖·优秀奖
General Prize Awarded · Honorable Mention Prize

注 册 号：100386
Registration No: 100386

项目名称：绿意·环生
Green Germination

作　　者：李莹莹、林绮琪、陈伟霖、谈浩然
Authors：Li Yingying, Lin Qiqi, Chen Weilin, Tan Haoran

参赛单位：广州大学
Participating Unit：Guangzhou University

指导教师：席明波、万丰登
Instructors：Xi Mingbo, Wan Fengdeng

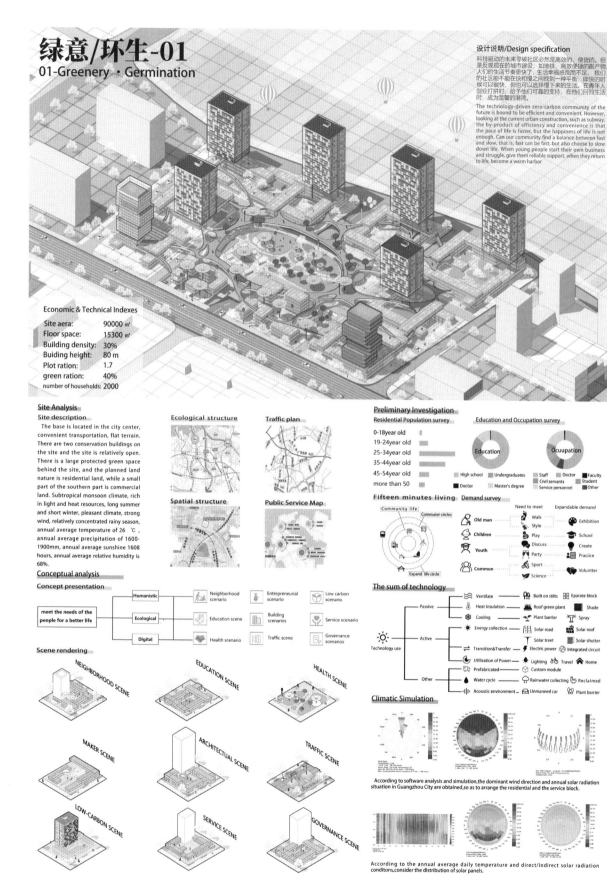

绿意/环生-02
02-Greenery · Germination

Site-plan 1:1000

Plane generation analysis

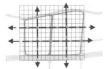

In the 30*30 grid system, each site is divided into twelve blocks.

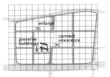

According to the actual situation of the site, adjust the road network to suit the site.

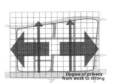

According to the nature of the surrounding land and road, determine the different degree of privacy of the block.

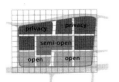

Finally, the division of site privacy degree is obtained.

Based on the degree of privacy, the main function of each block is obtained according to the nature of the surrounding land.

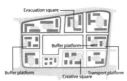

The layout of the first floor creates a rich square space. (Unmarked squares are activity squares)

Through the layout of the ground floor, the surrounding landscape is connected with the site by sight, so that there is no obstruction to the green landscape in the block.

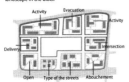

According to the flow direction and functional attributes create different spaces

Network analysis

Air corridors: Designed for walking and cycling traffic speed, the core is an integral part of the walking and cycling system.

only for unmanned vehicles, bicycle, walking.
first grade road (public): 40km/h, no motor vehicles other than unmanned vehicles are allowed
second grade road(transition): 30km/h
second grade road(residents): 20km/h

The city trunk road sinks, enters the parking lot from both sides.

Traffic pattern
Traditional TOD mode of transportation:

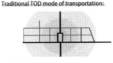

New TOD mode of transportation:

Autonomous vehicles can reach any place, so that community transportation from central to traffic mode to multi-point community activities

Technical support

Modularization   Heatable

The precast board process enables rapid repair and replacement of road pavement

Conductive concrete materials can heat the pavement

Heatable   Heatable

LED lights can change the right of way by changing color.

Paving allows for exclusive space for landscaped areas.

Streets that dynamically allocate space

Ground traffic control system

Technical support
Energy recovery technology of piezoelectric materials

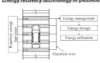

The mechanical energy produced by vehicle dynamic load on road surface is converted into electric energy by piezoelectric effect. It can generate electricity to meet the power supply demand of microelectronic devices and LED lights.

The thermoelectric model

A thermoelectric mode of converting thermal energy into electrical energy by means of thermoelectric conversion materials in a thermal gradient space. Black asphalt pavement has a strong solar energy absorption capacity, its absorption coefficient can reach 90%.

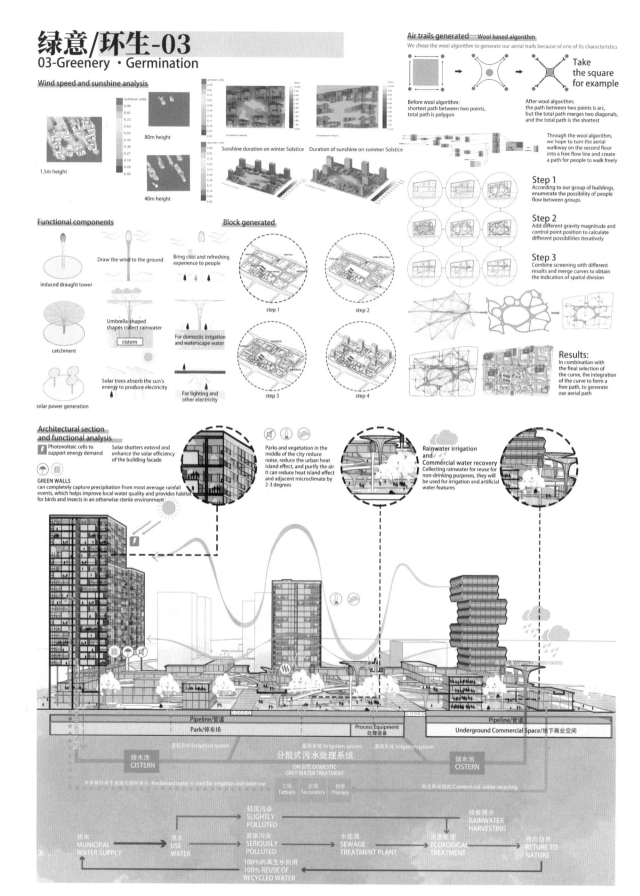

绿意/环生-04
04-Greenery · Germination

Zero carbon design

Passive energy saving technology

Natural ventilation under wind pressure

Plane optimization

Control the window opening direction to form a wind deflector to optimize indoor ventilation

Active energy saving technology
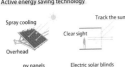
Spray cooling / Overhead / pv panels / Track the sun / Clear sight / Electric solar blinds

Residential concept

Space requirement

single 30㎡ / 2 people 80㎡ / 3 people 100㎡ / more people 120+㎡

Different numbers of users have different requirements for the size of the space

Functional requirement

artist - Exhibition / Net Red - Studio / dancer - Dance room / other occupations - other room

Different users have different requirements for the function of the space

Traditional house

change house or using one house

Traditional residential space has a fixed size, and there is insufficient or surplus space during use

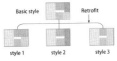
Basic style / Retrofit / style 1 / style 2 / style 3

Residents transform their basic apartment types to get a space that suits them, and the process wastes resources

DIY House

We want a house like this. → DIY on the app → transport → build
5 years later
We want to expand our house. → add modules → transport → build
Good!

Information of DIY House

flat
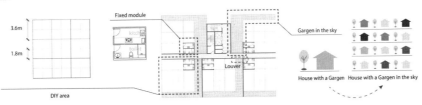

Fixed module / DIY area / Louver / Gargen in the sky / House with a Gargen / House with a Gargen in the sky

Module

Fixed module (3.6m × 3.6m)
Including kitchen and toilet. Each layer is fixed for piping layout.
 Toilet / Kitchen

Outdoor module
Outdoor use

Living module (3.6m × 3.6m)
Meet basic life needs.
 Baby room / Dinning room / Bedroom / Living room / more

Add-on module (3.6m × 1.8m)
Enrich living space.
 Studio / Bar / Tea room / Cloakroom / Studyrom

Module combination

add module → add module → ...

Examples of floor plans

Different stages of construction
1. Initial state
2. Remove floor slab
3. Early state
4. Mid-term state
5. Saturation state
6. Updatable modules

7. Prefabricated modules

Straw building materials: Heat insulation, Carbon fixation, Recyclable, Sustainable / Steel

纳兰住集·向阳而居
Living in the Naran Resistance towards the sun

01

本次设计位于内蒙古巴彦淖尔市，"太阳"在蒙古语中意为"纳兰"，纳兰住集由此而来。方案从地域文化和气候出发，结合低碳理念、太阳能技术，旨在设计出符合地域特色的满足低收入人群需求的低碳、绿色、生态社区。

针对"零碳社区"，方案主要从场地设计和住宅设计两个方面做出回应。社区规划以慢行交通和电动车为主，利用当地太阳能和风能两大优势，设置太阳能充电桩，四周设置风光互补路灯。

建筑设计：一是设置共享服务空间，节约用地，减少碳排放；二是针对内蒙古巴彦淖尔气候特征设计集合住宅，以冬季保暖为主，夏季隔热为辅。方案设计通过技术可以达到夏季建筑内部不使用空调制冷，冬季减少采暖期时长。

在地域性的回应中，选取具有当地代表性的向日葵和黄河几字湾两个文化形象。设计与文化形象相呼应，生成"黄河边上的向日葵"的艺术形态。

This design is located in Inner Mongolia Bayannaoer City, ' Sun ' in Mongolian meaning ' Nalan ', Nalan living set from this. Starting from regional culture and climate, combined with low-carbon concept and solar technology, the scheme aims to design low-carbon, green and ecological communities that meet the needs of low-income people with regional characteristics.

For ' zero carbon community ', the program mainly responds from two aspects site design and residential design.The community planning is mainly based on slow traffic and electric vehicles.Using the two advantages of local solar and wind energy, solar charging piles are set up, and complementary street lights are set around.

Architectural design : one is to set up shared service space, save land, reduce carbon emissions. The second is to design a set of residential buildings according to the climatic characteristics of Bayannaoer in Inner Mongolia, which is mainly warm in winter and supplemented by heat insulation in summer. Scheme design through technology can achieve the summer building without air conditioning cooling, reduce the heating period in winter.

In the regional response, two representative cultural images of sunflower and Yellow River Chiziwan are selected. The design corresponds to the cultural image and forms the artistic form of ' sunflowers on the Yellow River '.

综合奖·优秀奖
General Prize Awarded·
Honorable Mention Prize

注　册　号：100406
Registration No：100406

项目名称：纳兰住集·向阳而居
Living in the Naran Resistance Towards the Sun

作　　者：夏　凡、王璐瑶、王端阳、
　　　　　蒲云云、孙子璇、乌　润
Authors：Xia Fan，Wang Luyao，
　　　　 Wang Duanyang，Pu Yunyun，
　　　　 Sun Zixuan，Wu Run

参赛单位：内蒙古工业大学
Participating Unit：Inner Mongolia University of Technology

指导教师：杨春虹
Instructor：Yang Chunhong

Location

The project base is located in Linhe District, Bayannaoer City, Inner Mongolia Autonomous Region. The annual average temperature in Bayannur is not more than 10 degrees, and the sunshine in Bayannur is sufficient, which is one of the areas with the most sunshine in China. The growth of crops in sufficient sunshine time is helpful. The south of Yinshan is Hetao plain, with the name of ' Yellow River peril, only one set of rich '. Water resources are extremely excellent, and it is one of the important commodity grain production bases in China.

Diagram of Design Process

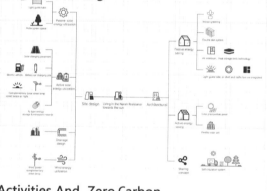

Regional Culture

Sunflower : Bayannur is one of the most abundant areas of light energy resources in China. Therefore, there are various vegetables and fruits in Bayannur, and sunflower is the ' golden card ' of the city.

The Yellow River : Bayannur is located in the Hetao Plain, with the reputation of ' outside the Yangtze River south '.

Sandstorm : the dominant wind direction in Bayannaoer is northwest wind. Sandstorm weather has been a special memory of local residents in Inner Mongolia.

Climatic Simulation

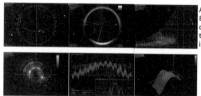

According to the analysis, Bayannaoer area is relatively dry, rich in solar resources, the dominant wind direction is northwest wind,

Activities And Zero Carbon

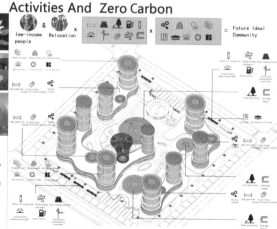

纳兰住集·向阳而居
Living in the Naran Resistance towards the sun

02

Logical Generation

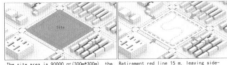

The site area is 90000 m² (300m*300m), the north side is the school, the west side is the hospital, and the east and west sides are adjacent to the community.

Retirement red line 15 m, leaving sidewalks and green belts for cities.

An annular fire lane is arranged inside the community, with entrances and exits on both sides of the east and west, and a winding fire lane shuttles among them, like the Yellow River Cross Bay.

Layout of high-rise residential buildings along internal lanes to save land.

Introducing Sunflower and Yellow River Culture, Low Carbon Concept and Solar Energy Technology into Bayannur Community to Achieve Design Balance.

Solar photovoltaic panels are placed above the building, like sunflowers, and street lights with complementary scenery are arranged around the community, which can not only illuminate the interior of the community, but also illuminate the urban roads.

Electric vehicle charging piles and a large number of bicycle parking spaces in the community.

Connect the bottom of each assembly house to form a shared space for work, entertainment, rest and communication.

Combined with the principle of ground source heat pump, a large area of greening is set up in the community to collect rainwater. Green below combined with ground source heat pump water cycle utilization, to achieve more than 80 % of the water cycle.

Local mainly northwest wind, so the northwest side of the building elevation, shelter northwest wind.

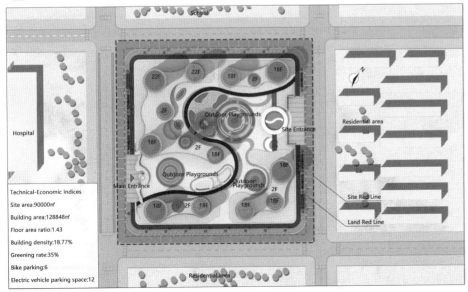

Technical-Economic Indices
Site area:90000㎡
Building area:128848㎡
Floor area ratio:1.43
Building density:18.77%
Greening rate:35%
Bike parking:6
Electric vehicle parking space:12

Site Plan 1:1500

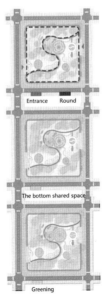

Analysis Chart

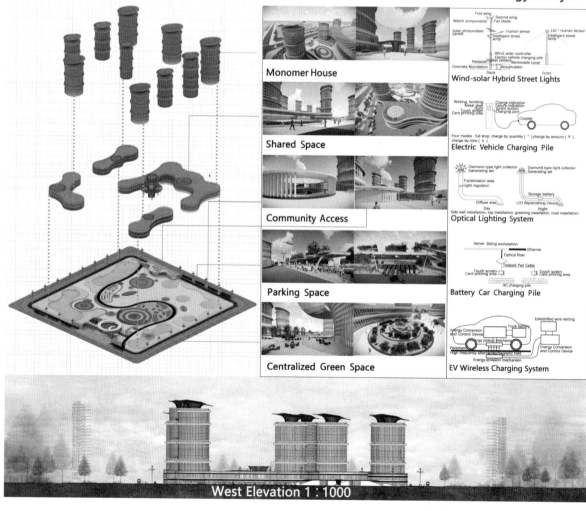

纳兰住集·向阳而居
Living in the Naran Resistance Towords the Sun

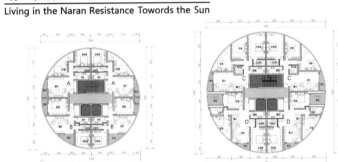
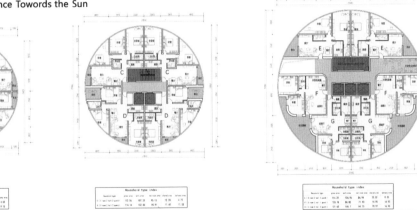

Unit Plan 1:200

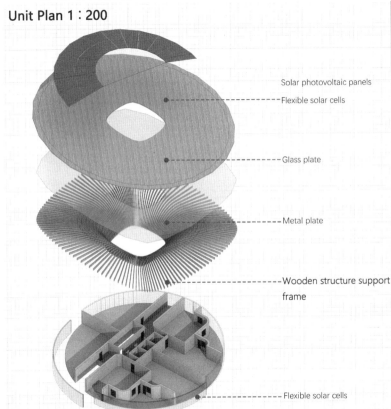

- Solar photovoltaic panels
- Flexible solar cells
- Glass plate
- Metal plate
- Wooden structure support frame
- Flexible solar cells

Unit Explosion Diagram

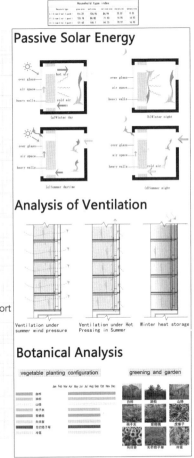

Passive Solar Energy

(a) Winter day — (b) Winter night
(c) Summer daytime — (d) Summer night

Analysis of Ventilation

Ventilation under summer wind pressure | Ventilation under Hot Pressing in Summer | Winter heat storage

Botanical Analysis

vegetable planting configuration | greening and garden

East Elevation 1:1000

综合奖·优秀奖
General Prize Awarded · Honorable Mention Prize

注　册　号：100416
Registration No：100416

项目名称：叠榭·回环——基于空间重组理念的能量自循环低碳社区设计
Pavilion Loop—Design of Energy Self-Circulation Zero Carbon Community Based on the Concept of Spatial Reorganization

作　　者：殷一心、李思静、郑宣亮、徐雯芳
Authors：Yin Yixin, Li Sijing, Zheng Xuanliang, Xu Wenfang

参赛单位：福州大学
Participating Unit：Fuzhou University

指导教师：吴木生
Instructor：Wu Musheng

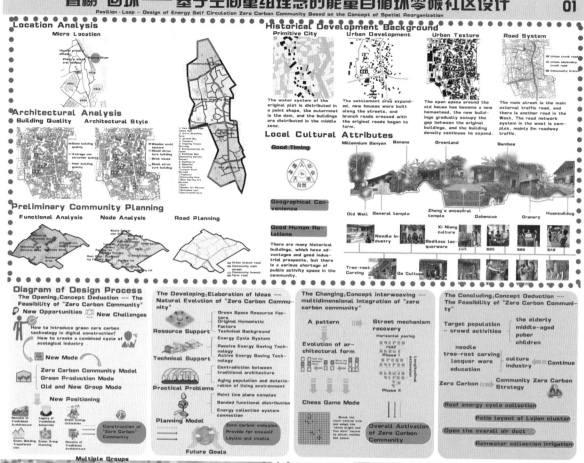

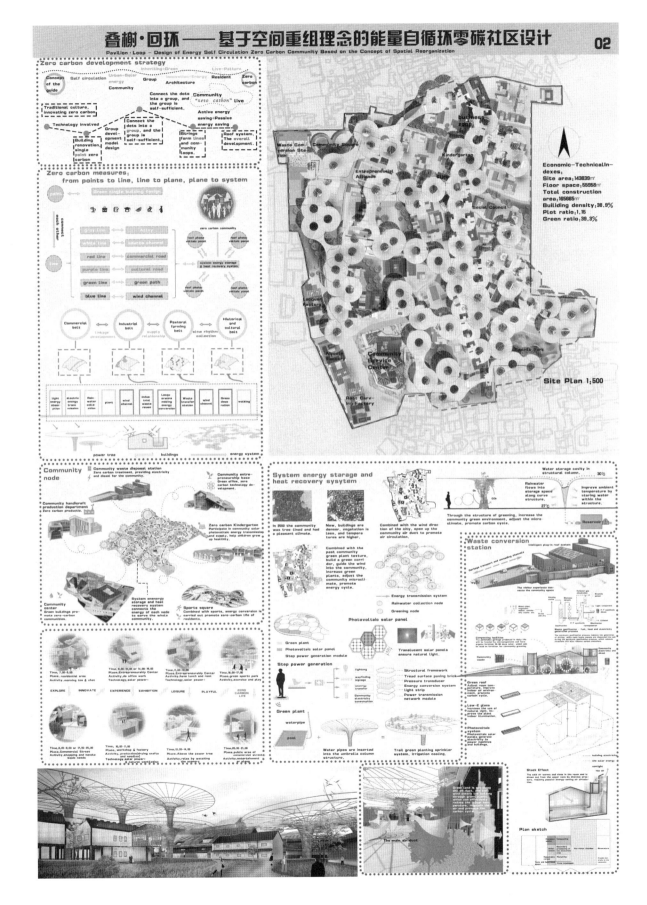

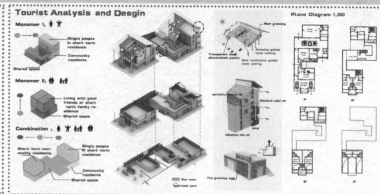
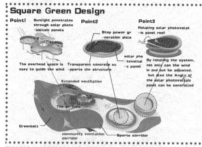

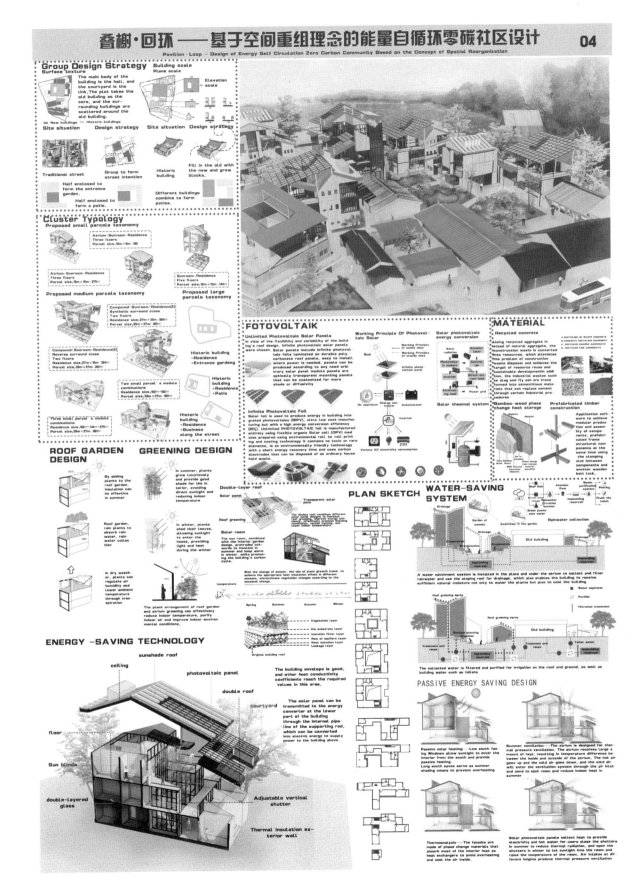

综合奖·优秀奖
General Prize Awarded·
Honorable Mention Prize

注 册 号：100463
Registration No：100463
项目名称：生长工厂
　　　　　Growing Factory
作　　者：黄灿光、张逸凡、魏　然、
　　　　　张华振
Authors：Huang Canguang, Zhang Yifan,
　　　　　Wei Ran，Zhang Huazhen
参赛单位：南京工业大学
Participating Unit：Nanjing University of
　　　　　Technology
指导教师：林杰文
Instructor：Lin Jiewen

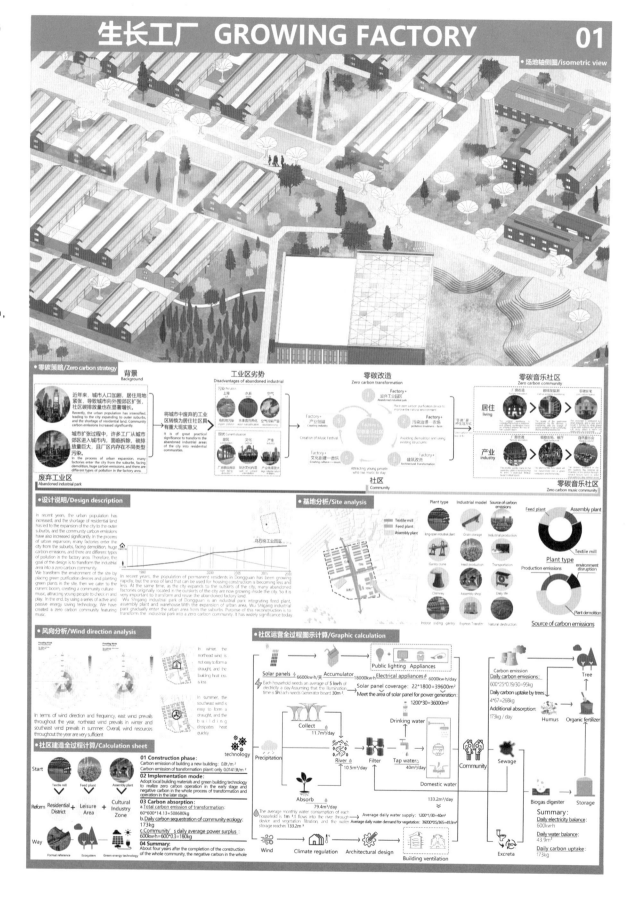

生长工厂 GROWING FACTORY　　02

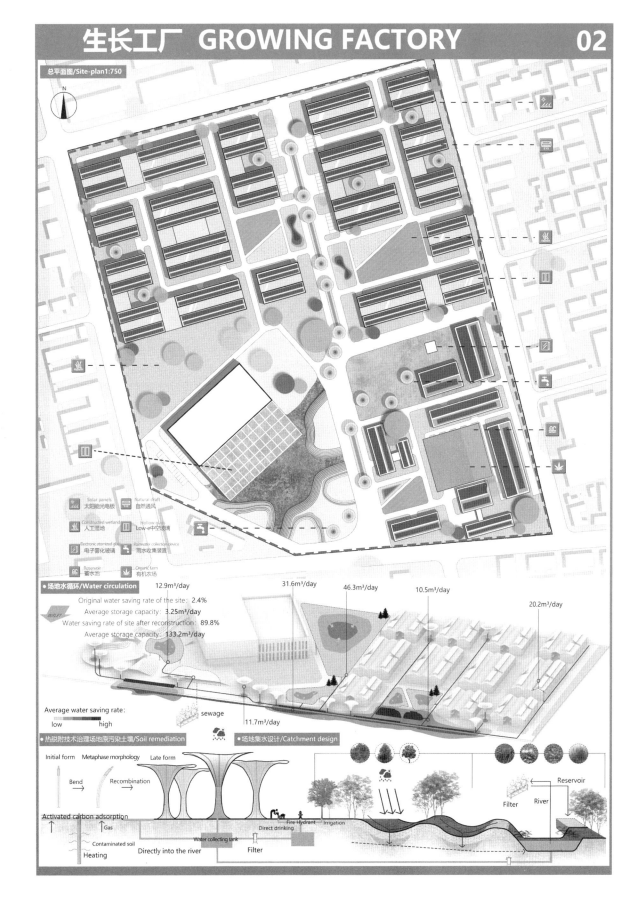

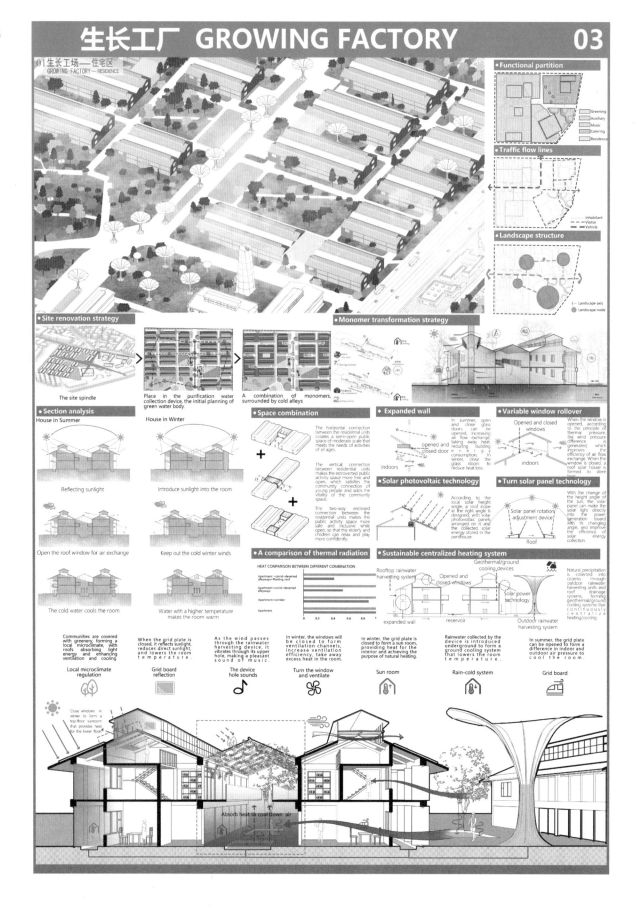

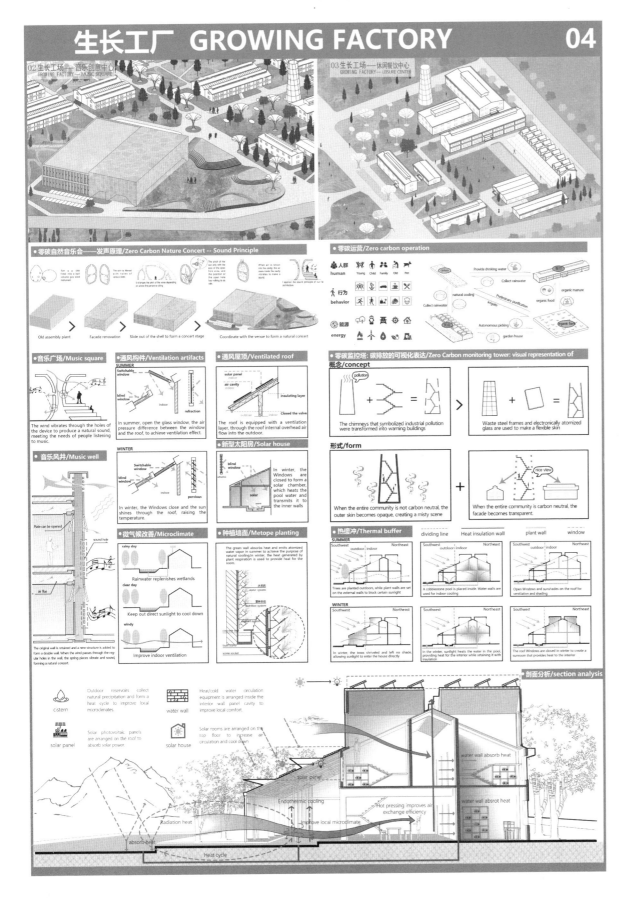

综合奖・优秀奖
General Prize Awarded・
Honorable Mention Prize

注 册 号：100480
Registration No：100480

项目名称：旧城新衣——基于岭南湿热气候的广州老旧小区零碳改造探索

Exploration of Zero Carbon Transformation of Old Communities in Guanghzhou Based on Lingnan'S Hot and Humid Climate

作　者：庄　霖、岑劲衡、倪立巧、王小瑜

Authors：Zhuang Lin, Cen Jinheng, Ni Liqiao, Wang Xiaoyu

参赛单位：华南理工大学

Participating Unit：South China University of Technology

指导教师：林正豪

Instructor：Lin Zhenghao

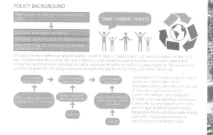

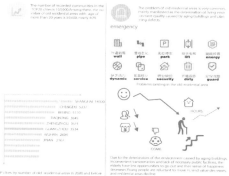

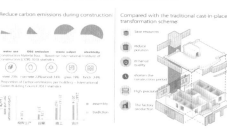

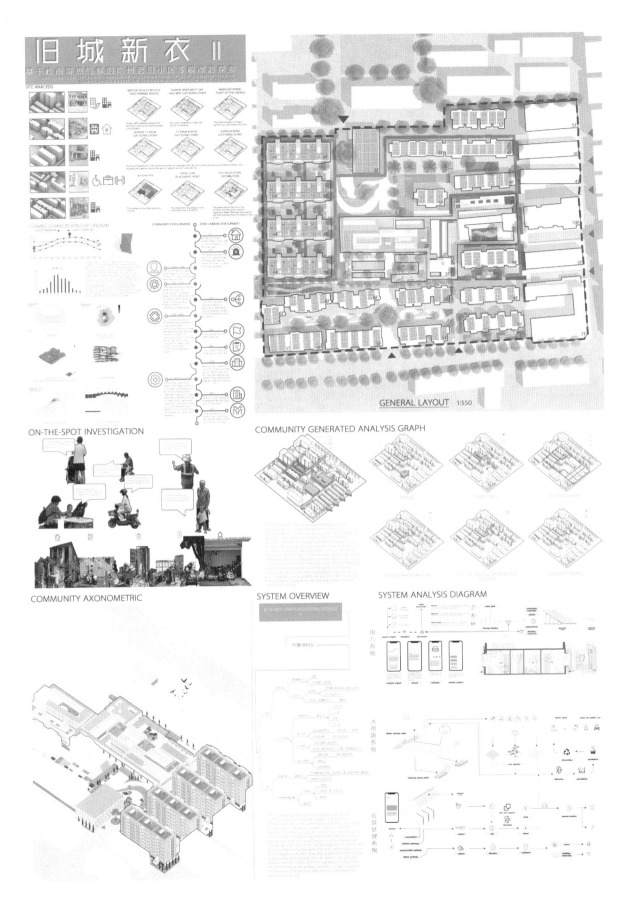

旧城新衣 III
基于岭南湿热气候的广州老旧小区零碳改造探索
EXPLORATION OF ZERO-CARBON TRANSFORMATION OF OLD COMMUNITIES IN GUANGZHOU BASED ON LINGNAN HOT AND HUMID CLIMATE

WET, HOT AND RAINY WEATHER IN GUANGZHOU

LANDSCAPE GREENING SYSTEM

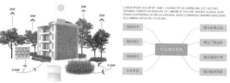

PLANT SELECTION

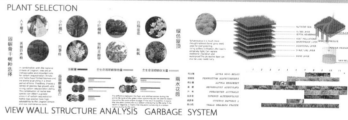

RAINWATER COLLECTION SYSTEM

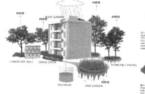

VIEW WALL STRUCTURE ANALYSIS   **GARBAGE SYSTEM**

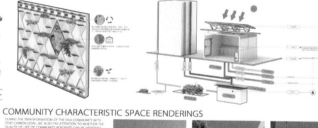

EXPLOSION ANALYSIS   **ACTIVITY ANALYSIS CHART**   **COMMUNITY CHARACTERISTIC SPACE RENDERINGS**

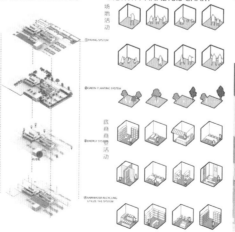

旧城新衣 IV
基于岭南湿热气候的广州老旧小区零碳改造探索
EXPLORATION OF ZERO-CARBON TRANSFORMATION OF OLD COMMUNITIES IN GUANGZHOU BASED ON LINGNAN HOT AND HUMID CLIMATE

ANALYSIS OF WALL INSULATION PERFORMANCE

energy saving & thermal insulation | sound proofing | water protection | exceptional results in renovation works | healthier environment | advanced

WALL EXPLODED VIEW
CLOSED EXPLODED VIEW
EXPLODED VIEW OF AIR INLET
VERTICAL GREENING ANALYSIS
WALL INSULATION
ROOF STRUCTURE

SCHEMATIC DIAGRAM OF PHOTOVOLTAIC PANEL SETUP
SCHEMATIC DIAGRAM OF ASSEMBLY PROCESS
BUILDING EXPLOSION ANALYSIS DIAGRAM

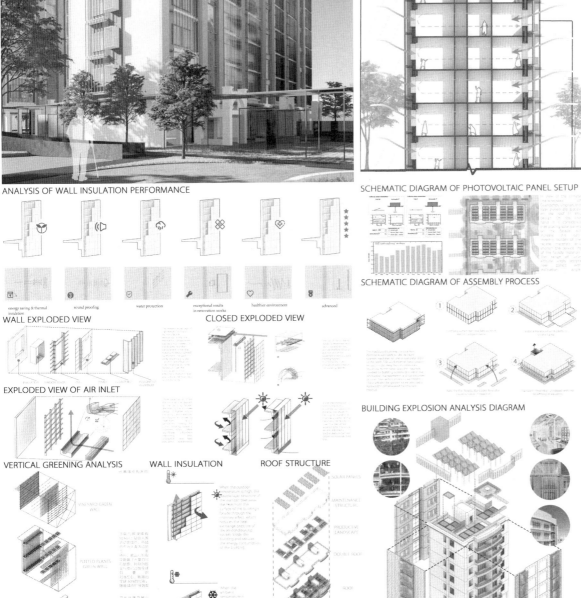

综合奖·优秀奖
General Prize Awarded · Honorable Mention Prize

注 册 号：100504
Registration No：100504
项目名称：何不归故里
        Why Not Return Home
作　　者：孙培竣、吕希萌
Authors：Sun Peijun, Lü Ximeng
参赛单位：山东科技大学
Participating Unit：Shandong University of Science and Technology
指导教师：王雅坤、郭清华、夏　斐
Instructors：Wang Yakun, Guo Qinghua, Xia Fei

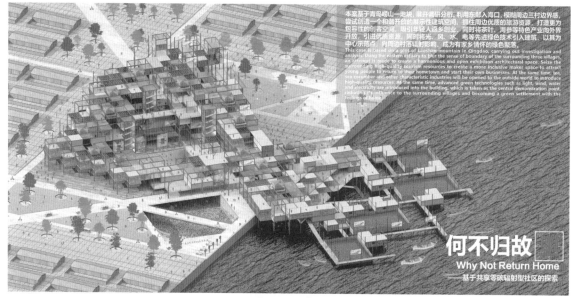

Pre-research

Conceptual Analysis

Four Sharing

01 The Community Share
During the field survey, it was found that the communication between the three surrounding villages was not smooth. The building also played a role in improving the intimacy and atmosphere of the block, and became a landmark to the outside world.

02 Residential Share
The residential area is for young people returning to their hometown and people who come to travel for a short period of time. Through the hollowing out of the central courtyard, the estrangement is broken and a warm shared space is created.

03 Business Share
In the eastern part of the building, commercial sharing space will be created, where emerging technologies, tea planting, aquaculture and other industries can tap value and create vitality, so that more young people can invest in the maker space.

04 Vertical Greening Sharing
The roof and the west of the building create vertical farms and vertical greening to improve the building's internal environment, while allowing people in the building to actively communicate and understand the green, healthy and ecological building operation system.

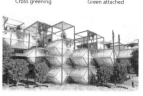

Elevation

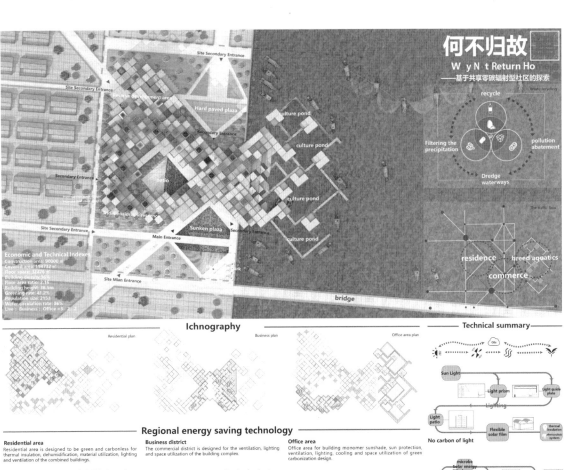

Ichnography

Residential plan | Business plan | Office area plan

Technical summary

Regional energy saving technology

Residential area
Residential area is designed to be green and carbonless for thermal insulation, dehumidification, material utilization, lighting and ventilation of the combined buildings.

Business district
The commercial district is designed for the ventilation, lighting and space utilization of the building complex.

Office area
Office area for building monomer sunshade, sun protection, ventilation, lighting, cooling and space utilization of green carbonization design.

The building's envelope utilizes local materials and adapts to local conditions
The 5mm glass cavity is filled with local ground granite and recycled waste from the river, which serves as the main frame of the 20mm inner wall. Its performance is similar to the concrete wall and can be dismantled.

Overhead and cut holes are set up in the commercial area to form grey space
On the one hand, it provides stalls for merchants to form a gathering point for people and create a local rural market atmosphere; On the other hand, strengthen internal ventilation and lighting, save energy.

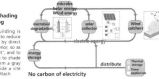

Roof grille for shading and cooling
The roof of the building is set with a grid grid to reduce the heat generated by direct overhead to the interior, so as to form a "loft effect", and to form a shadow area to shade roof activities, to form a gray space, and to provide a site for green plants to attach.

Use solar wall system instead of air conditioning and other equipment

Vertical traffic and wind traps are used for ventilation and lighting
When the wind speed above is large, the use of wind trap to drive the air flow; When the wind speed below is high, the air will flow upward using the principle of pulling out the wind, so that the living space is ventilated smoothly, and the light can illuminate into.

The outer corridor is enclosed to form the sun room
The solar room is insulated in winter, so that commercial activities can work normally under bad weather conditions. The space is arranged vertically, and the vertical space is utilized to save space and form a multi-dimensional shared business area.

Set louver grille for shading and ventilation
The west side of the louver grid, and can adjust the indoor ventilation and lighting, the blade has 0 – 105 degrees of turning Angle, can be adjusted at will, 90 degrees, can obtain the maximum ventilation effect, 15 – 25 degrees, can effectively prevent outdoor peep.

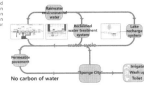

Cross-section drawn & water cycle

综合奖·优秀奖
General Prize Awarded · Honorable Mention Prize

注　册　号：100550
Registration No：100550

项目名称：阡陌交通
　　　　　Crisscross of Paths in Fields

作　　者：刘时羽、覃泷莹、陈香合、
　　　　　孙钰洁
Authors：Liu Shiyu, Tan Longying,
　　　　　Chen Xianghe, Sun Yujie

参赛单位：北京交通大学
Participating Unit：Beijing Jiaotong
　　　　　　　　　 University

指导教师：周艺南、董玉香
Instructors：Zhou Yinan, Dong Yuxiang

设计说明

场地位于北方某高校附近居住区境内，地处北三环以内寸土寸金的市中心位置。周边区域设施老旧，绿化稀少，居住空间拥挤，停车车位不足且占用公共活动空间。不仅如此，居民适应着城市快节奏的生活，无暇与他人、与自然享受片刻宁静。基于此，我们旨在给人们创造一个更加舒适的且供求平衡的零碳社区生活空间。第一大特色是阡陌交通的垂直立体式农园，便于太阳能光伏集热的同时也满足社区平衡的住水系统。多方位的流线丰富居民的休闲体验，让快节奏的都市人体验种植丰收的慢节奏生活。不仅如此，如果疫情突然来临，集中隔离时粮食储备也能满足居民一段时间内的刚需。
第二大特色是将伞亭作为景观置入小区，增加垂直绿化面积。白天乘凉避风供人休憩，晚上照明指引回家的道路。同时，伞亭的高可以产生瀑布，作为景观的同时也调节了周边的局部微气候。地块周边是千篇一律的住宅布局和冰冷的钢筋水泥。本设计在延续周边道路肌理的前提下，将各种有趣的元素融入，新居住区的布置延续了整体肌理，但是局部增加了新的有人情关怀的元素—农场伞亭、瀑布、数目、座椅、运动场等。让生活更加美好。

The site is located in a residential area near a university. The surrounding area has old facilities, scarce greenery, crowded living space, insufficient parking spaces, and occupying public space; besides, residents have adapted to the city's fast-paced life and have no time to enjoy a moment of tranquility with others and nature. So, we aim to create a comfortable, zero-carbon community living space with a balanced supply and demand for people.
The first significant feature is the vertical three-dimensional farm garden with terraces and moss traffic, which is convenient for solar photovoltaic heat collection and, at the same time, satisfies the community to balance the community water filtration system. The multi-directional streamlines enrich the leisure experience of residents, allowing fast-paced urbanites to experience the slow-paced life of planting and harvesting. Not only that, if the epidemic comes suddenly, food reserves during centralized isolation can also meet the daily needs of residents for a while.
The second prominent feature is that the umbrella pavilion is placed in the community as a landscape to increase the vertical green area. In the daytime, people can take shelter from the wind to rest, and at night, the lighting guides the way home.
At the same time, the height difference of the umbrella pavilion can produce waterfalls, which serve as a landscape and regulates the local microclimate of the surrounding area.
Around the site is the exact layout of houses and cold steel and concrete. On the premise of continuing the surrounding road texture, this design incorporates various exciting elements. The layout of the new residential area continues the overall texture but locally adds new humane and caring elements-farms, umbrella pavilions, waterfalls, numbers, Seats, sports fields, which make life better.

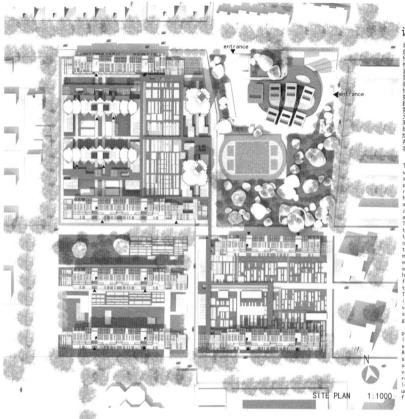

SITE PLAN 1:1000

Crisscross of Paths in Fields | 阡陌交通

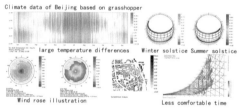

占地面积　　cover area: 90000m²
建筑基地面积 building area: 280226m²
总建筑面积　 overall floorage: 602975m²
容积率　　　 volume fraction: 0.67
建筑密度　　 building density: 31%
绿化率　　　 greening rate: 48%

Haidian district

Design Generation

A diverse community with a theme of Agriculture

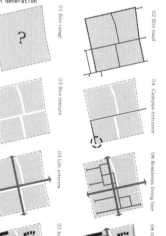

Crisscross of Paths in Fields | 阡陌交通

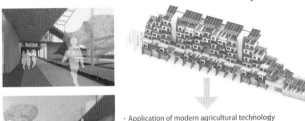
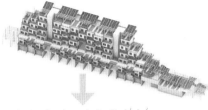
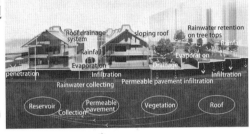
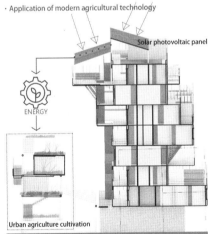
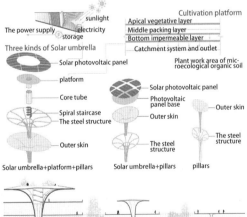

Agricultural garden rainwater collection and circulation system
Photovoltaic utilization system in agricultural garden

综合奖·优秀奖
General Prize Awarded·
Honorable Mention Prize

注 册 号：100563
Registration No: 100563

项目名称："厂·能"低碳社区
　　　　　Factory Energy

作　　者：刘和靖、吴达逊、罗雨晴、
　　　　　蔡晓莹
Authors: Liu Hejing, Wu Daxun,
　　　　　Luo Yuqing, Cai Xiaoying

参赛单位：广州大学
Participating Unit: Guangzhou University

指导教师：席明波、李　丽
Instructors: Xi Mingbo, Li Li

FACTORY·ENERGY
"厂·能"—零碳社区

Design Specification

设计说明：该"厂·能"零碳社区以厂房为改造对象，根据数据模拟等手段将场地优化处理。我们保留了部分厂房并将其移至中心区域，从而打造一个以工业文化核心区为中心的综合多元化零碳居住社区。

居住单体以6000mm×6000mm为桁架框体模块，加以独立的居住模块，形成一个产能的、学习式的工业、居住综合零碳社区。在技术方面，我们充分运用了太阳能光伏板、种植绿化、"海绵城市"式水系循环体系、立面与桁架相互结合设计。最后再将主动式、被动式的太阳能板与其他技术结合起来。

The "Factory·Energy" zero carbon community design takes the factory buildings as the reconstruction objects. Optimize the site according to data simulation and other measures. We retained part of the factory buildings and moved them to the central area in order to create a comprehensive and diversified zero carbon residential community centered on the core area of industrial culture.

The residential monomer takes 6000mm6000mm as truss frame module, adding individual residential blocks. Thus, an energy-producing, learning-type industrial residential comprehensive zero carbon community will be formed. In terms of technology, we fully use the solar photovoltaic panels、planting greenery、"Sponge City" water circulation system. The facade is tightly integrated with the trusses. Finally we combined active and passive solar energy using with other use of technology.

FACTORY·ENERGY
"厂·能"—零碳社区

ECONOMIC & TECHNICAL INDEXES:
Site area: 90000 m²
Floor space: 128000 m²
Plot ratio: 1.42
Green ratio: 68%

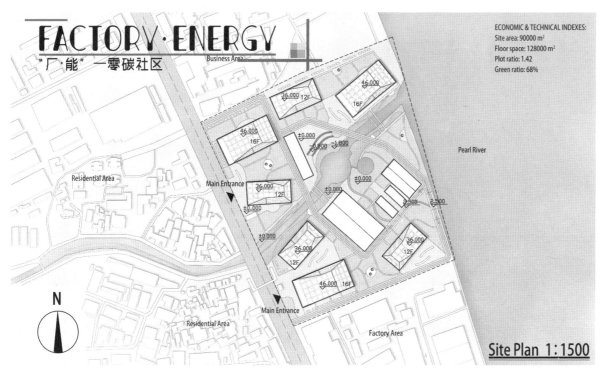

Site Plan 1:1500

Site Analysis

The site is located in Liwan District, Guangzhou, with convenient transportation. Besides, the land we selected is crossed by a river. To its east is the Pearl River. Considering the needs of zero carbon communities, we retained part of the original factory buidings, moving them to the center of the site in order to create an industrial and Cultural Zore.

The site has a subtropical monsoon climate, rich in light and heat resources, short winters and long summers, pleasant climate, smooth winds, and concentrated rainy seasons. According to the data provided by the Meteorological Bureau, the annual average temperature is 21.9℃, the average annual rainfall volume is 1806 mm, the annual sunshine hours average are 1608 hours and the average annual relative humidity is 77%.

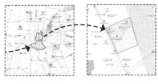

Liwan, Guangzhou   Industrial Park, Liwan

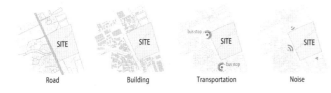

Road   Building   Transportation   Noise

Diagram of Design Strategy

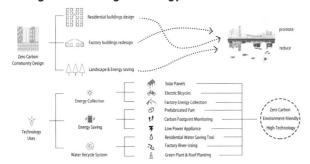

Climatic Analysis

According to software analysis and simulation, We chose to combine "Ladybug" and "Ecotect" to visualize the data. Based on the weather data of Guangzhou, this analysis analyzes the climate of the site from multiple angles.

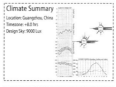
Climate Summary
Location: Guangzhou, China
Timezone: +8.0 hrs
Design Sky: 9000 Lux

Prevailing Winds
Wind Frequency (Hrs)
Location: Guangzhou, China
Date: 1st January-31st December
Time: 00:00-24:00

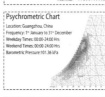
Psychrometric Chart
Location: Guangzhou, China
Frequency: 1st January to 31st December
Weekday Times: 00:00-24:00 Hrs
Weekend Times: 00:00-24:00 Hrs
Barometric Pressure:101.36 kPa

Stereographic Diagram
Location: Guangzhou, China
Time: 12:00
Date: 1st April
Dotted lines: July-December
Sun Position: 159.1°, 70.7°
HSA: 159.1° VSA: 108.1°

Weekly Summary
Average Temperature
Location: Guangzhou, China

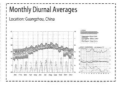
Monthly Diurnal Averages
Location: Guangzhou, China

Hourly Temperatures
All Visible Thermal Zones
(Values shown are environments, not air temperatures.)

Hourly Solar Exposure
Location: Guangzhou, China
Date: Monday, 1st January

FACTORY·ENERGY
"厂·能"—零碳社区

Body Generation

Select the existing plant size of 20m*80m*6m in the site as the base modulus to superposition the building volume

Cut the block according to the wind direction, while reducing the sense of clumsiness

Buildings echo each other, protecting surrounded by residential buildings, highlighting industrial and cultural areas

Single Spatial Structure

Keep the plant truss structure

Put in the prefabricated house type unit

Green plants grown in the truss produce oxygen through photosynthesis while filtering sunlight

Prefabricated Wood Structure

- First framework structure
- Forth framework structure
- Structure base
- Third framework structure
- Fill the cavity
- Second framework structure
- Fill the structure

- Assembly bit

The structural frame comprises interconnected first, second, third and fourth frame structures surrounding each other around a filling cavity for setting a filling structure

simple, stable, low cost
The structural connection ends are all inclined, fixed by inclined welding,

Prevent high temperature and fire prevention
Heat insulation, sound insulation
The upper and lower layer filling structure design

Renewable, and sustainable
After removal is easy to be handled, can be recycled, no need to landfill to occupy the cultivated land

First Floor Plan 1:350

Population Analysis

The elderly accounted for relatively few, the industrial zone environment is not suitable for pension
The community has only a small number of residential units for the elderly

Young and middle-aged people account for the largest proportion, most of them are office workers, married families, and single young people
Independent office space is needed

Primary and secondary school students account for a moderate proportion, because there are well-known primary and secondary schools around the area
The community should have a proper family of three

Population forecast:
According to the previous plan, there may be more young entrepreneurs in the future, and backpackers will come to the area in the future

Household Social Type

Live alone
Far spacing between cells. There are separate outdoor balconies. Not shared with others

Children social type
There is a small tree house between the units to provide interesting activity space for children

Pension social type
With a separate balcony and a public communication space, the elderly can choose quiet or lively

Work social type
Independent large balcony can be used as a new outdoor office space, where neighbors can see and communicate with each other

Farming social type
There is a common garden between the units. There can be increased neighborhood communication opportunities during agricultural activities

Entrepreneurship social type
Provide an interesting activity exchange platform for a group of young people

FACTORY·ENERGY
"厂·能"—零碳社区

◻ Solar Illumination Analysis

It can be seen from the different lighting conditions in specific months that the layout of buildings in the residential area meets the lighting requirements and the interaction between them is small.

Insolation Analysis:
Average Daily PAR
Value Range: 0.60-5.60 MJ/md

Insolation Analysis:
Total Sunlight Hours
Value Range: 0.0-10.0 Hrs

Insolation Analysis:
Average Daily PAR
Value Range: 0.80-7.60 MJ/md

Insolation Analysis:
Average Daily PAR
Value Range: 0.70-5.80 MJ/md

◻ Ventilation Analysis

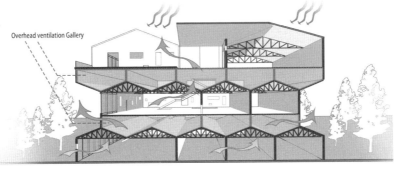

Overhead ventilation Gallery

◻ Plant reconstruction Analysis

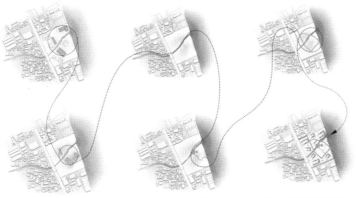

◻ Drainage Analysis

The height of the central plant area of the site is lower than the surrounding area, and the river is expanded in the center to form a transition pool area. When the rain is heavy, the water in the site will flow to the middle, converge to the pool area, and then drain to the river through the river.

The plant at the north corner of the site was demolished to the plot along the river and used to form an industrial and cultural center with the other two reserved buildings

Introduce the river flow into the site, on the one hand, the sponge city water resources management will be implemented in the residential areas. On the other hand, the biological energy in the river

In terms of traffic, the main road is set up around the industrial and cultural center to form a smooth visit streamline

◻ Plant Elevation

Northeast Elevation 1:300

Northeast Elevation 1:300

综合奖·优秀奖
General Prize Awarded·Honorable Mention Prize

注　册　号：100611
Registration No：100611

项目名称：告别孤独的低碳社区——基于"共享"理念的定制化社区
Join the We Share Say Goodbye to the Loneliness—Customized Community Based on "Sharing" Concept

作　　者：聂大为、杨喆雨
Authors：Nie Dawei, Yang Zheyu

参赛单位：华东建筑设计研究院总院、同济大学建筑设计研究院（集团）有限公司
Participating Units：East China Architectural Design and Research Institute General Institute, Tongji University Architectural Design and Research Institute (Group) Co., Ltd.

指导教师：陈　易
Instructor：Chen Yi

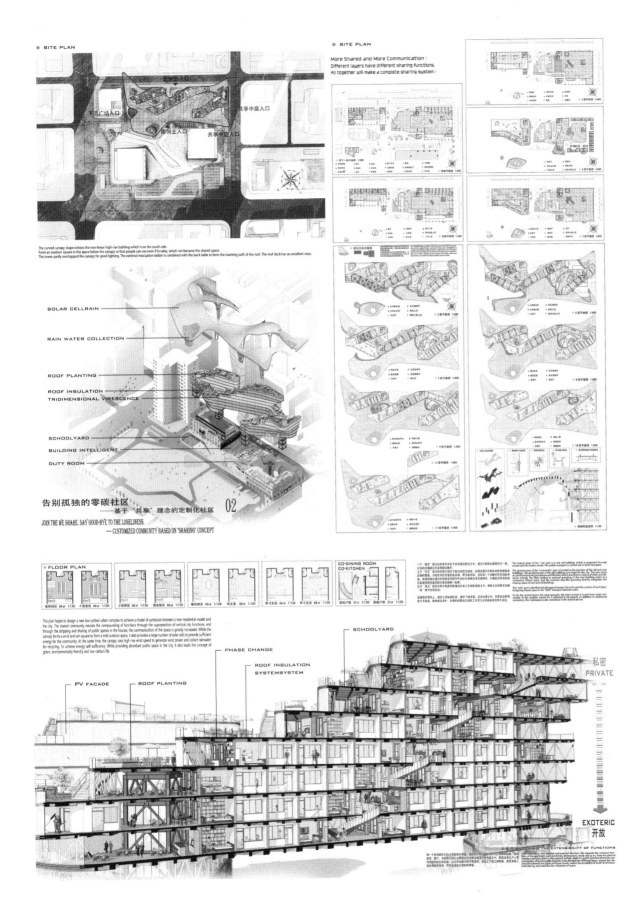

● **AXONOMETRIC DRAWING**

①于"城市"部分的共享空间处于旧有建筑部分之中,通过中庭和边庭组织在一起,公共的交通部分在这两部分展开;
②于"社区"部分的共享空间位于新旧建筑交接处,旧建筑部分外面布对城市的商业店铺覆盖,内建作为社区级的键身房、图书馆存在,也形成一个动静分区的功能布局。新建筑部分通向外部绿化的前厅作为社区级的共享空间存在,外部的台阶状绿化直接将新老建筑的共享空间统一起来;
③于"单元"部分分布于每层的转角处和单元之间的缝隙之中,将单元之间挤压出的"场"赋予共享空间;

在建筑的竖向上,越往上功能越私密,越往下越开放。在综合体之中,共享生活体现在方方面面,除传统商业外,商铺的经营也以创客工作室为主导结合在共享中庭边。

The shared space of the "city" part is in the historical building, which is organized by atrium and the border atrium. The public transport is carried out in these two parts.

The shared space of the "community" part is at the junction of the historical and present buildings. The peripheral part of the old buildings is covered by shops for the city. The core exists as community-level gymnasiums and libraries which also forms a layout of static and dynamic zoning. The lobby leading to external greening of the new building exists as a communal shared space. And the external step-like greening directly integrates the sharing space of new and old.

The "unit" part is distributed in the gaps between the units and the corners of each layer. Assigning shared space to the "field" extruded between units.

As for the vertical layout, the more upwards, the more private it is, and more open conversely. In the complex, shared life is reflected in all aspects. In addition to traditional commerce, the hackerspace is also combined in the shared atrium.

This zero-carbon community hopes to achieve a combination of low-carbon energy saving and symbiotic comfort. On the premise of creating a shared space that satisfies living comfort and enhances communication and exchanges, it will realize green, energy-saving, low-carbon and environmental protection. The building plans to renovate the Shengli Warehouse in Hankou District of Wuhan to realize the renewal and utilization of the city.

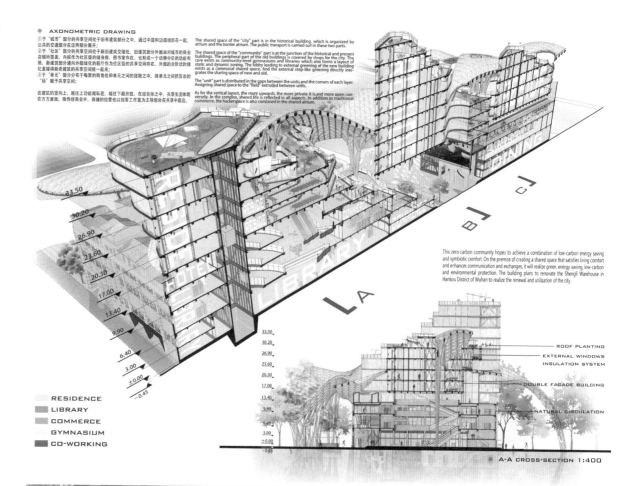

RESIDENCE
LIBRARY
COMMERCE
GYMNASIUM
CO-WORKING

ROOF PLANTING
EXTERNAL WINDOWS
INSULATION SYSTEM
DOUBLE FACADE BUILDING
NATURAL CIRCULATION

A-A CROSS-SECTION 1:400

告别孤独的零碳社区
——基于"共享"理念的定制化社区
JOIN THE WE SHARE, SAY GOOD-BYE TO THE LONELINESS
—CUSTOMIZED COMMUNITY BASED ON "SHARING" CONCEPT

告别孤独的零碳社区
——基于"共享"理念的定制化社区
JOIN THE WE SHARE, SAY GOOD-BYE TO THE LONELINESS.
—— CUSTOMIZED COMMUNITY BASED ON SHARING CONCEPT

GREEN TECHNOLOGY VERIFICATION ABOUT WIND

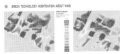 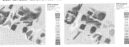   

When the canopy was not added, the wind speed was between 2.4m/s and 4m/s, and normal ventilation could not be achieved. The wind speed can be controlled between 0.3m/s and 1.2m/s after added the rain shelter, which is suitable for normal ventilation.

When the canopy is not added, the wind speed at the ground pedestrian level is between 1m/s and 2.4m/s, and pedestrians will feel uncomfortable. The wind environment of pedestrian height on the ground is improved obviously in winter after adding outdoor rain shelter. The wind speed in the simulated ground layer can be roughly controlled between 0.3m/s and 0.5m/s.

The vertical wind directly blows the ground, which has a great impact on pedestrians. After adding the building to the outdoor canopy, the vertical wind under the canopy will have less impact, thus creating a more ideal outdoor space.

When the canopy is not added, the wind speed of the roof layer of the old building is between 1.5m/s and 2.7m/s, and the tourists will feel uncomfortable. The wind environment in winter was improved after added the outdoor rain shelter.

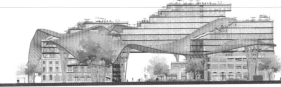

● South Elevation 1:300

● West Elevation 1:300

● East Elevation 1:300   ● North Elevation 1:500

This zero-carbon community hopes to achieve a combination of low-carbon energy saving and symbiotic comfort. On the premise of creating a shared space that satisfies living comfort and enhances communication and exchanges, it will realize green, energy-saving, low-carbon and environmental protection. The building plans to renovate the Shengli Warehouse in Hankou District of Wuhan to realize the renewal and utilization of the city. This plan hopes to design a new low-carbon urban complex to achieve a model of symbiosis between a new residential model and the city. The shared community realizes the compounding of functions through the superposition of vertical city functions, and through the stripping and sharing of public spaces in the houses, the communication of the space is greatly increased. While the canopy forms a wind and rain square to form a mild outdoor space, it also provides a large number of solar cells to provide sufficient energy for the community. At the same time, the canopy uses high-rise wind speed to generate wind power and collect rainwater for recycling. To achieve energy self-sufficiency. While providing abundant public space in the city, it also leads the concept of green, environmentally friendly and low-carbon life.

综合奖·优秀奖
General Prize Awarded·
Honorable Mention Prize

注 册 号：100632
Registration No：100632
项目名称："芯"院里——基于碳中和视角下的社区设计
Ismart Community
作　者：王子璇、张 敏
Authors：Wang Zixuan, Zhang Min
参赛单位：青岛理工大学
Participating Unit：Qingdao University of Technology
指导教师：李岩学、高伟俊、张洪恩
Instructors：Li Yanxue, Gao Weijun, Zhang Hongen

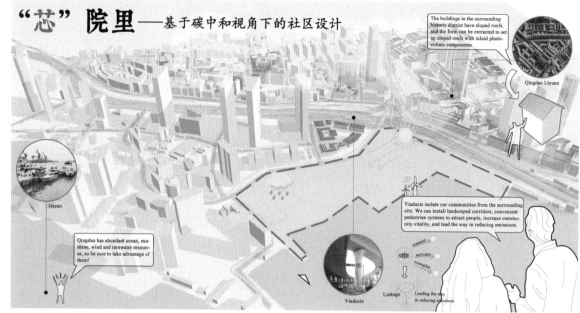

▋Site Analysis

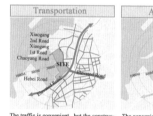

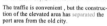

The traffic is convenient, but the construction of the elevated area has separated the port area from the old city.

The convenience of public transportation is conducive to low-carbon travel for community residents.

The convenience of public transportation is conducive to low-carbon travel for community residents.

▋Climate Simulation

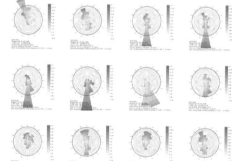

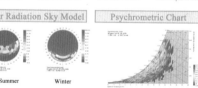

▋Culture & Features

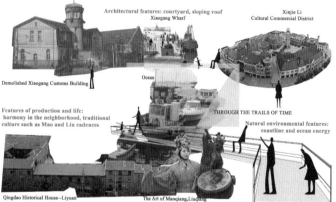

A brand new corridor is used to connect the broken stream of history, carrying memories and passing on culture through the combination of new ideas and old material forms.

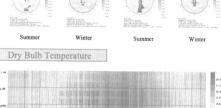

"芯"院里——基于碳中和视角下的社区设计

Design Background

Current urban operations already account for about 75% of global greenhouse gas emissions and create 80% of the world's pollution. The building sector produces about 40% of the world's CO_2 each year, with concrete accounting for 6%–11% and steel for 10%, so we are addressing this phenomenon by designing zero carbon communities.

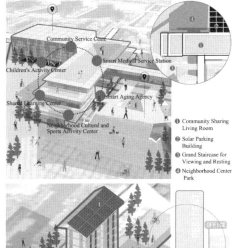

1. Community Sharing Living Room
2. Solar Parking Building
3. Grand Staircase for Viewing and Resting
4. Neighborhood Center Park

Vibrant Community Ring
1. Solar panels
2. Shared Office
3. Rooftop Garden
4. 7.5m ground
5. Coastal Business

Isometric View

设计说明
Design Description

本地块位于青岛市小港港口区域，方案在提取青岛文化属性的基础上，充分利用青岛自然资源。在规划层面，方案明确街区开放、组团封闭的空间组织模式，多功能融合，打造24小时活力社区圈。在技术层面，考虑到过度依赖光伏将造成电力供需关系失衡，由于太阳能有季节性和时间差距，26%的住宅能源需求将依赖非电池储能。因此本设计方案提出一种低碳、预制装配式重型木结构滨海住宅，同时配有新型集成太阳能发电系统。该系统包括光伏热（PV/T）面板组、光电化学（PEC）反应器组、热泵、吸收冷却系统（ACS）、质子交换膜（PEM）燃料电池。在白天利用太阳能提供电能和热能，PEC反应堆利用太阳能和PV/T的电力从水中产生氢气，在PEM燃料电池中用于夜间发电。冬季所需的加热和夏季所需的冷却分别通过热泵机组和吸收式制冷机提供。最后，所有能源系统的调控将依赖人工智能技术进行滨海社区智能调控，实现住宅区的零碳目标。

This site is located in the port area of Xiaogang, Qingdao, and the scheme makes full use of Qingdao's natural resources on the basis of extracting the cultural attributes of Qingdao. At the planning level, the scheme specifies the spatial organization model of open neighborhoods and closed clusters, with multi-functional integration to create a 24-hour vibrant community circle. At the technical level, considering that over-reliance on PV will cause an imbalance between electricity supply and demand, 26% of residential energy demand will rely on non-battery energy storage due to the seasonality and time gap of solar energy. This design proposal therefore proposes a low-carbon, prefabricated, heavy-duty wood-frame coastal residence with a new integrated solar power system. The system includes a photovoltaic thermal (PV/T) panel set, photovoltaic chemical (PEC) reactor set, heat pump, absorption cooling system (ACS), and proton exchange membrane (PEM) fuel cell. Solar energy is used to provide electricity and heat during the day. the PEC reactor uses solar and PV/T electricity to produce hydrogen from water in the PEM fuel cell for nighttime electricity generation. The heating required in winter and the cooling required in summer are provided by heat pump units and absorption cooling units, respectively. Finally, the regulation of all energy systems will rely on artificial intelligence technology for intelligent regulation of the coastal community to achieve a zero carbon goal for the residential area.

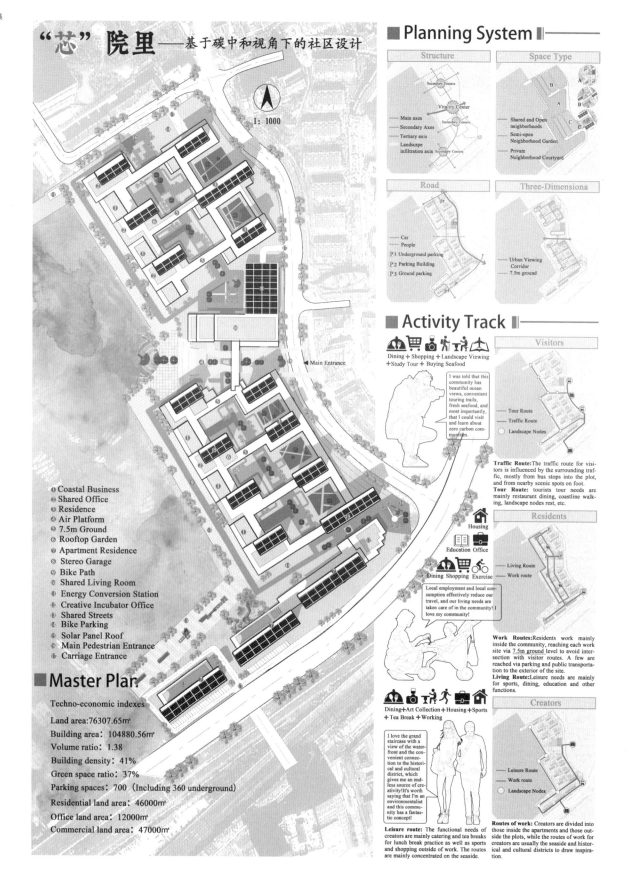

"芯"院里——基地碳中和视角下的社区设计

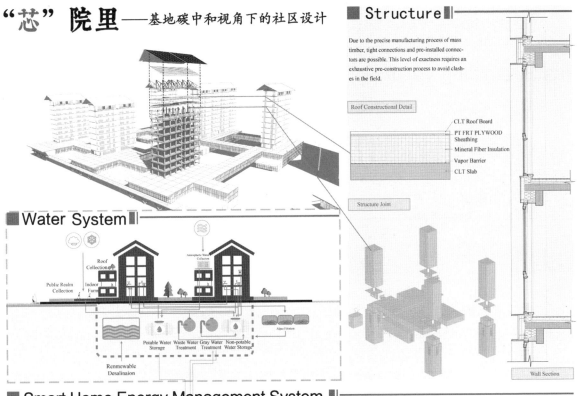

Artificial Intelligence Technology in Energy System

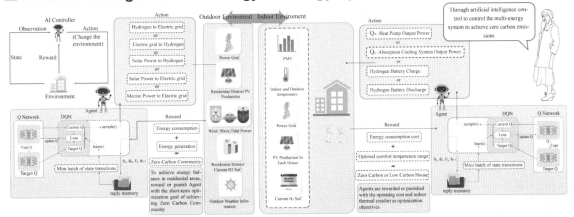

有效作品参赛团队名单
Name List of All Participants Submitting Valid Works

注册号	作者	单位名称	指导人	单位名称
100003	李亦枫、赵钰铭、高鸣飞、纪硕、唐润乾、谢文驰、王明宇、陈楠	天津大学、合肥工业大学、东南大学	冯刚、李早	天津大学、合肥工业大学
100015	武玉艳、高新萍、杨晴、巩博源、张莹	西安建筑科技大学	陈景衡	西安建筑科技大学
100046	武昕迪、夏仰星、李健豪	华东交通大学	彭小云	华东交通大学
100054	郭嘉钰、张玉琪、王松瑞	北京建筑大学	俞天琦、丁光辉	北京建筑大学
100061	陈春桦、邓忠鹏、梁迪生	广东工业大学	李超	广东工业大学
100069	马远林、阎丽伊、王慎薇、刘佳、张偌涵	石家庄铁道大学	高力强	石家庄铁道大学
100072	赵时、张玉琪、郭嘉钰、周文宇	天津大学、河北工程大学、哈尔滨工业大学	刘家韦华、张伟亮、侯万钧	中粮华商国际工程有限公司、河北工程大学
100099	江雨涵、查婉宁、杨乐	福州大学	邱文明	福州大学
100100	李婧、李玉盟、齐琪、蒋东霖、李敬良、桑弘翼	中国矿业大学（北京）	李晓丹	中国矿业大学（北京）
100118	张嘉倩、曾雅清、林凯思	福州大学	王炜、赵丽珍	福州大学
100120	方鸣、刘晓者	广东博意建筑设计院有限公司		
100129	郝旌潮、胡梦娇	河南工业大学	马静	河南工业大学
100134	王贝琳、蒋仁杰、梁璧选、冯宇欢	南京工业大学	郭兰	南京工业大学
100140	韩楚玉、王璐	河南工业大学	张华	河南工业大学
100145	林欣怡	三明学院	王琦	三明学院
100148	余宜冰、张颖晖、周益萍	福州大学	邱文明	福州大学
100159	柳雯瀚、范书玮、范吉彬、徐翔	山东建筑大学	薛一冰、管振忠	山东建筑大学
100166	许利、胡文韬	兰州交通大学	高发文	兰州交通大学
100172	徐明哲	间作设计		
100222	甄琪、杜一鸣	天津大学	张安晓	天津大学
100232	李金举、雷毕江、李增、覃燕玲、梁永琦	广西科技大学	叶雁冰	广西科技大学
100235	王晨雨、苗杰、陈骥、孙治宇	南阳理工学院	赵敬辛、高力强	南阳理工学院、石家庄铁道大学
100237	王彬屹、黄頔、郭世鹏、李浩铭、李自钊	广西科技大学	叶雁冰	广西科技大学
100238	王珂、黄志铮、张朔、朱紫悦	新疆大学	塞尔江·哈力克、艾斯卡尔·模拉克	新疆大学
100244	王晓艳、李少轩、张吉森	西北工业大学	王晋、邵腾	西北工业大学

续表

注册号	作者	单位名称	指导人	单位名称
100261	叶鑫、宋尧、杨赞峰、李铭泉	广西科技大学	叶雁冰	广西科技大学
100263	乔昕蕊	郑州轻工业大学	李晓阳	郑州轻工业大学
100265	李晓鸥、郎一郎、闫孟委	中国矿业大学（北京）	郑利军	中国矿业大学（北京）
100270	罗峻杉、刘玲玲、陆玉婷、黎惠	广西科技大学	叶雁冰	广西科技大学
100276	张浩然、刘梓涵	昆明理工大学	谭良斌	昆明理工大学
100288	李秋儒、张岩、陈文霜、杨晋	西安建筑科技大学	何泉	西安建筑科技大学
100291	王远航、席斌、蔡天舒、潘鹏宇	南京工业大学	薛春霖、罗靖	南京工业大学
100294	孙佳奇、杨向婷、黄先奇	重庆大学	张海滨、周铁军	重庆大学
100295	许其弘、丁明琦、张涵琪、崔娅茹、程明、童晓泉	同济大学	宋德萱、叶海	同济大学
100311	桂熙、杨雅雯、杨玲钰、蔺富荣	广西科技大学	叶雁冰	广西科技大学
100320	丁阳权、冯磊磊、龙泓昊、冉大林	重庆大学	周铁军、张海滨	重庆大学
100325	张磊、张芷宁、郭振辉、史皓冉、杨逸轩	吉林建筑大学	常悦	吉林建筑大学
100328	程凯、毛长远、刘佳妹	中国矿业大学（北京）	贺丽洁	中国矿业大学（北京）
100329	席巧、龙伟凡、段跃臣、许文亚	青岛理工大学	马青松	青岛理工大学
100330	张骁、李皓妍、郭昕晨	北京交通大学	杜晓辉、胡映东	北京交通大学
100331	黄富庚、张承博、常海佳、徐一丹	厦门大学	石峰	厦门大学
100335	李彩琼、谢明轩	昆明理工大学	李莉萍	昆明理工大学
100336	李舒颖、张小雨、蔡诺、蔡润栩	西安科技大学	孙倩倩	西安科技大学
100339	田若入、宋蔚青、司圆梦、刘淑娟、刘尚阳、马昊、李秀秀	潍坊科技学院	周超、武海泉	潍坊科技学院
100340	安馨悦、郑园梦	河南工业大学	张华	河南工业大学
100349	甘一新、金朝阳、赵轩	中国矿业大学（北京）	郑利军	中国矿业大学（北京）
100352	张子裕、许碧康、宁娇荣、蔡雨辰	南京工业大学	胡振宇、王一丁、薛春霖	南京工业大学
100354	徐昕宇、李玥、王梦婷、夏斌	南京工业大学	叶起瑾	南京工业大学
100358	杜舰、姜雨、邓立瑞、赵雨濛、林子琪、葛祎明	东南大学	沈宇驰、王伟	东南大学
100367	王珍珍、田蓓茹、陈团、郭大民	安阳工学院	张仲军	安阳工学院
100371	王妍淇、张雯、凌嘉敏	重庆大学	周铁军、张海滨	重庆大学
100376	杨新哲、姚其郁	河南工业大学	张华	河南工业大学

续表

注册号	作者	单位名称	指导人	单位名称
100377	范凯峰、王逸甫	河北大学	李纪伟	河北大学
100382	陈彦儒、陈潇语、李匀思	福州大学	邱文明	福州大学
100385	刘人龙	宇朔建筑设计咨询（上海）有限公司	张椀晴	宇朔建筑设计咨询（上海）有限公司
100388	刘锐捷、郑仲意、杨雪娴	山东建筑大学	房涛	山东建筑大学
100389	孙之桐、肖瑶、沈冰珏、励姿玮	南京工业大学	薛洁、刘强	南京工业大学
100390	尚超、常志韦、田春兰、宋之涵	南阳理工学院	赵敬辛、高力强	南阳理工学院、石家庄铁道大学
100391	唐梦	西藏大学	索朗白姆	西藏大学
100394	胡昕月、郭承跞、句泽宇	西安建筑科技大学	罗智星、朱新荣、刘月超	西安建筑科技大学
100396	唐中博、廖洁、伍泓杰	北京交通大学	张文、王鑫	北京交通大学
100400	杨华婷、吴相礼、顾迦艺、字月婷	南京工业大学	林杰文	南京工业大学
100405	宋优、杨颖汐、奚秩华	苏州科技大学	刘长春、陈守恭、高姗	苏州科技大学
100407	冯慧玉、宋俊慷、李砖砖、刘西南	新疆大学	滕树勤、王万江、潘永刚	新疆大学
100410	张宇、陈勃翰、唐存彬	中国矿业大学（北京）	冯萤雪	中国矿业大学（北京）
100412	李若天、李金洋、金在纯、邹庆泽	苏州科技大学	刘长春、陈守恭、高姗	苏州科技大学
100414	朱耘业、赵永青	湖南大学	田真	湖南大学
100424	严凡辉、高兴、唐超、邓棋	河南工业大学	刘强	河南工业大学
100432	余庭瑶、赵文杰、周晴、嵇昕睿	南京工业大学	刘强	南京工业大学
100436	杨瑞航、曾俊鸿、蒋铁	重庆大学	张海滨、周铁军	重庆大学
100439	赵倩文	北京建筑大学	欧阳文	北京建筑大学
100451	苏泽勇、唐沛、张朝臣	深圳大学	夏珩	深圳大学
100453	康丽颖、杲宏月、何良耀、宋慧华、李程	西藏大学、西藏昂彼特堡能源科技有限公司	索朗白姆、魏琴	西藏大学
100454	郭士宸、晁龙、姬文哲	石家庄铁道大学	高力强	石家庄铁道大学
100464	胡思怡、曹茜、朗潇月	昆明理工大学	谭良斌	昆明理工大学
100465	吴祥泽、严军涛、林秋洁	福州大学	崔育新	福州大学
100467	吴浩、赵思佳、于越	北京交通大学	杜晓辉、胡映东	北京交通大学
100476	李世萍、李舒祺、卓金明、许浩川、周超超、苏湘茗	西安建筑科技大学	李涛、张斌、朱新荣	西安建筑科技大学、西耶建筑设计工作室

续表

注册号	作者	单位名称	指导人	单位名称
100478	张一鸣、杨若玲、奚秩华、邹庆泽	苏州科技大学	刘长春、陈守恭、高姗	苏州科技大学
100479	李文强、张一鸣、杨敏敏、谭景柏、张入川	西北工业大学	刘煜	西北工业大学
100485	田彬、吴凡、张栩锋	重庆大学	周铁军、张海滨	重庆大学
100487	宋家雪、牛延延、王纬乾、王勇超	西安建筑科技大学	朱新荣、李涛、罗智星	西安建筑科技大学
100489	孙一宸、朱潇凡、孙雅萌、薛浩然、周浩、王文鑫	河北建筑工程学院	武欣	河北建筑工程学院
100491	杨乐融、祝一佳、聂凯	福州大学	邱文明	福州大学
100498	孙思雨、邓凯元、刘乃歌、李祐平、张淑婧	吉林建筑大学	赫双龄、周春艳	吉林建筑大学
100506	李睿晨、何嘉琦、谢冲、冉铖李、王文轩	重庆大学	张海滨、李臻赜	重庆大学
100514	罗雪松、王珍	河南工业大学	张华	河南工业大学
100520	李昊辰、芦天怡、王运生、金雨晨	南京工业大学	刘强	南京工业大学
100534	詹进、高雪梅、王艳廷、霍然	山东建筑大学	何文晶、杨倩苗	山东建筑大学
100544	陈欣怡、金天、刘艾佳、祁思语	南京工业大学	薛春霖	南京工业大学
100555	白苏日吐	天津大学	朱丽	天津大学
100560	曹瑞	西北工业大学	刘煜	西北工业大学
100569	高博、朱可迪、罗犇	重庆大学	张海滨、周铁军	重庆大学
100574	兰良建、梁今浒、张俊哲	华南理工大学	王静	华南理工大学
100586	刘奕辰、徐子牧、邓紫岚、康旭东、吴伯超、周浩	湖南大学	徐峰、严湘琦、彭晋卿、李洪强、周晋、肖坚	湖南大学
100591	李小转、邵红、王奇鹏	西安建筑科技大学	李涛	西安建筑科技大学
100596	宁洁、陈佳文、瞿可心、吴雯静	同济大学浙江学院	司舵	同济大学浙江学院
100606	王宇清、朱瑞东、沈志成、秦浩之、王鑫、刘书东、田昭源、张豪	山东建筑大学	郑斐、王月涛	山东建筑大学
100611	聂大为、杨喆雨	华东建筑设计研究总院、同济大学建筑设计研究院（集团）有限公司	陈易	同济大学
100613	赵玉晗、李梦楠	河南工业大学	张华	河南工业大学
100620	吴静文、李成成、卢易、赵鹏飞	安徽建筑大学	桂汪洋	安徽建筑大学
100631	罗昕钰、陈芊、仇梦佳、陈泽涛、刘永建	中国矿业大学	邵泽彪、孙良	中国矿业大学
100634	雷珊珊、梁欢、何秀敏、郑何山、雷宸骁	西安建筑科技大学	何泉	西安建筑科技大学

2021台达杯国际太阳能建筑设计竞赛办法
Competition Brief on International Solar Building Design Competition 2021

竞赛宗旨：

围绕我国2030碳达峰和2060碳中和的战略目标，贯彻落实国家实现能源利用转型，推动低碳社会发展、建设绿色健康的生活环境，将低碳、绿色、可持续理念融入社区建设中。

竞赛主题：阳光·低碳社区
竞赛题目：实地建设项目：西藏班戈县青龙乡东嘎村低碳社区
　　　　　概念设计项目：零碳社区
主办单位：国际太阳能学会
　　　　　中国建筑设计研究院有限公司
承办单位：国家住宅与居住环境工程技术研究中心
冠名单位：台达集团
评委会专家：**Deo Prasad**：澳大利亚科技与工程院院士、澳大利亚勋章获得者、澳大利亚新南威尔士大学教授。
　　　　　杨经文：马来西亚汉沙杨建筑师事务所创始人、2016年梁思成建筑奖获得者。
　　　　　Peter Luscuere：荷兰代尔伏特大学建筑系教授。
　　　　　崔愷：中国工程院院士、全国工程勘察设计大师、中国建筑设计研究院有限公司总建筑师。
　　　　　王建国：中国工程院院士、教育部高等学校建筑类专业指导委员会主任委员、东南大学建筑学院教授。
　　　　　庄惟敏：中国工程院院士、全国工程勘察设计大师、2019梁思成建筑奖获得者、清华大学建筑学院院长。
　　　　　林宪德：中国台湾绿色建筑委员会主席、中国台湾成功大学建筑系教授。
　　　　　宋晔皓：清华大学建筑学院建筑与技术研究所所长、教授、博士生导师，清华大学建筑设计研究院副总建筑师。
　　　　　钱锋：全国工程勘察设计大师、同济大学建筑与城市规划学院教授，

Goal of Competition:

Centered on China's strategic goals to have carbon dioxide emissions peak before 2030 and achieve carbon neutrality before 2060, the low-carbon, green and sustainable concepts should be integrated into community construction for the transformation of energy use, a low-carbon society, and green and healthy living environment.

Theme of Competition:

Sunshine & Low-Carbon Community

Subject of Competition:

Field Construction Project: Donggar Village Low-Carbon Community in Qinglong Township, Baingoin County, southwest China's Tibet Autonomous Region
Concept Design Project: Zero-Carbon Community

Host:

International Solar Energy Society (ISES)
China Renewable Energy Society (CRES)
China Architecture Design & Research Group (CAG)

Organizer:

China National Engineering Research Center for Human Settlements (CNERCHS)
Special Committee of Solar Buildings, CRES

Title Sponsor:

Delta Electronics

Experts of Judging Panel:

Mr. Deo Prasad: Academician of Academy of Technological Sciences and

博士生导师，高密度人居环境生态与节能教育部重点实验室主任。

仲继寿：中国可再生能源学会太阳能建筑专业委员会主任委员、中国建筑设计研究院有限公司副总建筑师。

黄秋平：华东建筑设计研究总院总建筑师。

冯雅：中国建筑西南设计研究院顾问总工程师、中国建筑学会建筑热工与节能专业委员会副主任。

刘泓志：AECOM亚洲区高级副总裁，中国区战略与发展负责人，城市设计策略团队负责人。

组委会成员：由主办单位、承办单位及冠名单位相关人员组成。办事机构设在国家住宅与居住环境工程技术研究中心。

设计任务书及专业术语等附件：（见资料下载）

附件1：实地建设项目：西藏班戈县青龙乡东嘎村低碳社区

附件2：概念设计项目：低碳社区

附件3：专业术语

附件4：参赛者信息表

奖项设置及奖励形式：

实地建设项目：

综合奖：

一等奖作品：1名 颁发奖杯、证书及人民币100000元奖金（税前）；

二等奖作品：2名 颁发奖杯、证书及人民币30000元奖金（税前）；

三等奖作品：4名 颁发奖杯、证书及人民币10000元奖金（税前）。

优秀奖作品：30名 颁发证书。

技术专项奖：名额不限，颁发证书。

概念设计项目：

一等奖作品：1名 颁发奖杯、证书及人民币30000元奖金（税前）；

二等奖作品：2名 颁发奖杯、证书及人民币10000元奖金（税前）；

三等奖作品：4名 颁发奖杯、证书及人民币5000元奖金（税前）。

优秀奖作品：30名 颁发证书。

设计创意奖：名额不限，颁发证书。

Engineering, Winner of the Order of Australia, and Professor of University of New South Wales, Sydney, Australia

Mr. King Mun YEANG: President of T. R. Hamzah & Yeang Sdn. Bhd, 2016 Liang Sicheng Architecture Prize Winner

Mr. Peter Luscuere: Professor of Department of Architecture, Delft University of Technology

Mr. Cui Kai: Academician of China Academy of Engineering, National Engineering Survey and Design Master and Chief Architect of China Architecture Design & Research Group (CAG)

Mr. Wang Jianguo: Academician of China Academy of Engineering; Director Academic Council of School of Architecture. Professor and Doctoral Supervisor of School of Architecture, Southeast University

Mr. Zhuang Weimin: Academician of Chinese Academy of Engineering, National Engineering Survey and Design Master, 2019 Liang Sicheng Architecture Prize Winner, and currently Dean, Professor and Doctoral Supervisor of School of Architecture, Tsinghua University

Mr. Lin Xiande: President of Taiwan Green Building Committee, and Professor of Faculty of Architecture of Cheng Kung University, China

Mr. Song Yehao: Director, Professor and Doctoral Supervisor of Institute of Architecture and Technology, School of Architecture, Tsinghua University, and Deputy Chief Architect of Architectural Design and Research Institute of Tsinghua University

Mr. Qian Feng: Professor and Doctoral Supervisor of College of Architecture and Urban Planning Tongji University (CAUP), Director of Key Laboratory of Ecology and Energy-saving Study of Dense Habitat (Tongji University), Ministry of Education

Mr. Zhong Jishou: Deputy Chief Architect of CAG, and Chief Commissioner of Special Committee of Solar Buildings, CRES

Mr. Huang Qiuping: Chief Architect of East China Architectural Design & Research Institute (ECADI)

Mr. Feng Ya: Chief Engineer of China Southwest Architectural Design and Research Institute Corp. Ltd., and Deputy Director of Special Committee of Building Thermal and Energy Efficiency, Architectural Society of China (ASC)

Mr. Liu Hongzhi: AECOM Asia Senior Vice President, Leader for Strategy & Development Cities Market Sector Leader, China

Members of the Organizing Committee:

It is composed of the competition host, organizer and title sponsor. The administration office is a standing body in Special Committee of Solar Buildings, CRES.

The Design Specifications and Professional Glossary (Found in Annex)

Annex 1: Donggar Village Low-Carbon Community in Qinglong Township, Baingoin County, Tibet Autonomous Region

Annex 2: Zero-Carbon Community

参赛要求：

1. 欢迎建筑设计院、高等院校、研究机构、绿色建筑部品研发生产企业等单位，组织专业人员组成竞赛小组参加竞赛。

2. 请参赛者访问 www.isbdc.cn，按照规定步骤填写注册表，提交后会得到唯一的注册号，即作品编号，一个作品对应一个注册号。提交作品时把注册号标注在每副作品的左上角，字高6mm。注册时间2021年3月30日～2021年8月15日。

3. 参赛者同意组委会公开刊登、出版、展览、应用其作品。

4. 被编入获奖作品集的作者，应配合组委会，按照出版要求对作品进行相应调整。

注意事项：

1. 参赛作品电子文件须在2021年9月30日前提交组委会，请参赛者访问www.isbdc.cn，并上传文件，不接受其他递交方式。

2. 作品中不能出现任何与作者信息有关的标记内容，否则将视其为无效作品。

3. 组委会将及时在网上公布入选结果及评比情况，将获奖作品整理出版，并对获奖者予以表彰和奖励。

4. 获奖作品集首次出版后30日内，组委会向获奖作品的创作团队赠样书2册。

5. 竞赛活动消息发布、竞赛问题解答均可登陆竞赛网站查询。

所有权及版权声明：

参赛者提交作品之前，请详细阅读以下条款，充分理解并表示同意。

依据中国有关法律法规，凡主动提交作品的"参赛者"或"作者"，主办方认为其已经对所提交的作品版权归属作如下不可撤销声明：

1. 原创声明

参赛作品是参赛者原创作品，未侵犯任何他人的任何专利、著作权、商标权及其他知识产权；该作品未在报纸、杂志、网站及其他媒体公开发表，未申请专利或进行版权登记，未参加过其他比赛，未以任何形式进入商业渠道。参赛者保证参赛作品终身不以同一作品形式参加其他的设计比赛或转让给他方。否则，主办单位将取消其参赛、入围与获奖资格，收回奖金、奖品及并保留追究法律责任的权利。

2. 参赛作品知识产权归属

为了更广泛地推广竞赛成果，所有参赛作品除作者署名权以外的全部著作权归竞赛承办单位及冠名单位所有，包括但不限于以下方式行使著作权：享有对所

Annex 3: Professional Glossary
Annex 4: Information Table

Award Setting and Award Form:

Field Construction Project:
General Prize:
First Prize: 1 winner
The Trophy Cup, Certificate and Bonus RMB 100,000 (before tax) will be awarded.
Second Prize: 2 winners
The Trophy Cup, Certificate and Bonus RMB 30,000 (before tax) will be awarded.
Third Prize: 4 winners
The Trophy Cup, Certificate and Bonus RMB 10,000 (before tax) will be awarded.
Honorable Mention Prize: 30 winners
The Certificate will be awarded.
Prize for Technical Excellence Works:
The quota is open-ended. The Certificate will be awarded.
Concept Design Project:
First Prize: 1 winner
The Trophy Cup, Certificate and Bonus RMB 30,000 (before tax) will be awarded.
Second Prize: 2 winners
The Trophy Cup, Certificate and Bonus RMB 10,000 (before tax) will be awarded.
Third Prize: 4 winners
The Trophy Cup, Certificate and Bonus RMB 5,000 (before tax) will be awarded.
Honorable Mention Prize: 30 winners
The Certificate will be awarded.
Prize for Design Originality:
The quota is open-ended. The Certificate will be awarded.

Participation Requirements:

1. Professionals from institutes of architectural design, colleges and universities, research institutions and green building product development and manufacturing enterprises are welcomed to take part in the competition in the form of competition groups.

2. Please visit www.isbdc.cn. You may fill in the registry according to the instruction and gain an ID of your work after submitting the registry. That's the number of your work, and one work only has one ID. The number should be indicated in the top left corner of each submitted work with word height of 6mm. Registration time: March 30, 2021 - July 30, 2021.

3. Participants must agree that the Organizing Committee may publish, print, exhibit and apply their works in public.

4. The authors whose works are edited into the publication should cooperate with the Organizing Committee to adjust their works according to the requirements of press.

属竞赛作品方案进行再设计、生产、销售、展示、出版和宣传的权利；享有自行使用、授权他人使用参赛作品用于实地建设的权利。竞赛主办方对所有参赛作品拥有展示和宣传等权利。其他任何单位和个人（包括参赛者本人）未经授权不得以任何形式对作品转让、复制、转载、传播、摘编、出版、发行、许可使用等。参赛者同意竞赛承办单位及冠名单位在使用参赛作品时将对其作者予以署名，同时对作品将按出版或建设的要求做技术性处理。参赛作品均不退还。

3. 参赛者应对所提交作品的著作权承担责任，凡由于参赛作品而引发的著作权属纠纷均应由作者本人负责。

声明：

1. 参与本次竞赛的活动各方（包括参赛者、评委和组委），即表明已接受上述要求。

2. 本次竞赛的参赛者，须接受评委会的评审决定作为最终竞赛结果。

3. 组委会对竞赛活动具有最终的解释权。

4. 为维护参赛者的合法权益，主办方特提请参赛者对本办法的全部条款、特别是"所有权及版权"声明部分予以充分注意。

Important Consideration:

1. Participant's digital file and report on works' design method must be uploaded to the Organizing Committee's FTP site (www.isbdc.cn) before September 30, 2021. Other ways will not be accepted.

2. Any mark, sign or name related to participant's identity should not appear in, on or included within the submitted files, otherwise the submission will be deemed invalid.

3. The Organizing Committee will publicize the process and result of the appraisal online in a timely manner, compile and publish the awarded works. The winners will be honored and awarded.

4. In 30 days after the collection of works being published, two books of award works will be freely presented by the Organizing Committee to the competition teams who are awarded.

5. The information concerning the competition as well as explanation about all activities may be checked and inquired on the website of the competition.

Announcement about Ownership and Copyright:

Before submitting the works, participants should carefully read following clauses, fully understand and agree with them.

According to relevant national laws and regulations, it is confirmed by the competition hosts that all "participants" or "authors" who have submitted their works on their own initiative have received following irrevocable announcement concerning the ownership and copyright of their works submitted:

1. Announcement of originality

The entry work of the participant is original, which does not infringe any patent, copyright, trademark and other intellectual property; it has not been published in any newspapers, periodicals, magazines, webs or other media, has not been applied for any patent or copyright, not been involved in any other competition, and not been put in any commercial channels. The participant should assure that the work has not been put in any other competition by the same work form in its whole life or legally transferred to others, otherwise, the competition hosts will cancel the qualification of participation, being shortlisted and being awarded of the participant, call back the prize and award and reserve the right of legal liability.

2. The ownership of intellectual property of the works

In order to promote competition results, the participants should relinquish copyright of all works to competition organizers and titled sponsors except authorship. It includes but is not limited to the exercise of copyright as follows: benefit from the right of the works on redesigning, production, selling, exhibition, publishing and publicity; benefit from the right of the works on construction for self use or accrediting to others for use. Hosts of the competition have such rights to display and publicize all the works. Without accreditation, any organizations and individuals (including authors themselves) cannot transfer, copy, reprint, promulgate, extract and edit, publish and admit others to use the works by any way. Participants have to agree that competition organizers and titled sponsors will sign the name of authors when their works are used and the works will be treated for technical processing according to the requirements of publication and construction. All works are not returned to the author.

3. All authors must take responsibility for their copyrights of the works including all disputes of copyright caused by the works.

Announcement:

1. It implies that everybody who has attended the competition activities including participants, jury members and members of the Organizing Committee has accepted all requirements mentioned above.

2. All participants must accept the appraisal of the jury as the final result of the competition.

3. The Organizing Committee reserves final rights to interpret the competition activities.

4. In order to safeguard the legitimate rights and interests of the participants, the organizers ask participants to fully pay attention to all clauses in this document, especially the Announcement about Ownership and Copyright.

附件1：
西藏班戈县青龙乡东嘎村低碳社区赛题（实地建设项目）

1. 项目背景

项目地位于班戈县青龙乡东嘎村，北纬31.09190.801°，东经90.801°，距纳木措景区约30公里，距青龙乡16公里，距班戈县82公里。省道S206班洛线于基地东侧南北向穿过。

该项目将计划建设为集居住、旅游接待为一体化的低碳社区，目前已建成纳木错国家公园露营区游客中心、畜牧品交易中心、合作社产业基地等建筑。一期计划建设包括商业街区、牧民住宅、居住区配套服务设施等。

2. 自然条件

班戈县青龙乡东嘎村位于西藏自治区那曲市，属高原亚寒带半干旱季风气候区。气候寒冷，空气稀薄，四季不分明，冬长夏短，多风雪天气，年温差相对大

Annex1: Design Specifications of Donggar Village Low-Carbon Community in Qinglong Township, Baingoin County, southwest China's Tibet Autonomous Region (Concept Design Project)

I. Project Background

Located in Donggar Village, Qinglong Township, Baingoin County, southwest China's Tibet Autonomous Region, with a latitude of 31.091°N and a longitude of 90.801°E, the project site is about 30km away from Namtso Lake Scenic Area, 16km away from Qinglong Township, and 82km away from Baingoin County. The Provincial Highway S206 Bange-Luozha Line passes north-south on the east side of the site.

The project is scheduled to be built into a low-carbon community with the aim of residence and tourism reception, and buildings such as the visitor center of the Namtso Lake National Park camping area, livestock trading center and cooperative industrial base have been built currently. In the first phase, such facilities as commercial blocks,

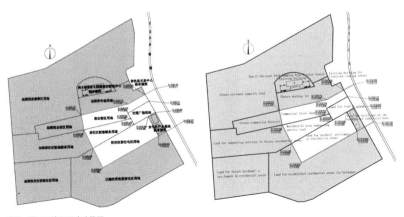

图2 项目用地及周边功能图
Figure 2 Project Site and Surrounding Land

图1 项目所在地俯视图
Figure 1 Vertical View of the Project Site

图3 项目用地平面图
Figure 3 Plan of the Project Site (North Arrow Presented)

于日温差。班戈县境由于群山隔断，印度洋潮湿空气难以进入，属高原亚寒带季风半干旱气候区，空气稀薄，寒冷干燥，气候变幻无常，昼夜温差大，无绝对无霜期。年平均气温零度左右，年日照时间为2850多个小时，年最高气温21.9℃，最低气温 –28.6℃。一月份平均气温为 –17.1℃，七月份平均气温 16.5℃，冻土深度3m。班戈县年降水量为289~390mm，主要集中在6~9月份，占全年降水总量的80%。年蒸发量为 1993.4~2104.1mm，为降水量的6.9~7.3倍，年相对湿度为41%，年径流量为59.6mm。

3. 基础设施

基地内基础设施较差，有自来水系统，供电稳定性不佳、无排水措施，通信不稳定。

4. 竞赛场地

本项目为西藏班戈县青龙乡东嘎村低碳社区，项目用地较为平坦。用地东侧S206省道（班洛线）临街处已建成畜牧品交易中心用房和合作社产业基地用房，北侧为远期规划停车场用地及纳木措国家公园露营区游客中心，西侧为远期规划露营区用地、远期规划商业街区用地、远期规划居住区配套服务用地，南侧为已建牧民定居居住区用地。用地面积 75391.5m²。建设用地面积 62892.3m²，其中广场用地 6157.9m²，居住区配套服务用地 12632.3m²，牧民定居住宅区用地 27122.2m²，商业街区用地 19703.5m²。

5. 设计要求

1）在给定的竞赛用地范围内设计文化广场、商业街区、牧民定居住宅及居住区配套服务设施，总建筑面积 12000m² 左右。

2）在牧民定居住宅区用地内布置独立住宅，单户建筑面积不大于120m²，建筑不高于2层，总建筑面积约为5000m²。

3）在本期居住区配套服务用地内布置服务于牧民定居住宅区的卫生院、文

residential areas for herdsmen, and supporting service facilities for residential areas are planned to be constructed.

II. Natural Conditions

Located in Nagqu Prefecture of Tibet Autonomous Region, categorized as the semi-arid monsoon climate region of plateau subfrigid zone, Donggar Village, Qinglong Township, Baingoin County is cold, windy and snowy, with thin air, indistinctive seasons, as well as long winter and short summer. Besides, the annual temperature difference is relatively larger than the daily temperature difference. Separated by mountains, humid air of the Indian Ocean is hardly accessible, so Baingoin County develops the semi-arid monsoon climate of plateau subfrigid zone, which is cold, dry and changeable, with thin air, large diurnal temperature variation, and no absolute frost-free period. Here are some parameters of natural conditions in Baingoin County: annual average temperature: around 0℃; annual sunshine duration: over 2,850h; annual maximum temperature: 21.9 ℃; minimum temperature: -28.6 ℃; average temperature in January: -17.1 ℃; average temperature in July: 16.5℃; depth of frozen soil: 3m; annual precipitation: 289-390mm, mainly from June to September, accounting for 80% of the annual precipitation; annual evaporation: 1,993.4-2,104.1mm, 6.9-7.3 times of the precipitation; annual relative humidity: 41%; and annual runoff: 59.6mm.

III. Infrastructure

The infrastructure in the base is backward, with poor power supply stability, no drainage measures and unstable communications, in addition to a tap water system.

IV. Competition Venue

The project is a low-carbon community in Donggar Village, Qinglong Township, Baingoin County, Tibet Autonomous Region, with relatively flat land. On the east side of the site, the area near the Provincial Highway S206 (Bange-Luozha Line) has been built into the livestock trading center and cooperative industrial base. In the long term, the north side is planned to be built into the parking lot and visitor center of Namtso Lake National Park camping area, and the west side is planned to be built into the camping land, commercial blocks and supporting service land for residential areas. The south side has been built into residential areas for herdsmen. The site area is 75,391.5m², and the construction area is 62,892.3m², including 6,157.9m² for squares, 12,632.3m² for supporting services of residential areas, 27,122.2m² for herdsmen's

图4 已建纳木措国家公园露营区游客中心照片
Figure 4 Constructed Visitor Center of the Namtso Lake National Park Camping Area

图5 在建牧民定居项目照片
Figure 5 Residential Areas for Herdsmen under Construction

图6 已建畜牧品交易中心
Figure 6 Constructed Livestock Trading Center

Figure 7　Cooperative Industrial Base

residential areas, and 19,703.5m² for commercial blocks.

V. Requirements for Design

1) Design cultural squares, commercial blocks, herdsmen's residential areas and supporting service facilities for residential areas within the given scope of competition land, with a total construction area of about 15,000m².

2) Arrange independent houses in herdsmen's residential areas, with the single-family construction area no more than 120m², no more than 2 floors, and the total construction area of about 5,000m².

3) Arrange such service facilities as health centers, cultural activity stations and village committee offices in supporting service land for current residential areas to serve herdsmen, and to provide basic community services and concentrated exchange activity space for villages around the ranch, with no more than 2 floors, the total construction area no more than 1,500m², and equipped with the parking lot, traffic square, etc.

4) Set up the image entrance of the project and cultural square near the Provincial Highway S206 in the traffic square site.

5) Build shopping and catering facilities and other commercial facilities in commercial blocks, with no more than 3 floors, the total construction area no more than 6,000m², and the single building area no more than 600m².

6) Organize the relationship and streamline among the usable land in the community for convenient contact and less interference.

7) Consider low-carbon emission reduction for implementation from community construction, operation to the whole process.

8) Solve the problem of building insulation and heating in winter, and apply active and passive solar technology and other renewable energy technologies, taking renewable energy transmission and distribution into account, based on local climate characteristics and natural environment, local building features and materials, as well as energy characteristics of different types of buildings.

9) Set up corresponding measures such as garbage disposal and sewage harmless treatment as there is a lack of municipal supporting measures in the project site.

10) Consider the economy and popularization of the project.

11) Function settings of the land in the community are shown in the table below:

Functional Space	Function	Quantity	Total Usable Area (m²)	Note
Land for Herdsmen's Residential Areas	Herdsmen's residential areas	42 sets	5,000	No more than 2 floors of the building, with independent courtyards, the single building area no more than 120m², and internal functional division of the building not required
Supporting Service Land for Residential Areas	Village committee	1	400	No more than 2 floors of the building, with convenient transportation, easy access to residential areas and commercial blocks, no less than 6 parking lots, and internal functional division of the building not required

续表

功能空间	功能要求	数量	总使用面积（m²）	备注
居住区配套服务用地	村委会	1	400	建筑不高于2层，要求交通便利，与居住区、商业街区联系便捷，不要求建筑内部功能划分，设置不少于6个停车位
	卫生院	1	300	建筑不高于2层，要求与居住区联系便捷，不要求建筑内部功能划分，设置不少于4个停车位
	文化活动站	1	300	建筑为1层，室外设置600m²以上的活动场地。与居住区联系便捷，不要求建筑内部功能划分设置
	警务处	1	100	建筑为1层，与居住区、商业街区联系便捷，不要求建筑内部功能划分设置不少于6个停车位
	垃圾中转站	1	200	建筑为1~2层，建筑入口处设置200m²以上空地，与其他建筑保持15m以上距离，不要求建筑内部功能划分
	超市	1	200	建筑为1层，主要服务与居住区，不要求建筑内部功能划分
交通广场用地	形象入口	1		社区主入口
	文化广场	1		社区交通中心，可便捷通往各个分区，可设置主题性景观
商业街区用地	特产售卖	不限	2000	包括文化商品、手工商品藏药等，宜设置在1~2层，独立经营单元面积50~200m²，要求进行建筑内部功能划分
	民俗体验	不限	1000	包括民族习俗体验、农牧产品体验、藏医体验等，宜设置在1~2层，独立经营单元面积50~200m²，要求进行建筑内部功能划分
	餐饮服务	不限	2000	宜设置在1~3层，独立经营单元面积100~500m²
	住宿服务	不限	1000	为旅游人群提供住宿服务，宜设置在1~3层，独立经面积300~600m²

Continue Table

Functional Space	Function	Quantity	Total Usable Area (m²)	Note
Supporting Service Land for Residential Areas	Health center	1	300	No more than 2 floors of the building, with easy access to residential areas, no less than 4 parking lots, and internal functional division of the building not required
	Cultural activity station	1	300	1 floor of the building, with the outdoor venue area of over 600m², easy access to residential areas, and internal functional division of the building not required
	Police station	1	100	1 floor of the building, with easy access to residential areas and commercial blocks, no less than 6 parking lots, and internal functional division of the building not required
	Waste transfer station	1	200	1-2 floor(s) of the building, with an open space of over 200m² at the entrance, a distance of over 15m from other buildings, and internal functional division of the building not required
	Supermarket	1	200	1 floor of the building, serving residence areas, and internal functional division of the building not required
Land for Traffic Squares	Image entrance	1		Main entrance of the community
	Cultural square	1		Transportation center of the community, easily accessible to various partitions and themed landscapes can be set up
Land for Commercial Blocks	Special local product selling	Open-ended	2,000	Including cultural goods, handmade goods and Tibetan medicine, set on 1-2F appropriately, with an area of 50-200m² of independent business units, and internal functional division of the building required
	Folk custom experience	Open-ended	1,000	Including folk custom experience, farm product and pasture product and Tibetan medicine experience, set on 1-2F appropriately, with an area of 50-200m² of independent business units, and internal functional division of the building required

6. 评比办法

1）由组委会审查参赛资格，并确定入围作品。

2）由评委会评选出竞赛获奖作品。

7. 评比标准

1）参赛作品须符合本竞赛"作品要求"的内容。

2）作品应具有原创性，鼓励创新。

3）作品应满足使用功能、绿色低碳、安全健康的要求，建筑技术与太阳能利用技术具有适配性。

4）作品应充分体现太阳能利用技术对降低建筑使用能耗的作用，在经济、技术层面具有可实施性。

评比指标	指标说明	分值
规划与建筑设计	规划布局、建筑空间组合、功能流线组织、建筑艺术	50
被动太阳能采暖技术	利用建筑设计与建筑构造实现建筑采暖与蓄热	25
主动太阳能利用技术	利用太阳能光伏系统、集热器等实现太阳能的利用及传输	10
采用的其他技术	社区建造与运行过程中的绿色、低碳、安全、健康技术	15
可操作性（加分项）	作品的可实施性，技术的经济性和普适性	10

8. 作品要求

1）建筑设计方面应达到方案设计深度，技术应用方面应有相关的技术图纸和指标。

2）作品图面、文字表达清楚，数据准确。

3）作品基本内容包括：

（1）简要建筑方案设计说明（限200字以内），包括方案构思、太阳能技术、低碳技术与设计创新、技术经济指标表等。

（2）项目的竞赛作品需进行竞赛用地范围内的规划设计，总平面图（含活动场地及环境设计）。

（3）单体设计：

a. 商业街区地块中能充分表达建筑与室内外环境关系的各功能典型建筑各层平面图、外立面图、剖面图，比例不小于1：300。

b. 牧民定居住宅区用地、居住区配套服务用地、交通广场用地内的建筑、设施及景观布局及交通流线组织。

Continue Table

Functional Space	Function	Quantity	Total Usable Area (m²)	Note
Land for Commercial Blocks	Food & Beverage	Open-ended	2,000	Set on 1-3F appropriately, with an area of 100-500m² of independent business units
	Accommodation	Open-ended	1,000	Provide accommodation services for tourists, set on 1-3F appropriately, with an area of 300-600m² of independent business units

Ⅵ. **Appraisal Methods**

1. Organizing Committee will check up eligible entries and confirm shortlist entries.

2. Judging Panel will appraise and select out the awarded works.

Ⅶ. **Appraisal Standards**

1) The entries must meet the demands of the "Requirements for Works".

2) The entries should be original to encourage innovation.

3) The submitted works should meet the demands for use, green and low-carbon concept, safety and health. And the building technology and solar energy utilization technology should be compatible.

4) The submitted works should play the role of reducing building energy consumption by using solar energy technology and are feasible in economy and technology.

Appraisal Indicator	Explanation	Score
Planning and architecture design	Planning design, architectural space combination, functional division and streamline organization, and architectural art.	50
Passive solar heating technology	Realize the heating and thermal storage of buildings by architecture and construction design.	25
Active solar utilization technology	Realize the utilization and transmission of solar energy by solar photovoltaic systems and collectors.	10
Other technologies	Green, low-carbon, safe and healthy technologies during the community construction and its running process.	15
Operability of the technology (scores additionally added)	Feasibility, economy and popularity of works	10

Ⅷ. **Requirements for Works**

1) The submitted drawing sheets should meet the requirements of scheme design level and should be accompanied with relevant technical drawings and technology data.

2) Drawings and text should be expressed in a clear and readable way.

b. 能表现出技术与建筑结合的重点部位、局部详图及节点大样，比例自定；低碳及其他相关的技术图、分析图。

（4）整体社区、地块、建筑单体效果表现图。

（5）参赛者须将作品文件编排在 A1 展板 (594mm×841mm) 区域内（统一采用竖向构图），作品张数应为 4 张或 6 张。中文字体不小于 6mm，英文字体不小于 4mm，文件分辨率 300 dpi，格式为 JPG 或 PDF 文件，提交参赛者信息表，格式为 JPG 或 PDF 文件。

（6）参赛者通过竞赛网页上传功能将作品递交竞赛组委会，入围作品由组委会统一编辑板眉、出图、制作展板。

（7）作品文字要求："建筑方案设计说明"采用中英文外，其他为英文；建议使用附件 3 中提供的专业术语。

Mentioned data should be accurate.

3) The submitted work should include:

(1) A project description (not exceeding 200 words) including the following elements: schematic design concept; solar energy technology; low-carbon technology and innovative design; technical and economic indicators.

(2) Participants should provide a planning design within the outline of the competition site, and provide a site plan (including the venue / environment design).

(3) Monomer Design:

a. Participants should provide floor plans, elevations and sections of commercial blocks with the scale not less than 1:300, which can fully express the relationship between architecture and indoor and outdoor environment.

b. Buildings, facilities and landscape layout, and the traffic streamline organization in the land for herdsmen's residential areas, supporting service land for residential areas and land for traffic squares.

c. Participants should provide detailed drawings (without limitation of scale) that illustrate the integration of technology in the architectural project, as well as low-carbon and other relevant elements, such as technical charts and analysis diagram.

(4) Rendering perspective drawing of community, land and single building

(5) Participants should arrange the submission into four or six exhibition panels, each 594 mm × 841 mm (A1 format) in size (arranged vertically). Word height of Chinese is not less than 6mm and that of English is not less than 4mm. File resolution: 300 dpi in JPG or PDF format. And the information form of participants should also be submitted in JPG or PDF format.

(6) Participants should upload a digital version of submission via FTP to the Organizing Committee, who will compile, print and make exhibition panels for shortlist works.

(7) Text requirement: The submission should be in English, in addition to "architectural design description" in English and Chinese. Participants should use the words from the Professional Glossary in Annex 3.

附件2：
零碳社区设计任务书（概念设计项目）

1. 项目背景

当前气候变暖有 90% 以上是由人类活动造成的，而城市作为人类活动的主要场所，其运行过程中消耗了大量的化石能源，排放的温室气体已占到全球总量的 75% 左右，制造出全球 80% 的污染。随着城市扩张速度越来越快，城市也因此变得越来越脆弱，频繁发生的气候灾害威胁到了城市居民正常的生产生活。社区作为城市碳减排的重要单元和主体，是实现全球减碳和低碳城市化的关键所在。

2. 设计要求

1）功能要求

参赛者在世界范围内任意城市，在城市内选定一块 300m×300m 的场地（可真实存在或虚拟假定，但应符合物理规律），打造为零碳社区，能够满足 1000～3000 人在社区中的居住与工作生活，需满足居住、综合服务办公、商业、交通等需求，并与邻近社区及城市有紧密联系，各功能分区的面积比例合理。

2）技术要求

社区内实现零碳排放，社区内充分利用太阳能等实现能源自给甚至向外能源输送，水资源可 80% 以上循环利用，绿地率≥35%；设计应体现当地文化属性以及对所在地区自然环境的适应与利用并应描述人在其中的活动轨迹与状态。

3. 评比办法

1）由组委会审查参赛资格，并确定入围作品。

2）由评委会评选出竞赛获奖作品。

4. 评比标准

1）参赛作品须符合本竞赛"作品要求"的内容。

2）作品应具有前瞻性、原创性和创新性。

3）作品应注重绿色低碳、安全健康、气候适应性以及可再生能源应用。

评比指标	指标说明	分值
规划与建筑设计	规划理念、布局、交通流线组织、建筑空间组合	40
地域及气候适应性	体现当地文化属性、对所在地区自然环境的适应与利用	20
低碳及能源自给	建造及运行过程实现碳中和、社区内用能与产能平衡	20
创新与可实施性	设计及技术应用具有创新性及可实施性	20

Annex2: Design Specifications of Zero-Carbon Community (Field Construction Project)

I. Project Background

It is human activities that account for more than 90% of the current climate warming. As the main place for human activities, cities consume much fossil energy during the operation, emitting about 75% of the global greenhouse gases, and causing 80% of the global pollution. As cities expand faster and faster, they have become increasingly vulnerable. Frequent climatic disasters have posed threats to normal production and life of urban residents. Communities, an important role and subject in urban carbon emission reduction, constitute a key to achieving global carbon reduction and low-carbon urbanization.

II. Requirements for Design

Conceptual design scheme

Participants choose a 300m×300m site in any city around the world (it can be real or virtual, but should conform to natural and physical principles) to build a zero-carbon community, which can meet the demands of living and working for 1,000-3,000 people, satisfy people's requirements for housing, comprehensive services (education, medical care, etc.), working, business, transportation, etc., and have close connections with neighboring communities and cities, with reasonable ratio of each functional zone.

In the zero-carbon community, solar energy can be fully utilized to achieve self-sufficiency and even transmit energy to the external world. The RRUP of water resources is more than 80%, and the greening rate is not less than 35%. The design should reflect the local cultural attributes and the adaptation to and utilization of natural environment of the area, and describe people's activities and state.

III. Appraisal Methods

1. Organizing Committee will check up eligible entries and confirm shortlist entries.
2. Judging Panel will appraise and select out the awarded works.

IV. Appraisal Standards

1. The entries must meet the demands of the "Requirements for Works".
2. The entries should be forward-looking, original and innovative.
3. The submitted works should focus on the green and low-carbon concept, safety, health, climatic adaptability and application of renewable energy.

Appraisal Indicator	Explanation	Score
Planning and architecture design	Planning concept, layout, traffic streamline organization, architectural space combination	40
Regional and climatic adaptability	Reflect the local cultural attributes and the adaptation to and utilization of natural environment of the area	20
Low carbon and energy self-sufficiency	Achieve carbon neutrality and balance between energy consumption and production capacity within the community during the construction and operation	20
Innovation and operability	Design and technology application are innovative and operable	20

5. 作品要求

1) 简要建筑方案设计说明（限400字以内），包括方案构思、设计理念等。

2) 能充分表达作品创作意的总平面图、功能分区图、建筑表现图、透视图、分析图等，不要求建筑内部功能划分。

3) 能表达实现能源自给及零碳排放的相关设计图纸、技术分析等；

4) 效果表现图1~4个。

参赛者须将作品文件编排在A1展板(594mm×841mm)区域内（统一采用竖向构图），作品张数应为2张或4张。中文字体不小于6mm，英文字体不小于4mm，文件分辨率300 dpi，格式为JPG或PDF文件，提交参赛者信息表，格式为JPG或PDF文件。

5) 参赛者通过竞赛网页上传功能将作品递交竞赛组委会，入围作品由组委会统一编辑板眉、出图、制作展板。

6) 作品文字要求："建筑方案设计说明"采用中英文外，其他为英文；建议使用附件3中提供的专业术语。

V. Requirements for Works

1) A project description (not exceeding 400 words) including the schematic design concept and design concept.

2) Participants should provide the general site plan, functional zoning diagram, architectural performance diagram, perspective diagram, analysis diagram, etc. of the work, and does not require the internal function division of the building.

3) Participants should provide related design drawings, technical analysis, etc., which can express the energy self-sufficiency and zero-carbon emission.

4) Rendering perspective drawing (1-4)

Participants should arrange the submission into four or six exhibition panels, each 594 mm × 841 mm (A1 format) in size (arranged vertically), the number of works should be 2 or 4. Word height of Chinese is not less than 6mm and that of English is not less than 4mm. File resolution: 300 dpi in JPG or PDF format. And the information form of participants should also be submitted in JPG or PDF format.

5) Participants should upload a digital version of submission via FTP to the Organizing Committee, who will compile, print and make exhibition panels for shortlist works.

6) Text requirement: The submission should be in English, in addition to "architectural design description" in English and Chinese. Participants should use the words from the Professional Glossary in Annex 3.

附件3：专业术语
Annex 3: Professional Glossary

百叶通风	— shutter ventilation
保温	— thermal insulation
被动太阳能利用	— passive solar energy utilization
敞开系统	— open system
除湿系统	— dehumidification system
储热器	— thermal storage
储水量	— water storage capacity
穿堂风	— through-draught
窗墙面积比	— area ratio of window to wall
次入口	— secondary entrance
导热系数	— thermal conductivity
低能耗	— lower energy consumption
低温热水地板辐射供暖	— low temperature hot water floor radiant heating
地板辐射采暖	— floor panel heating
地面层	— ground layer
额定工作压力	— nominal working pressure
防潮层	— wetproof layer
防冻	— freeze protection
防水层	— waterproof layer
分户热计量	— household-based heat metering
分离式系统	— remote storage system
风速分布	— wind speed distribution
封闭系统	— closed system
辅助热源	— auxiliary thermal source
辅助入口	— accessory entrance
隔热层	— heat insulating layer
隔热窗户	— heat insulation window
跟踪集热器	— tracking collector
光伏发电系统	— photovoltaic system
光伏幕墙	— PV façade
回流系统	— drainback system
回收年限	— payback time
集热器瞬时效率	— instantaneous collector efficiency
集热器阵列	— collector array
集中供暖	— central heating
间接系统	— indirect system
建筑节能率	— building energy saving rate
建筑密度	— building density
建筑面积	— building area
建筑物耗热量指标	— index of building heat loss
节能措施	— energy saving method
节能量	— quantity of energy saving
紧凑式太阳热水器	— close-coupled solar water heater
经济分析	— economic analysis
卷帘外遮阳系统	— roller shutter sun shading system
空气集热器	— air collector
空气质量检测	— air quality test (AQT)
立体绿化	— tridimensional virescence
绿地率	— greening rate
毛细管辐射	— capillary radiation
木工修理室	— repairing room for woodworker
耐用指标	— permanent index
能量储存和回收系统	— energy storage & heat recovery system
平屋面	— plane roof
坡屋面	— sloping roof
强制循环系统	— forced circulation system
热泵供暖	— heat pump heat supply
热量计量装置	— heat metering device
热稳定性	— thermal stability
热效率曲线	— thermal efficiency curve
热压	— thermal pressure
人工湿地效应	— artificial marsh effect
日照标准	— insolation standard
容积率	— floor area ratio
三联供	— triple co-generation
设计使用年限	— design working life
使用面积	— usable area
室内舒适度	— indoor comfort level
双层幕墙	— double facade building

太阳方位角	– solar azimuth
太阳房	– solar house
太阳辐射热	– solar radiant heat
太阳辐射热吸收系数	– absorptance for solar radiation
太阳高度角	– solar altitude
太阳能保证率	– solar fraction
太阳能带辅助热源系统	– solar plus supplementary system
太阳能电池	– solar cell
太阳能集热器	– solar collector
太阳能驱动吸附式制冷	– solar driven desiccant evaporative cooling
太阳能驱动吸收式制冷	– solar driven absorption cooling
太阳能热水器	– solar water heating
太阳能烟囱	– solar chimney
太阳能预热系统	– solar preheat system
太阳墙	– solar wall
填充层	– fill up layer
通风模拟	– ventilation simulation
外窗隔热系统	– external windows insulation system
温差控制器	– differential temperature controller
屋顶植被	– roof planting
屋面隔热系统	– roof insulation system
相变材料	– phase change material (PCM)
相变太阳能系统	– phase change solar system
相变蓄热	– phase change thermal storage
蓄热特性	– thermal storage characteristic
雨水收集	– rain water collection
运动场地	– schoolyard
遮阳系数	– sunshading coefficient
直接系统	– direct system
值班室	– duty room
智能建筑控制系统	– building intelligent control system
中庭采光	– atrium lighting
主入口	– main entrance
贮热水箱	– heat storage tank
准备室	– preparation room
准稳态	– quasi-steady state
自然通风	– natural ventilation
自然循环系统	– natural circulation system
自行车棚	– bike parking